# INFINITE ABELIAN GROUPS

*Volume I*

This is Volume 36 in
PURE AND APPLIED MATHEMATICS
A Series of Monographs and Textbooks
Edited by PAUL SMITH AND SAMUEL EILENBERG
A complete list of titles in this series appears at the end of this volume

# INFINITE ABELIAN GROUPS

*László Fuchs*

*Tulane University*
*New Orleans, Louisiana*

*VOLUME I*

*ACADEMIC PRESS*    *NEW YORK AND LONDON*    *1970*

ACADEMIC, PRESS, INC.
111 Fifth Avenue, New York, New York 10003

*United Kingdom Edition published by*
ACADEMIC PRESS, INC. (LONDON) LTD.
Berkeley Square House, London W1X 6BA

LIBRARY OF CONGRESS CATALOG CARD NUMBER: 78–97479
AMS 1968 SUBJECT CLASSIFICATION 2030

PRINTED IN THE UNITED STATES OF AMERICA

# PREFACE

The theory of abelian groups is a branch of algebra which deals with commutative groups. Curiously enough, it is rather independent of general group theory: its basic ideas and methods bear only a slight resemblance to the noncommutative case, and there are reasons to believe that no other condition on groups is more decisive for the group structure than commutativity.

The present book is devoted to the theory of abelian groups. The study of abelian groups may be recommended for two principal reasons: in the first place, because of the beauty of the results which include some of the best examples of what is called algebraic structure theory; in the second place, it is one of the principal motives of new research in module theory (e.g., for every particular theorem on abelian groups one can ask over what rings the same result holds) and there are other areas of mathematics in which extensive use of abelian group theory might be very fruitful (structure of homology groups, etc.).

It was the author's original intention to write a second edition of his book "Abelian Groups" (Budapest, 1958). However, it soon became evident that in the last decade the theory of abelian groups has moved too rapidly for a mere revised edition, and consequently, a completely new book has been written which reflects the new aspects of the theory. Some topics (lattice of subgroups, direct decompositions into subsets, etc.) which were treated in "Abelian Groups" will not be touched upon here.

The twin aims of this book are to introduce graduate students to the theory of abelian groups and to provide a young algebraist with a reasonably comprehensive summary of the material on which research in abelian groups can be based. The treatment is by no means intended to be exhaustive or even to yield a complete record of the present status of the theory—this would have been a Sisyphean task, since the subject has become so extensive and is growing almost from day to day. But the author has tried to be fairly complete in what he considers as the main body of up-to-date abelian group theory, and the reader should get a considerable amount of knowledge of the central ideas, the basic results, and the fundamental methods. To assist the reader in this, numerous exercises accompany the text; some of them are straightforward, others serve as additional theory or contain various complements. The exercises are not used in the text except for other exercises, but the reader is advised to attempt some exercises to get a better understanding of the theory. No mathematical knowledge is presupposed beyond the rudiments of abstract algebra, set theory, and topology; however, a certain maturity in mathematical reasoning is required.

The selection of material is unavoidably somewhat subjective. The main emphasis is on structural problems, and proper place is given to homological questions and to some topological considerations. A serious attempt has been made to unify methods, to simplify presentation, and to make the treatment as self-contained as possible. The author has tried to avoid making the discussion too abstract or too technical. With this view in mind, some significant results could not be treated here and maximum generality has not been achieved in those places where this would entail a loss of clarity or a lot of technicalities.

Volume I presents what is fundamental in abelian groups together with the homological aspects of the theory, while Volume II is devoted to the structure theory and to applications. Each volume has a Bibliography listing those works on abelian groups which are referred to in the text. The author has tried to give credit wherever it belongs. In some instances, however, especially in the exercises, it was nearly impossible to credit ideas to their original discoverers. At the end of each chapter, some comments are made on the topics of the chapter, and some further results and generalizations (also to modules) are mentioned which a reader may wish to pursue. Also, research problems are listed which the author thought interesting.

The system of cross-references is self-explanatory. The end of a proof is marked with the symbol □. Problems which, for some reason or other, seemed to be difficult are often marked by an asterisk, as are some sections which a beginning reader may find it wise to skip.

The author is indebted to a number of group theorists for comments and criticisms; sincere thanks are due to all of them. Special thanks go to B. Charles for his numerous helpful comments. The author would like to express his gratitude to the Mathematics Departments of University of Miami, Coral Gables, Florida, and Tulane University, New Orleans, Louisiana, for their assistance in the preparation of the manuscript, and to Academic Press, Inc., for the publication of this book in their prestigious series.

# CONTENTS

## III.   Direct Sums of Cyclic Groups

## IV.   Divisible Groups

## V.   Pure Subgroups

## VI.   Basic Subgroups

## VII.   Algebraically Compact Groups

# I

## PRELIMINARIES

The principal purpose of this introductory chapter is to acquaint the reader with the terminology and basic facts of abelian groups which will be used throughout the text. Some of the proofs will be omitted as they are standard and can be found in textbooks on algebra or on group theory. The fundamental types of groups, together with their main properties, are briefly discussed here. We shall save numerous repetitions by the adoption of their conventional notations. Maps, diagrams, categories, and functors are also presented; they will play an important role in our developments. Some of the most useful topologies in abelian groups will also be surveyed.

A reader not familiar with the subject treated here is advised to read this chapter most carefully.

## 1. DEFINITIONS

Abelian groups, like other algebraic systems, are defined on sets. In abelian group theory, however, certain set-theoretical features of the underlying sets seem to play a much more important role than in other parts of algebra. Therefore, we shall frequently have occasion to refer to cardinal and ordinal numbers, and to some results in set theory. In spite of this, we are not going to discuss the set-theoretical backgrounds of abelian groups. We accept the Gödel–Bernays axioms of set theory, including the *Axiom of Choice* which we use mainly in the equivalent form called *Zorn's lemma*.

Let $P$ be a partially ordered set, i.e., a set with a binary relation $\leq$ such that $a \leq a$; $a \leq b$ and $b \leq a$ imply $a = b$; $a \leq b$ and $b \leq c$ imply $a \leq c$, for all $a, b, c \in P$. A subset $C$ of $P$ is a chain, if $a, b \in C$ implies either $a \leq b$ or $b \leq a$. The element $u \in P$ is an upper bound for $C$, if $c \leq u$, for all $c \in C$, and $P$ is said to be *inductive*, if every chain in $P$ has an upper bound in $P$. A $v \in P$ is maximal in $P$, if $v \leq a$ with $a \in P$ implies $a = v$.

**Zorn's Lemma.** *If a partially ordered set is inductive, then it contains a maximal element.*

Whenever necessary, we assume the Continuum Hypothesis, too; this fact will always be stated explicitly.

*Class* and *set* will be used as customary in set theory. If we say *family* or *system*, then we do not exclude the repeated use of the same element. We adapt the conventional notations of set theory [see the table of notations, p. 281 ] except for writing $\alpha : a \mapsto b$ to mean that $\alpha$ is a function that maps the element $a$ of some class [set] $A$ upon the element $b$ of a class [set] $B$, while $\alpha : A \to B$ denotes that $\alpha$ is a function mapping the class [set] $A$ into $B$.

The word "group" will mean, throughout, an additively written abelian [i.e., commutative] group. That is, by *group* is meant a set $A$ of elements, such that with every pair $a, b \in A$ there is associated an element $a + b$ of $A$, called the *sum* of $a$ and $b$; there is an element $0 \in A$, the *zero*, such that $a + 0 = a$ for every $a \in A$; to each $a \in A$ there exists an $x \in A$ with $a + x = 0$, this $x = -a$ is the *inverse* of $a$; finally, both the associative and the commutative laws hold:

$$(a + b) + c = a + (b + c), \qquad a + b = b + a,$$

for all $a, b, c \in A$.

Note that a group is never empty, because it contains a zero, and that in an equality $a + b = c$, any two of $a, b, c$ uniquely determine the third one. The associative law enables us to write a sum of more than two summands without parentheses, and due to commutativity, the terms of a sum can be permuted. For the sake of brevity, one writes $a - b$ for $a + (-b)$; thus $-a - b$ is the inverse of $a + b$. The sum $a + \cdots + a$ [$n$ summands] is abbreviated as $na$, and $-a - \cdots - a$ [$n$ summands] as $(-n)a$ or $-na$. A sum without terms is 0; accordingly, $0a = 0$ for all $a \in A$ [notice that we do not distinguish in notation between the integer 0 and the group element 0]. An element $na$, with $n$ an integer, is called a *multiple* of $a$.

We shall use the same symbol for a group and for the set of its elements. The *order* of a group $A$ is the cardinal number $|A|$ of the set of its elements. If $|A|$ is a finite [countable] cardinal, $A$ is called a *finite* [*countable*] group.

A subset $B$ of $A$ is a *subgroup* if the elements of $B$ form a group under the same rule of addition. If $A$ is finite, by Lagrange's theorem, $|B|$ is a divisor of $|A|$. A subgroup of $A$ always contains the zero of $A$, and a nonempty subset $B$ of $A$ is a subgroup of $A$ if and only if $a, b \in B$ implies $a + b \in B$, and $a \in B$ implies $-a \in B$, or, more simply if and only if $a, b \in B$ implies $a - b \in B$. The *trivial* subgroups of $A$ are $A$ and the subgroup consisting of 0 alone; there being no danger of confusion, the latter subgroup will also be denoted by 0.

A subgroup of $A$, different from $A$, is a *proper* subgroup of $A$. We shall write $B \leq A$ [$B < A$] to indicate that $B$ is a subgroup [a proper subgroup] of $A$.

If $B \leq A$ and $a \in A$, the set $a + B = \{a + b \mid b \in B\}$ is called a *coset of $A$ modulo $B$*. Recall that

(i) $b \mapsto a + b$ is a one-to-one correspondence between $B$ and $a + B$;

(ii) $a_1, a_2 \in A$ belong to the same coset mod $B$ if and only if $a_1 - a_2 \in B$; one may write then $a_1 \equiv a_2$ mod $B$ and say: $a_1, a_2$ are *congruent* mod $B$;

(iii) two cosets are either identical or disjoint;

(iv) $A$ is the set-theoretical union of pairwise disjoint cosets of $A$ mod $B$.

An element of a coset is called a *representative* of this coset. A set consisting of just one representative from each coset mod $B$ is a *complete set of representatives* mod $B$. Its cardinality, i.e., the cardinal number of the set of different cosets mod $B$, is the *index* of $B$ in $A$, denoted as $|A : B|$. This may be finite or infinite; in the first case, $B$ is *of finite index* in $A$. If $A$ is a finite group, then $|A : B| = |A|/|B|$.

The cosets of $A$ mod $B$ form a group $A/B$ known as the *quotient* or *factor group* of $A$ mod $B$. In $A/B$, the sum of two elements $C_1, C_2$ [which are cosets of $A$ mod $B$] is defined to be the coset $C$ containing the set $\{c_1 + c_2 \mid c_1 \in C_1, c_2 \in C_2\}$; actually, this set is itself a coset and thus it is identical with $C$. The zero element of $A/B$ is $B$ [*qua* its own coset], and the inverse of a coset $C_0$ is the coset $-C_0 = \{-c \mid c \in C_0\}$. $A/B$ is a *proper* quotient group of $A$ if $B \neq 0$.

We shall frequently refer to the natural one-to-one correspondence between the subgroups of the quotient group $A^* = A/B$ and the subgroups of $A$ containing $B$. The elements of $A$ contained in elements [i.e., cosets of $A$] of some subgroup $C^*$ of $A^*$ form a subgroup $C$ such that $B \leq C \leq A$. On the other hand, if $B \leq C \leq A$, then the cosets of $A$ mod $B$ containing at least one element from $C$ form a subgroup $C^*$ of $A^*$. In this way, $C$ and $C^*$ correspond to each other, and we may write $C^* = C/B$. Notice that $|C^*| = |C : B|$, and $|A^* : C^*| = |A : C|$.

The set-theoretic intersection $B \cap C$ of two subgroups $B, C$ of $A$ is again a subgroup of $A$. More generally, if $B_i$ is a family of subgroups of $A$, then their intersection $B = \bigcap_i B_i$ is likewise a subgroup of $A$. We agree to put $B = A$ if $i$ ranges over the empty set.

If $S$ is a subset of $A$, the symbol $\langle S \rangle$ will denote the subgroup of $A$ *generated by* $S$, i.e., the intersection of all subgroups of $A$ containing $S$. If $S$ consists of the elements $a_i$ ($i \in I$), we also write

$$\langle S \rangle = \langle \cdots, a_i, \cdots \rangle_{i \in I},$$

or simply $\langle S \rangle = \langle a_i \rangle_{i \in I}$. This $\langle S \rangle$ consists of all sums of the form $n_1 a_1 + \cdots + n_k a_k$ [this is called a *linear combination* of $a_1, \cdots, a_k$] with $a_i \in S$, $n_i$ integers, and $k$ a nonnegative integer. If $S$ is empty, then $\langle S \rangle = 0$.

If $\langle S \rangle = A$, $S$ is said to be a *generating system* of $A$; the elements of $S$ are *generators* of $A$. A *finitely generated* group is one which has a finite generating system. Notice that $\langle S \rangle$ is of the same power as $S$ unless $S$ is finite, in which case $\langle S \rangle$ may be finite or countably infinite.

If $B$ and $C$ are subgroups of $A$, then the subgroup $\langle B, C \rangle$ they generate consists of all elements of $A$ of the form $b + c$ with $b \in B$, $c \in C$. We may write, therefore, $\langle B, C \rangle = B + C$. For a possibly infinite collection of subgroups $B_i$ of $A$, the subgroup $B$ they generate consists of all finite sums $b_{i_1} + \cdots + b_{i_k}$ with $b_{i_j}$ belonging to some $B_{i_j}$; we shall then write $B = \sum_{i \in I} B_i$.

The group $\langle a \rangle$ is the *cyclic* group generated by $a$. The order of $\langle a \rangle$ is also called the *order* of the element $a$, in notation: $o(a)$. The order $o(a)$ is thus either a positive integer or the symbol $\infty$. If $o(a) = \infty$, all the multiples $na$ of $a$ ($n = 0, \pm 1, \pm 2, \cdots$) are different and exhaust $\langle a \rangle$, while if $o(a) = m$, a positive integer, then $0, a, \cdots, (m - 1)a$ are the different elements of $\langle a \rangle$, and $ra = sa$ if and only if $m \mid r - s$.

If every element of $A$ is of finite order, $A$ is called a *torsion* or *periodic group*, while $A$ is *torsion-free* if all its elements, except for 0, are of infinite order. *Mixed* groups contain both nonzero elements of finite order and elements of infinite order. A *primary group* or *p-group* is defined to be a group the orders of whose elements are powers of a fixed prime $p$.

**Theorem 1.1.** *The set $T$ of all elements of finite order in a group $A$ is a subgroup of $A$. $T$ is a torsion group and the quotient group $A/T$ is torsion-free.*

Since $0 \in T$, $T$ is not empty. If $a, b \in T$, i.e., $ma = 0$ and $nb = 0$ for some positive integers $m$, $n$, then $mn(a - b) = 0$, and so $a - b \in T$, $T$ is a subgroup. To show $A/T$ torsion-free, let $a + T$ be a coset of finite order, i.e., $m(a + T) \subseteq T$ for some $m > 0$. Then $ma \in T$, and there exists $n > 0$ with $n(ma) = 0$. Thus, $a$ is of finite order, $a \in T$, and $a + T = T$ is the zero of $A/T$. $\square$

We shall call $T$ the *maximal torsion subgroup* or the *torsion part* of $A$, and shall denote it by $T(A)$. Note that if $B$ is a torsion subgroup of $A$, then $B \leq T$, and if $C \leq A$ such that $A/C$ is torsion-free, then $T \leq C$.

For a group $A$ and an integer $n > 0$, let

$$nA = \{na \mid a \in A\}$$

and

$$A[n] = \{a \mid a \in A, na = 0\}.$$

Thus $g \in nA$ if and only if the equation $nx = g$ has a solution $x$ in $A$, and $g \in A[n]$ if and only if $o(g) \mid n$. Clearly, $nA$ and $A[n]$ are subgroups of $A$.

If $a$ is an element of order $p^k$, $p$ a prime, we call $k$ the *exponent* of $a$, and write $e(a) = k$. Given $a \in A$, the greatest nonnegative integer $r$ for which $p^r x = a$ is solvable for some $x \in A$, is called the *p-height* $h_p(a)$ of $a$. If $p^r x = a$ is solvable whatever $r$ is, $a$ is *of infinite p-height*, $h_p(a) = \infty$. The zero is of infinite height at every prime. If it is completely clear from the context which prime $p$ is meant, we call $h_p(a)$ simply the *height* of $a$ and write $h(a)$.

The *socle* $S(A)$ of a group $A$ consists of all $a \in A$ such that $o(a)$ is a square-free integer. $S(A)$ is a subgroup of $A$; it is 0 if and only if $A$ is torsion-free, and it is equal to $A$ if and only if $A$ is an *elementary* group in the sense that every element has a square-free order. For a $p$-group $A$, we have $S(A) = A[p]$.

The set of all subgroups of a group $A$ is partially ordered under the inclusion relation. It is, moreover, a lattice where $B \cap C$ and $B + C$ are the lattice operations "inf" and "sup" for subgroups $B$, $C$ of $A$. This lattice $\mathbf{L}(A)$ has a maximum and a minimum element ($A$ and 0), and it satisfies the *modular law*: if $B$, $C$, $D$ are subgroups of $A$ such that $B \leq D$, then

$$B + (C \cap D) = (B + C) \cap D.$$

In fact, the inclusion $\leq$ being evident, we need only prove that every $d \in (B + C) \cap D$ belongs to the subgroup on the left member. Write $d = b + c$ with $b \in B$, $c \in C$; thus $d - b = c$ belongs to $D$ and $C$. Hence $c \in C \cap D$, and $d = b + c \in B + (C \cap D)$, indeed.

EXERCISES

1. Prove that a finite group $A$ contains an element of order $p$ if and only if $p$ divides the order of $A$.

2. If $B < A$ and $|B| < |A|$, then $|A/B| = |A|$, provided $|A|$ is infinite.

3. Let $B$, $C$ be subgroups of $A$ such that $C \leq B$ and $|B : C|$ is finite. Then, for every subset $S$ of $A$, $\langle S, C \rangle$ is of finite index in $\langle S, B \rangle$, and this index divides $|B : C|$.

4. (a) (W. R. Scott) Let $B_i$ ($i \in I$) be subgroups of $A$, and let $B$ denote their intersection. Then the index $|A : B|$ is not larger than the product of the $|A : B_i|$, $i \in I$.
   (b) The intersection of a finite number of subgroups of finite index is of finite index.

5. Let $B$, $C$ be subgroups of $A$.
   (a) For every $a \in A$, $a + B$ and $a + (B + C)$ meet the same cosets mod $C$.
   (b) A coset mod $B$ contains $|B : (B \cap C)|$ pairwise incongruent elements mod $C$.

6. (O. Ore) $A$ has a common system of representatives mod two of its subgroups, $B$ and $C$, if and only if
$$|B : (B \cap C)| = |C : (B \cap C)|.$$

[*Hint*: for necessity, use Ex. 5; for sufficiency, divide the cosets mod $B$ into blocks mod $B + C$ and make one-to-one correspondences within the blocks.]

7.* (N. H. McCoy)

(a) If $B$, $C$, $G$ are subgroups of $A$ such that $G$ is contained in the set-union $B \cup C$, then either $G \leq B$ or $G \leq C$. [*Hint*: if $b \in (B \cap G)\backslash C$, then $c \in C \cap G$ implies $b + c \in B \cap G$, $c \in B \cap G$.]

(b) The same does not hold for the set-union of three subgroups.

(c) If $G \leq A$ is contained in the set-union of the subgroups $B_1, \cdots, B_n$ of $A$, but not in the union of any $n-1$ of the $B_i$, then

$$mG \leq B_1 \cap \cdots \cap B_n \qquad \text{for some integer } m > 0.$$

[*Hint*: apply an argument like the one in (7.3) *infra*.]

8. Let $B \leq A$, and let $S$ be a subset of $A$ disjoint from $B$. There exists a subgroup $C$ of $A$ such that: (i) $B \leq C$; (ii) $C$ does not intersect $S$; (iii) $C < C' \leq A$ implies that $C'$ does intersect $S$.

9. Let $B$, $X$ be subgroups of $A$. There exists a subgroup $C$ of $A$ such that: (i) $B \leq C$; (ii) $B \cap X = C \cap X$; (iii) $C < C' < A$ implies $B \cap X < C' \cap X$.

10. (Honda [1]) If $B \leq A$ and $m$ is a positive integer, define

$$m^{-1}B = \{a \mid a \in A, ma \in B\}.$$

Prove that (a) $m^{-1}B$ is a subgroup; (b) $m^{-1}0 = A[m]$; (c) $m^{-1}(mB) = B + A[m]$; (d) $m(m^{-1}B) = B \cap mA$; (e) $m^{-1}(n^{-1}B) = (mn)^{-1}B$.

11. Prove the "triangle inequality" for the heights:

$$h_p(a + b) \geq \min(h_p(a), h_p(b)),$$

and equality holds if $h_p(a) \neq h_p(b)$.

12. If $A$ contains elements of infinite order, then the set of all elements of infinite order in $A$ generates $A$.

13. If $B \leq A$, then $T(B) = T(A) \cap B$, and $S(B) = S(A) \cap B$.

14. For every integer $n > 0$, $T(nA) = nT(A)$.

15. If $B$, $C$, $D$ are subgroups of $A$, then

$$(B \cap C) + (B \cap D) \leq B \cap (C + D)$$

and

$$B + (C \cap D) \leq (B + C) \cap (B + D).$$

Find examples where proper inclusions hold.

## 2.  MAPS AND DIAGRAMS

Let $A$ and $B$ be arbitrary groups. A *map*

$$\alpha : A \to B$$

[often denoted as $A \xrightarrow{\alpha} B$] is a function that associates with each element $a \in A$ a unique element $b \in B$, $\alpha : a \mapsto b$. This $b$ is the *image* of $a$ under $\alpha$, $b = \alpha(a)$, or simply $b = \alpha a$. $A$ is called the *domain* and $B$ the *range* or *codomain* of $\alpha$. A map $\alpha : A \to B$ is a *homomorphism* [of $A$ into $B$] if it preserves addition, that is,

$$\alpha(a_1 + a_2) = \alpha a_1 + \alpha a_2 \qquad \text{for all} \quad a_1, a_2 \in A.$$

If there is no need to name the homomorphism, we write simply $A \to B$.

Every homomorphism $\alpha : A \to B$ gives rise to two subgroups: $\text{Ker } \alpha \leqq A$ and $\text{Im } \alpha \leqq B$. The *kernel* of $\alpha$, $\text{Ker } \alpha$, is the set of all $a \in A$ with $\alpha a = 0$, while the *image* of $\alpha$, $\text{Im } \alpha$, consists of all $b \in B$ such that some $a \in A$ satisfies $\alpha a = b$. One may write $\alpha A$ for $\text{Im } \alpha$. If $\text{Im } \alpha = B$, $\alpha$ is called *surjective* or *epic*; we also say that $\alpha$ is an *epimorphism*. If $\text{Ker } \alpha = 0$, $\alpha$ is said to be *injective* or *monic*; also, $\alpha$ is a *monomorphism*. If both $\text{Im } \alpha = B$ and $\text{Ker } \alpha = 0$, then $\alpha$ is one-to-one between $A$ and $B$ [i.e., it is *bijective*]; in this case it is called an *isomorphism*. The groups $A$, $B$ are *isomorphic* [denoted as $A \cong B$] if there is an isomorphism $\alpha : A \to B$; then the inverse map $\alpha^{-1} : B \to A$ exists and is again an isomorphism. As customary in algebra, we make no distinction between isomorphic groups, unless they are distinct subgroups of the same larger group considered. If $G$ is a subgroup both of $A$ and $B$, and if $\alpha : A \to B$ fixes the elements of $G$, then $\alpha$ is called a *homomorphism over G*.

A homomorphism with 0 image is referred to as a *zero homomorphism*; it will be denoted by 0. If $A \leqq B$, then the map that assigns every $a \in A$ to itself may be regarded as a homomorphism of $A$ into $B$; it is called an *injection* [or *inclusion*] *map*. The injection $0 \to A$ is the unique homomorphism of 0 into $A$. If $\alpha : A \to B$ and $C \leqq A$, then the *restriction* $\alpha | C$ of $\alpha$ to $C$ has the domain $C$ and range $B$, and coincides with $\alpha$ on $C$.

Let $\alpha : A \to B$ and $\beta : B \to C$ be homomorphisms; here the range of $\alpha$ is the same as the domain of $\beta$. The composite map $A \to B \to C$, called the *product* of $\alpha$ and $\beta$ and denoted by $\beta \circ \alpha$ or simply by $\beta\alpha$ [notice the order of factors], is again a homomorphism. Recall that $\beta\alpha$ acts according to the rule

$$(\beta\alpha)a = \beta(\alpha a) \qquad \text{for all} \quad a \in A.$$

We have the associative law

$$\gamma(\beta\alpha) = (\gamma\beta)\alpha$$

whenever the products $\beta\alpha$ and $\gamma\beta$ are defined. It follows easily that $\alpha$ is *right-cancellable* [i.e., $\beta\alpha = \gamma\alpha$ always implies $\beta = \gamma$] exactly if $\alpha$ is an epimorphism, and *left-cancellable* [$\alpha\beta = \alpha\gamma$ always implies $\beta = \gamma$] if and only if it is a monomorphism. The product of two epimorphisms [monomorphisms] is again one.

A homomorphism of $A$ into itself is called an *endomorphism*, an isomorphism of $A$ with itself an *automorphism*. The identity automorphism $1_A$ of $A$ satisfies

$$1_A \alpha = \alpha \qquad \text{and} \qquad \beta 1_A = \beta$$

whenever the left-hand products are defined. A subgroup $B$ of $A$ that is carried into itself by every endomorphism [automorphism] of $A$ is said to be a *fully invariant* [*characteristic*] subgroup of $A$.

Let $\alpha : A \to B$ be an epimorphism, and let Ker $\alpha = K$. The complete inverse image $\alpha^{-1}b = \{a \mid a \in A, \alpha a = b\}$ of an element $b \in B$ is a coset $a + K$ in $A$. It follows that the map $a + K \mapsto \alpha a$ [being independent of the special choice of the representative $a$ of the coset] induced by $\alpha$ is an isomorphism between $A/K$ and $B$. In the same way, every homomorphism $\alpha : A \to B$ induces an isomorphism between $A/\text{Ker } \alpha$ and Im $\alpha$. The mapping $a \mapsto a + K$ is the *canonical* or *natural* epimorphism of $A$ onto $A/K$. If $C \leqq B \leqq A$, then $1_A$ induces the epimorphism $a + C \mapsto a + B$ of $A/C$ onto $A/B$.

If the homomorphisms $\alpha$, $\beta$ have the same domain $A$ and the same range $B$, then their *sum* $\alpha + \beta$ can be defined by the formula

$$(\alpha + \beta)a = \alpha a + \beta a \qquad \text{for every} \quad a \in A.$$

It is readily checked that $\alpha + \beta : A \to B$ is likewise a homomorphism, and one has

$$\alpha + \beta = \beta + \alpha, \qquad \alpha + 0 = \alpha,$$

$$(\alpha + \beta) + \gamma = \alpha + (\beta + \gamma),$$

$$(\alpha + \beta)\gamma = \alpha\gamma + \beta\gamma, \qquad \delta(\alpha + \beta) = \delta\alpha + \delta\beta,$$

whenever the sums and products are defined.

A *sequence* of groups $A_i$ and homomorphisms $\alpha_i$

$$A_0 \xrightarrow{\ \alpha_1\ } A_1 \xrightarrow{\ \alpha_2\ } \cdots \xrightarrow{\ \alpha_k\ } A_k \qquad (k \geqq 2)$$

is *exact* if

$$\text{Im } \alpha_i = \text{Ker } \alpha_{i+1} \qquad \text{for} \quad i = 1, \cdots, k - 1.$$

In particular, $0 \to A \xrightarrow{\ \alpha\ } B$ is exact if and only if $\alpha$ is monic, while $B \xrightarrow{\ \beta\ } C \to 0$ is exact if and only if $\beta$ is epic. The exactness of $0 \to A \xrightarrow{\ \alpha\ } B \to 0$ is equivalent to the fact that $\alpha$ is an isomorphism. We call an exact sequence of the form

$$0 \to A \xrightarrow{\ \alpha\ } B \xrightarrow{\ \beta\ } C \to 0$$

a *short exact sequence*; here $\alpha$ is an injection of $A$ into $B$ such that $\beta$ is an epimorphism with Im $\alpha$ as kernel. [Notice that in this case $A$ can be identified with the subgroup Im $\alpha$ of $B$, and $C$ with the quotient group $B/A$.]

Roughly speaking, a *diagram* of groups and homomorphisms consists of capital letters representing groups, and arrows between certain pairs of capital letters representing homomorphisms between the indicated groups. A diagram is *commutative* if we get the same composite homomorphisms whenever we follow directed arrows along different paths from one group to another group in the diagram. For instance, the diagram

$$
\begin{array}{ccc}
A & \xrightarrow{\mu} & B & \xrightarrow{\nu} & C \\
\downarrow{\alpha} & & \downarrow{\beta} & & \downarrow{\gamma} \\
A' & \xrightarrow{\mu'} & B' & \xrightarrow{\nu'} & C'
\end{array}
$$

is commutative exactly if the homomorphisms $\beta\mu$ and $\mu'\alpha$ of $A$ into $B'$ coincide, and the same holds for the homomorphisms $\gamma\nu$ and $\nu'\beta$ of $B$ into $C'$; then the equality of the homomorphisms $\gamma\nu\mu$, $\nu'\beta\mu$, $\nu'\mu'\alpha$ follows. In diagrams, the identity map will often be denoted by the sign of equality, as, e.g., in

$$
\begin{array}{ccc}
A & \xrightarrow{\alpha} & B \\
\| & & \downarrow{\beta} \\
A & \xrightarrow{\gamma} & C
\end{array}
$$

This diagram is essentially the same as

If this is commutative, we shall say that $\gamma$ *factors through* $B \to C$.

The following two lemmas are rather elementary.

**Lemma 2.1.** *A diagram*

(1)
$$
\begin{array}{c}
G \\
\downarrow{\eta} \\
0 \to A \xrightarrow{\alpha} B \xrightarrow{\beta} C
\end{array}
$$

*with exact row can be embedded in a commutative diagram*

(2)
$$
\begin{array}{c}
G \\
{}^{\phi}\swarrow \quad \downarrow{\eta} \\
0 \to A \xrightarrow{\alpha} B \xrightarrow{\beta} C
\end{array}
$$

*if and only if $\beta\eta = 0$. Moreover, $\phi : G \to A$ is unique.*

If such a $\phi$ exists, then $\eta = \alpha\phi$ implies $\beta\eta = \beta\alpha\phi = 0\phi = 0$. Thus the stated condition is necessary. Conversely, if $\beta\eta = 0$ in (1), then Im $\eta \leq$ Ker $\beta$. By the exactness of the row, Ker $\beta =$ Im $\alpha$, and $\alpha$ is a monomorphism; hence the map $\phi = \alpha^{-1}\eta$ of $G$ into $A$ is well defined. It is readily shown to be a homomorphism that makes (2) commutative. If $\phi' : G \to A$ does the same, then $\alpha\phi' = \eta = \alpha\phi$, whence $\phi' = \phi$, $\alpha$ being a monomorphism. $\square$

In order to save space, diagrams (1) and (2) will be replaced in the future by a single diagram

$$0 \to A \xrightarrow{\ \alpha\ } B \xrightarrow{\ \beta\ } C;$$

with $G$ above, $\phi$ mapping into $A$ and $\eta$ mapping into $B$.

thus dotted arrows will denote homomorphisms to be "filled in."

**Lemma 2.2.** *A diagram*

(3)
$$A \xrightarrow{\ \alpha\ } B \xrightarrow{\ \beta\ } C \to 0$$

with $\eta : B \to G$ and $\phi : C \to G$ below.

*with exact row can be filled in by a $\phi : C \to G$ so as to get a commutative diagram if, and only if, $\eta\alpha = 0$. Moreover, $\phi$ is unique.*

If such a $\phi$ exists, $\eta = \phi\beta$, then $\eta\alpha = \phi\beta\alpha = \phi0 = 0$, and the necessity is clear. Conversely, assume $\eta\alpha = 0$ and define $\phi : C \to G$ as follows, let $\phi c = \eta b$ if $b \in B$ satisfies $\beta b = c$. This is a good definition, for if $b' \in B$ also satisfies $\beta b' = c$, then $b' - b \in \operatorname{Ker} \beta = \operatorname{Im} \alpha \le \operatorname{Ker} \eta$, and so $\eta b' = \eta b$. It is readily seen that $\phi$ is a homomorphism making (3) commutative. Finally, if $\phi' : C \to G$ also satisfies $\phi'\beta = \eta$, then $\phi'\beta = \phi\beta$ and the epimorphic character of $\beta$ imply $\phi' = \phi$. $\square$

In the proof of the next lemma, we use the procedure of "chasing" elements around diagrams.

**Lemma 2.3** (the 5-lemma). *Let the diagram*

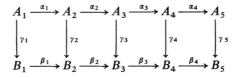

*be commutative with exact rows. Then*

    (a) *if $\gamma_1$ is epic and $\gamma_2$, $\gamma_4$ are monic, then $\gamma_3$ is monic;*
    (b) *if $\gamma_5$ is monic and $\gamma_2$, $\gamma_4$ are epic, then $\gamma_3$ is epic;*
    (c) *if $\gamma_1$ is epic, if $\gamma_5$ is monic, and if $\gamma_2$, $\gamma_4$ are isomorphisms, then $\gamma_3$ is an isomorphism.*

Assume the hypotheses of (a) and let $a_3 \in \operatorname{Ker} \gamma_3$. By the commutativity of the third square, $\gamma_4 \alpha_3 a_3 = \beta_3 \gamma_3 a_3 = 0$, whence $\alpha_3 a_3 = 0$ because $\gamma_4$ is monic. By the exactness of the top row, some $a_2 \in A_2$ satisfies $\alpha_2 a_2 = a_3$, and in view of the commutativity of the second square, $\beta_2 \gamma_2 a_2 = \gamma_3 \alpha_2 a_2 = \gamma_3 a_3 = 0$. The bottom row is exact, so $\beta_1 b_1 = \gamma_2 a_2$ for some $b_1 \in B_1$, and $\gamma_1$ being epic, $\gamma_1 a_1 = b_1$ for some $a_1 \in A_1$. Thus $\gamma_2 \alpha_1 a_1 = \beta_1 \gamma_1 a_1 = \beta_1 b_1 = \gamma_2 a_2$, and so $\alpha_1 a_1 = a_2$, for $\gamma_2$ is monic. This shows $a_3 = \alpha_2 a_2 = \alpha_2 \alpha_1 a_1 = 0$, i.e., $\gamma_3$ is a monomorphism.

Next start with the hypotheses of (b), and let $b_3 \in B_3$. The map $\gamma_4$ is epic, so $\beta_3 b_3 = \gamma_4 a_4$ for some $a_4 \in A_4$. Hence, $\gamma_5 \alpha_4 a_4 = \beta_4 \gamma_4 a_4 = \beta_4 \beta_3 b_3 = 0$, and so $\alpha_4 a_4 = 0$ because $\gamma_5$ is monic. The top row is exact, therefore $\alpha_3 a_3 = a_4$ for some $a_3 \in A_3$, whence $\beta_3 \gamma_3 a_3 = \gamma_4 \alpha_3 a_3 = \gamma_4 a_4 = \beta_3 b_3$. The exactness of the bottom row and the epic character of $\gamma_2$ imply $\beta_2 b_2 = \gamma_3 a_3 - b_3$ for some $b_2 \in B$, and $\gamma_2 a_2 = b_2$ for some $a_2 \in A_2$. Hence $\gamma_3 \alpha_2 a_2 = \beta_2 \gamma_2 a_2 = \gamma_3 a_3 - b_3$, $b_3 = \gamma_3(a_3 - \alpha_2 a_2) \in \operatorname{Im} \gamma_3$, and consequently, $\gamma_3$ is an epimorphism.

Combining (a) and (b), (c) follows readily.$\square$

Another noteworthy lemma of a somewhat different nature is the following, whose proof is again a routine element-chasing.

**Lemma 2.4** (the $3 \times 3$-lemma). *Assume that the diagram*

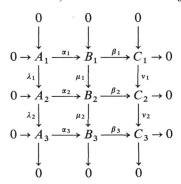

*is commutative and all three columns are exact. If the first two or the last two rows are exact, then the remaining row is exact.*

We prove only that the exactness of two first rows implies that the last is exact, while the proof of the other part [that runs dually] will be left to the reader.

Assume $a_3 \in \operatorname{Ker} \alpha_3$. Since $\lambda_2$ is epic, some $a_2 \in A_2$ satisfies $\lambda_2 a_2 = a_3$. From $\mu_2 \alpha_2 a_2 = \alpha_3 \lambda_2 a_2 = 0$, and the exactness of the middle column, follows the existence of a $b_1 \in B_1$ with $\mu_1 b_1 = \alpha_2 a_2$. From $\nu_1 \beta_1 b_1 = \beta_2 \mu_1 b_1 = \beta_2 \alpha_2 a_2 = 0$, we get $\beta_1 b_1 = 0$, since $\nu_1$ is monic, and so by the exactness of the first row, some $a_1 \in A_1$ satisfies $\alpha_1 a_1 = b_1$. Hence $\alpha_2 a_2 = \mu_1 b_1 = \mu_1 \alpha_1 a_1 = \alpha_2 \lambda_1 a_1$, and so, $\alpha_2$ being monic, $a_2 = \lambda_1 a_1$, whence $a_3 = \lambda_2 a_2 = \lambda_2 \lambda_1 a_1 = 0$, and $\alpha_3$ is a monomorphism.

Since $\beta_3 \alpha_3 \lambda_2 = \beta_3 \mu_2 \alpha_2 = \nu_2 \beta_2 \alpha_2 = 0$, and $\lambda_2$ is epic, we have $\beta_3 \alpha_3 = 0$. To show that $\operatorname{Ker} \beta_3 \leq \operatorname{Im} \alpha_3$, assume $b_3 \in \operatorname{Ker} \beta_3$. We know that some $b_2 \in B_2$ satisfies $\mu_2 b_2 = b_3$. Thus $\nu_2 \beta_2 b_2 = \beta_3 \mu_2 b_2 = 0$, and the exactness of the third column ensures that some $c_1 \in C_1$ satisfies $\nu_1 c_1 = \beta_2 b_2$. By the exactness of the first row, for some $b_1 \in B_1$, $\beta_1 b_1 = c_1$, whence $\beta_2 (b_2 - \mu_1 b_1) = \beta_2 b_2 - \nu_1 \beta_1 b_1 = 0$. From the second row, some $a_2 \in A_2$ satisfies $\alpha_2 a_2 = b_2 - \mu_1 b_1$, whence $\alpha_3 \lambda_2 a_2 = \mu_2 \alpha_2 a_2 = \mu_2 b_2 = b_3$, and $b_3 \in \operatorname{Im} \alpha_3$.

Finally,  $\operatorname{Im} \beta_3 \geq \operatorname{Im} \beta_3 \mu_2 = \operatorname{Im} \nu_2 \beta_2 = C_3$ shows $\beta_3$ is an epimorphism.$\square$

Let us recall the *isomorphism theorems* of E. Noether which are used often:

(i) If $B$ and $C$ are subgroups of $A$ such that $C \leq B$, then

$$A/B \cong (A/C)/(B/C),$$

where the natural isomorphism maps $a + B$ upon the coset containing $a + C$ $(a \in A)$.

(ii) If $B$ and $C$ are again subgroups of $A$, then

$$B/(B \cap C) \cong (B + C)/C,$$

the natural isomorphism being given by $\phi : b + (B \cap C) \mapsto b + C$ $(b \in B)$. [This $\phi$ makes the diagram

$$
\begin{array}{ccccccc}
0 \to B \cap C \to & B & \longrightarrow & B/(B \cap C) & \longrightarrow 0 \\
& \downarrow & \downarrow & \downarrow{\scriptstyle\phi} & \\
0 \longrightarrow C & \longrightarrow & B + C \to & (B + C)/C & \to 0
\end{array}
$$

commute, the first two vertical maps being inclusion maps.]

EXERCISES

1.  Let $\alpha : A \to B$ and $\beta : B \to C$. Prove that
    (a) $\operatorname{Ker} \beta\alpha \geq \operatorname{Ker} \alpha$, and equality holds if $\beta$ is a monomorphism;
    (b) $\operatorname{Im} \beta\alpha \leq \operatorname{Im} \beta$, and equality holds if $\alpha$ is an epimorphism.
2.  Let again $\alpha : A \to B$ and $\beta : B \to C$.
    (a) If $\beta\alpha$ is a monomorphism, then $\alpha$ is monic [but $\beta$ need not be a monomorphism].
    (b) If $\beta\alpha$ is an epimorphism, then $\beta$ is epic [but $\alpha$ is not necessarily].

3. For every homomorphism $\alpha : A \to B$ there is an exact sequence

$$0 \to \operatorname{Ker} \alpha \to A \xrightarrow{\ \alpha\ } B \to B/\operatorname{Im} \alpha \to 0.$$

4. If $\mu$ is the multiplication by the positive integer $m$, and if $\rho$ denotes the inclusion map, then the sequence

$$0 \to A[m] \xrightarrow{\ \rho\ } A \xrightarrow{\ \mu\ } mA \to 0$$

is exact.

5. Let $\alpha : A \to B$ and $B' \leqq B$.
   (a) If we write $\alpha^{-1}B' = \{a \mid a \in A, \alpha a \in B'\}$, then $\alpha(\alpha^{-1}B') \leqq B'$.
   (b) For $A' \leqq A$ we have $A' \leqq \alpha^{-1}(\alpha A')$, where equality does not hold in general.

6. If $0 \to A \xrightarrow{\ \alpha\ } B \xrightarrow{\ \beta\ } C \to 0$ is an exact sequence, and $B' \leqq B$, then there exist $A' \leqq A$ and $C' \leqq C$ such that the sequence $0 \to A' \xrightarrow{\ \alpha'\ } B' \xrightarrow{\ \beta'\ } C' \to 0$ is exact where $\alpha' = \alpha \mid A'$ and $\beta' = \beta \mid B'$.

7. In a diagram

$$
\begin{array}{ccc}
A & \xrightarrow{\ \alpha\ } & B \\
\downarrow{\scriptstyle \phi} & & \downarrow{\scriptstyle \eta} \\
0 \to C & \xrightarrow{\ \gamma\ } & D
\end{array}
$$

   with exact bottom row, the dotted arrow can be filled in to make the diagram commutative exactly if $\operatorname{Im} \eta\alpha \leqq \operatorname{Im} \gamma$. Moreover, $\phi$ is unique.

8. Formulate and prove the dual of Ex. 7 [all the arrows are reversed].

9. There is a homomorphism $A \to A'$ which makes the diagram

$$
\begin{array}{c}
0 \to A \to B \to C \to 0 \\
\downarrow \\
0 \to A' \to B' \to C' \to 0
\end{array}
$$

   commutative if and only if there is a homomorphism $C \to C'$ making it commutative, where both rows are assumed exact.

10. Let $\alpha, \beta$ be two homomorphisms $A \to B$, and assume the existence of a group $K$ and a homomorphism $\gamma : K \to A$ such that $\alpha\gamma = \beta\gamma$, and if $\gamma' : K' \to A$ satisfies $\alpha\gamma' = \beta\gamma'$, then there is a unique homomorphism $\phi : K' \to K$ with $\gamma' = \gamma\phi$. Prove that $\gamma$ is a monomorphism and $\operatorname{Im} \gamma = \operatorname{Ker}(\alpha - \beta)$.

11. Let $B$ and $C$ be subgroups of $A$ such that $A = B + C$, and $\beta : B \to X$, $\gamma : C \to X$ homomorphisms into the same group $X$. There is an $\alpha : A \to X$ with $\alpha \mid B = \beta$ and $\alpha \mid C = \gamma$ if and only if $\beta \mid B \cap C = \gamma \mid B \cap C$.

12. Prove the second part of (2.4).

13. For every positive integer $m$ and for every fully invariant subgroup $B$ of $A$, the subgroups $mB$ and $B[m]$ are fully invariant in $A$.

14.  A fully invariant subgroup of a fully invariant subgroup of $A$ is fully invariant in $A$.

15.  (a) If $B$ is a fully invariant subgroup of $A$, and $\eta$ is an endomorphism of $A$, then $a + B \mapsto \eta a + B$ is an endomorphism of $A/B$.
     (b) If $B$ is fully invariant in $A$, and $C/B$ in $A/B$, then $C$ is fully invariant in $A$.

16.  (a) If $B_i$ $(i \in I)$ are fully invariant [characteristic] subgroups of $A$, then so are $\bigcap B_i$ and $\sum B_i$.
     (b) Given a subset $S$ of $A$, there exists a unique minimal fully invariant [characteristic] subgroup of $A$ containing $S$. This is $\langle \cdots, \phi S, \cdots \rangle$ with $\phi$ running over all endomorphisms [automorphisms] of $A$.

17.  If $B$ is a fully invariant [characteristic] subgroup of $A$, and $S$ is a subset of $A$, such that $B \cap S = \varnothing$, then there exists a fully invariant [characteristic] subgroup $C$ of $A$ such that: (i) $B \leq C$, (ii) $C \cap S = \varnothing$, (iii) if $C'$ is fully invariant [characteristic] in $A$, and if $C < C'$, then $C' \cap S \neq \varnothing$.

18.  Let $\eta$ be an endomorphism of $A$ and $m$ an integer $>0$. Then $m^{-1} \operatorname{Ker} \eta = \operatorname{Ker} m\eta$.

## 3.   THE MOST IMPORTANT TYPES OF GROUPS

**Cyclic groups.** They were defined as groups that can be generated by a single element, i.e., they are of the form $\langle a \rangle$.

If $A = \langle a \rangle$ is an infinite cyclic group, then it is isomorphic to the additive group $Z$ of the rational integers $0, \pm 1, \pm 2, \cdots$, an isomorphism being given by the correspondence $na \mapsto n$. Thus all infinite cyclic groups are isomorphic; we denote them by the same symbol $Z$. Together with $a$, $-a$ is also a generator for $A$, but no other $na$ generates $A$.

A finite cyclic group $A = \langle a \rangle$ of order $m$ consists of the elements $0, a, 2a, \cdots, (m-1)a$. Because of $ma = 0$, we compute just as with the integers mod $m$; thus $A$ is isomorphic to the additive group $Z(m)$ of residue classes of the rational integers mod $m$. All finite cyclic groups of the same order $m$ are thus isomorphic; we shall use the notation $Z(m)$ for them.

Again, let $A = \langle a \rangle$ be cyclic of finite order $m$. Along with $a$, every $ka$ with $(k, m) = 1$ generates $A$. In fact, if $n > 0$ is an integer with $n(ka) = 0$, then $m \mid nk$ whence by hypothesis on $k$, we get $m \mid n$. This shows $o(ka) = m$, and thus $\langle ka \rangle = \langle a \rangle$. Conversely, if $ka$ generates $\langle a \rangle$, then $o(ka) = m$, and if we write $(k, m) = d$, then $md^{-1}ka = kd^{-1}ma = 0$ whence $o(ka) \leq md^{-1}$ and so $d = 1$. Therefore $\langle ka \rangle = \langle a \rangle$ if, and only if, $(k, m) = 1$. It follows that $Z(m)$ can be generated by a single element in $\phi(m)$ ways; here $\phi$ is Euler's function.

In connection with our notation for cyclic groups, it should always be kept in mind that neither $Z$ nor $Z(m)$ is an abstract group: while it is impossible to distinguish between the two generators of an abstract infinite cyclic group or between the $\phi(m)$ generators of an abstract cyclic group of order $m$, $Z$ and $Z(m)$ have distinguished generators, namely, 1 and the residue class of 1, respectively. [This will turn out to be of importance in certain natural isomorphisms.]

*Subgroups of cyclic groups are likewise cyclic.* In order to verify this, let $B$ be a nonzero subgroup of $\langle a \rangle$, and let $n$ be the smallest positive integer with $na \in B$. Then all the multiples of $na$ belong to $B$, and if $sa \in B$ with an integer $s = qn + r$ ($0 \leqq r < n$), then $ra = sa - q(na) \in B$ implies $r = 0$, showing that $B = \langle na \rangle$. If $a$ is of finite order $m$, then $n \mid m$. In fact, if $u, v$ are integers such that $mu + nv = (m, n)$, then $(m, n)a = mua + nva = v(na) \in B$, and so $n \leqq (m, n)$. For different divisors $n$ ($>0$) of $m$, the subgroups $\langle na \rangle$ are different; thus $Z(m)$ has as many subgroups as $m$ has divisors. Note that, of two subgroups of $Z(m)$, one contains the other if and only if the corresponding divisor of $m$ divides the other one. If $a$ is of infinite order, then so is $na$ ($n > 0$), and every nonzero subgroup of $Z$ is an infinite cyclic group. $\langle na \rangle$ is of index $n$ in $\langle a \rangle$, and it is the only subgroup of index $n$.

Let $A = \langle a \rangle$ and $B = \langle na \rangle$, with $n > 0$ a divisor of the order of $a$ if this is finite. Then the quotient group $A/B$ may be generated by the coset $a + B$ which is evidently of order $n$; thus $A/B \cong Z(n)$. Consequently, all proper quotient groups [epimorphic images] of a cyclic group are finite cyclic groups.

**Cocyclic groups.** A cyclic group can be characterized as a group $A$ containing an element $a$ such that any homomorphism $\phi : B \to A$ with $a \in \text{Im } \phi$ is epic. Dualizing this concept, we shall call a group $C$ *cocyclic* if there is an element $c \in C$ such that $\phi : C \to B$ and $c \notin \text{Ker } \phi$ imply that $\phi$ is monic. In this case, $c$ may be called a *cogenerator* of $C$. Since every subgroup is a kernel of a homomorphism, a cogenerator $c$ must belong to all nonzero subgroups of $C$. Hence the intersection of all nonzero subgroups of a cocyclic group $C$ is not zero; this is the smallest subgroup $\neq 0$ of $C$. Conversely, if a group has a smallest subgroup $\neq 0$, then the group is cocyclic, and any element $\neq 0$ in the smallest subgroup is a cogenerator.

A cyclic group $\langle a \rangle$ of prime power order $p^k$ is cocyclic where any element of order $p$ is a cogenerator. This follows from the simple fact that

$$0 < \langle p^{k-1}a \rangle < \cdots < \langle pa \rangle < \langle a \rangle$$

are the only subgroups of $\langle a \rangle$. Another type of cocyclic group was discovered by Prüfer [1].

Let $p$ denote a prime. The $p^n$th complex roots of unity, with $n$ running over all integers $\geqq 0$, form an infinite multiplicative group; in accordance with our convention, we switch to the additive notation. This group, called a *quasicyclic* group or a *group of type* $p^\infty$ [notation: $Z(p^\infty)$], can be defined as follows: it is generated by elements $c_1, c_2, \cdots, c_n, \cdots$, such that

(1) $$pc_1 = 0, \quad pc_2 = c_1, \cdots, pc_{n+1} = c_n, \cdots.$$

Here $o(c_n) = p^n$, and every element of $Z(p^\infty)$ is a multiple of some $c_n$.

In order to show $Z(p^\infty)$ cocyclic, let us choose a proper subgroup $B$ of $Z(p^\infty)$. There is a generator $c_{n+1}$ of a smallest possible index $n + 1$ which does not belong to $B$. We claim $B = \langle c_n \rangle$ (if $n = 0$, $B = 0$). Clearly $c_n \in B$. Furthermore, every $b \in B$ may be written in the form $b = kc_m$ for some $k$ and $m$, where $k$ may be assumed not to be divisible by $p$. If $r$, $s$ are integers such that $kr + p^m s = 1$, then $c_m = krc_m + p^m sc_m = rb \in B$, and thus $m \leq n$, $b \in \langle c_n \rangle$, establishing $B = \langle c_n \rangle$. Consequently all proper subgroups of $Z(p^\infty)$ are finite cyclic groups of order $p^n$ ($n = 0, 1, 2, \cdots$). These form a chain with respect to inclusion:

$$0 < \langle c_1 \rangle < \cdots < \langle c_n \rangle < \cdots,$$

since to a given $n$ there exists one and only one subgroup of order $p^n$, namely, that generated by $c_n$.

**Theorem 3.1.**  *A group $C$ is cocyclic if and only if $C \cong Z(p^k)$ with $k = 1, 2, \cdots$ or $\infty$.*

Let $c \in C$ denote a cogenerator of $C$. Then $\langle c \rangle$ is the smallest subgroup $\neq 0$ of $C$, and therefore $c$ has to be of prime order $p$. Since $c$ lies in every nonzero subgroup of $C$, $C$ contains neither elements of infinite order nor elements whose order is divisible by a prime $\neq p$, i.e., $C$ is a $p$-group. As a basis of induction, assume that $C$ contains at most one subgroup $C_n$ of order $p^n$ and this is cyclic, $C_n = \langle c_n \rangle$. If $A$, $B$ are subgroups of $C$ of order $p^{n+1}$, and if $a \in A \backslash C_n$, $b \in B \backslash C_n$, then $a$, $b \notin C_n$ implies $o(a) = p^{n+1} = o(b)$. Thus $pa = rc_n$, $pb = sc_n$ for suitable integers $r$, $s$ prime to $p$. If $r'$, $s'$ are such that $rr' \equiv 1 \equiv ss'$ mod $p^n$, then $r'$, $s'$ are prime to $p$, and $a' = r'a$, $b' = s'b$ satisfy: $\langle a' \rangle = \langle a \rangle$, $pa' = c_n$, $\langle b' \rangle = \langle b \rangle$, $pb' = c_n$. Hence $p(a' - b') = 0$, $a' - b' = tc_n$ for some integer $t$, that is, $a' = b' + tpb'$, $b' = a' - tpa'$, and $a'$, $b'$ generate the same cyclic group $\langle a' \rangle = \langle b' \rangle$. Consequently, $A = \langle a \rangle = \langle b \rangle = B$, and so $C$ is the union of a finite or infinite ascending chain of subgroups of orders $p^n$, i.e., $C$ is of the form $Z(p^k)$. □

Evidently, all the quasicyclic groups belonging to the same prime $p$ are isomorphic. Since the subgroups of $Z(p^\infty)$ are of type $Z(p^k)$, the quotient groups $\neq 0$ of $Z(p^\infty)$ are seen to be again $Z(p^\infty)$.

The group of all complex roots of unity, i.e., the group of all rotations of finite order of the circle, obviously contains $Z(p^\infty)$ for every prime $p$ as a subgroup. It has the remarkable property of being *locally cyclic* in the sense that all of its finitely generated subgroups are cyclic. It contains all finite cyclic groups as subgroups.

**Rational groups.**  Under addition the rational numbers form a group called the *full rational group*, denoted by $Q$. Like $Z(p^\infty)$, $Q$ can also be obtained as a

union of an infinite ascending chain of cyclic subgroups:
$$Z = \langle 1 \rangle < \langle 2!^{-1} \rangle < \cdots < \langle n!^{-1} \rangle < \cdots.$$
Thus $Q$ has a generating system $c_1, \cdots, c_n, \cdots$ satisfying

(2) $$2c_2 = c_1, \quad 3c_3 = c_2, \cdots, (n+1)c_{n+1} = c_n, \cdots.$$

It is easy to see that $Q$ is locally cyclic, too: every finite set of elements is contained in some $\langle n!^{-1} \rangle$; therefore, the subgroup they generate is a subgroup of a cyclic group, and so is itself cyclic. $Q$ contains numerous proper subgroups which are not finitely generated, as the group $Q_p$ of all rational numbers with denominators prime to $p$, or the group $Q^{(p)}$ of all rational numbers whose denominators are powers of $p$. The subgroups of $Q$, called *rational groups*, are of fundamental importance in the theory of torsion-free groups.

Every proper quotient group $Q/A$ of $Q$ (i.e., $A \neq 0$) is readily seen to be a torsion group, since every rational number has a multiple in $A$. In particular, $Q/Z$ is isomorphic to the group $C$ of all complex roots of unity, an isomorphism being induced by the epimorphism $r \mapsto e^{2i r \pi}$ [where $r \in Q$, $i = \sqrt{-1}$, and $e$ is the base of natural logarithms] of $Q$ onto $C$, whose kernel is $Z$. More generally, $Q/\langle r \rangle \cong Q/Z$ for every rational $r \neq 0$, while, e.g., $Q/Q_p \cong Z(p^\infty)$.

**p-adic integers.** The $p$-adic integers have many applications in various branches of abelian group theory. Let us sketch a method of introducing the $p$-adic integers.

Let $p$ be a prime and $Q_p$ the ring of rational numbers whose denominators are prime to $p$. The nonzero ideals of $Q_p$ are principal ideals generated by $p^k$ with $k = 0, 1, \cdots$ [i.e., it is a discrete valuation ring]. If the ideals $(p^k)$ are considered as a fundamental system of neighborhoods of 0, then $Q_p$ becomes a topological ring, and we may form the completion $Q_p^*$ of $Q_p$ in this topology [this completion process is described in detail for groups in **13**]. $Q_p^*$ is again a ring whose ideals are $(p^k)$ with $k = 0, 1, \cdots$, and which is complete [i.e., every Cauchy sequence in $Q_p^*$ is convergent] in the topology defined by its ideals.

The elements of $Q_p^*$ may be represented as follows: let $\{t_0, t_1, \cdots, t_{p-1}\}$ be a complete set of representatives of $Q_p$ mod $pQ_p$, e.g., $\{0, 1, \cdots, p-1\}$; then $\{p^k t_0, p^k t_1, \cdots, p^k t_{p-1}\}$ is one of $p^k Q_p$ mod $p^{k+1} Q_p$. Let $\pi \in Q_p^*$, and let $a_n \in Q_p$ be a sequence tending to $\pi$. Owing to the definition of Cauchy sequences, almost all $a_n$ [i.e., all with a finite number of exceptions] belong to the same coset mod $pQ_p$, e.g., to the one represented by $s_0$. Almost all differences $a_n - s_0$ belonging to $pQ_p$ belong to the same coset of $pQ_p$ mod $p^2 Q_p$, say, to that represented by $ps_1$. So proceeding, $\pi$ uniquely defines a sequence $s_0, s_1 p, s_2 p^2, \cdots$, and we assign to $\pi$ the formal infinite series $s_0 + s_1 p + s_2 p^2 + \cdots$. Its partial sums
$$b_n = s_0 + s_1 p + \cdots + s_n p^n \qquad (n = 1, 2, \cdots)$$

form a Cauchy sequence in $Q_p$ which converges in $Q_p^*$ to $\pi$, in view of $\pi - b_n \in p^k Q_p^*$ (for $n \geq k$). From the uniqueness of limits, it follows that, in this way, different elements of $Q_p^*$ are associated with different series, and since every series $s_0 + s_1 p + s_2 p^2 + \cdots$ with coefficients in a fixed system of representatives defines an element of $Q_p^*$, we may identify the elements $\pi$ of $Q_p^*$ with the formal series $s_0 + s_1 p + s_2 p^2 + \cdots$, with coefficients from $\{t_0, t_1, \cdots, t_{p-1}\}$, preferably from $\{0, 1, \cdots, p-1\}$, and write

$$(3) \qquad \pi = s_0 + s_1 p + \cdots + s_n p^n + \cdots \qquad (\text{with} \quad s_n = 0, 1, \cdots, p-1).$$

The arising ring $Q_p^*$ is a commutative domain [where *domain* is a ring without divisors of zero] called the ring of *p-adic integers*; its cardinality is the power of the continuum.

Notice that if $\rho = r_0 + r_1 p + \cdots + r_n p^n + \cdots$ ($r_n = 0, 1, \cdots, p-1$) is another *p*-adic integer, then the sum $\pi + \rho = q_0 + q_1 p + \cdots + q_n p^n + \cdots$ and the product $\pi \rho = q_1' + q_1' p + \cdots + q_n' p^n + \cdots$ are as follows: $q_0 = s_0 + r_0 - k_0 p$, $q_n = s_n + r_n + k_{n-1} - k_n p$, $q_0' = s_0 r_0 - m_0 p$, $q_n' = s_0 r_n + s_1 r_{n-1} + \cdots + s_n r_0 + m_{n-1} - m_n p$ ($n = 1, 2, \cdots$), where the integers $k_0, k_n, m_0, m_n$ are uniquely determined by the fact that all of $q_n$ and $q_n'$ are between 0 and $p-1$. As to subtraction and division, note that the negative of $\pi = s_n p^n + s_{n+1} p^{n+1} + \cdots$ ($s_n \neq 0$) is $-\pi = (p - s_n)p^n + (p - s_{n+1} - 1)p^{n+1} + \cdots$, and the inverse $\pi^{-1}$ of (3) exists if and only if $s_0 \neq 0$; it may be found by using the inverse rule to multiplication.

For the additive group of $Q_p^*$ we shall use the symbol $J_p$.

EXERCISES

1.  A simple (abelian) group is isomorphic to $Z(p)$ for some prime $p$.
2.  $A$ has a composition series if and only if $A$ is finite.
3.  (a) A subgroup $M$ of $A$ is called *maximal* if $M < A$ and $M \leq B < A$ implies $M = B$. Show that $M$ is maximal if and only if it is of prime index.
    (b) Prove that $Z(p^\infty)$, $Q$ have no maximal subgroups; $Z(p^k)$ ($k = 1, 2, \cdots$), $J_p$ have exactly one maximal subgroup; and $Z$ has infinitely many maximal subgroups.
4.  (a) The intersection of all maximal subgroups of $A$ of the same prime index $p$ is $pA$.
    (b) The *Frattini subgroup* of $A$ [i.e., the intersection of all maximal subgroups of $A$] is the intersection of all $pA$ with $p$ running over all primes $p$.
    (c) What are the Frattini subgroups of $Z(n)$, $Z$, $Z(p^\infty)$, $Q$, $Q_p$, $J_p$?
5.  Prove that neither $Z(p^\infty)$ nor $Q$ can be finitely generated.
6.  (a) Show that the group of all complex roots of unity is locally cyclic.

(b) Every subgroup and every quotient group of a locally cyclic group is again locally cyclic.

7. Prove the isomorphisms:

$$Q/Q_p \cong Z(p^\infty), \qquad Q^{(p)}/Z \cong Z(p^\infty), \qquad J_p/p^k J_p \cong Z(p^k)$$

for $k = 1, 2, \cdots$.

8. (L. Rédei) If a group $A$ contains subgroups isomorphic to any one of $Z(p^k)$ with a fixed prime $p$ and $k = 1, 2, \cdots$, but no proper subgroup of $A$ has this property, then $A \cong Z(p^\infty)$. [*Hint*: $pA = A$ and select generators.]

9. In a cyclic group $A$, two subgroups $B$ and $C$ coincide if $A/B \cong A/C$.

10. (a) Prove that a $p$-adic integer $\pi$ of the form (3) is a $p$-adic unit if and only if $s_0 \neq 0$.
    (b) The field of quotients of $Q_p^*$ consists of all elements of the form $\pi p^{-n}$ with $\pi \in Q_p^*$ and $n$ a nonnegative integer.

## 4. MODULES

Most of the theorems in abelian group theory can be generalized *mutatis mutandis* to unital modules over a principal ideal domain R with identity, and everything can be carried over—without any modification in the proofs—if R has the additional property that all quotient rings $R/(\alpha)$ with $0 \neq \alpha \in R$ are finite. It is, however, a delicate question to find the natural boundaries of a particular theorem in abelian groups, i.e., to describe the class of rings such that, for the modules over these rings, the theorem in question holds. A discussion of problems like this is beyond our present subject, and therefore we shall restrict ourselves to abelian groups only, i.e., modules over the ring Z of integers. Occasionally, however, we have to consider modules, since they yield a natural method of discussion. Therefore, let us recall the definition of modules.

Let R be an associative ring and $M$ an abelian group such that

(i) with $\alpha \in R$ and $a \in M$ there is associated an element of $M$, called the product of $\alpha$ and $a$, and denoted by $\alpha a$;

(ii) $(\alpha\beta)a = \alpha(\beta a)$ for all $\alpha, \beta \in R$ and $a \in M$;

(iii) $\alpha(a + b) = \alpha a + \alpha b$ for all $\alpha \in R$, $a, b \in M$;

(iv) $(\alpha + \beta)a = \alpha a + \beta a$ for all $\alpha, \beta \in R$, $a \in M$.

In this case, $M$ is said to be a *left* R-*module* or a *left module over* R. If R has a unit element $\varepsilon$, then it is in most cases assumed that $\varepsilon$ acts as the identity operator on $M$:

(v) $\varepsilon a = a$ for all $a \in M$.

Such R-modules $M$ are called *unital*. In our discussions, we shall only consider unital R-modules where R will always be commutative, in which case there is no need to distinguish between left and right R-modules. [In the *Notes*, modules are unital left modules.]

Recall that a *submodule* $N$ of an R-module $M$ is defined to be a subset of $M$ which is an R-module under the same operations, i.e., it is a subgroup of $M$ such that $\alpha N \subseteq N$ for all $\alpha \in R$. In this case, the quotient group $M/N$ becomes an R-module, the *quotient module*, where $\alpha(a + N) = \alpha a + N$ for all cosets $a + N$ and $\alpha \in R$.

If $M$, $N$ are R-modules, then a group homomorphism $\phi : M \to N$ is said to be an R-*homomorphism* if it satisfies

$$\phi(\alpha a) = \alpha\phi(a) \qquad \text{for all} \quad a \in M, \quad \alpha \in R.$$

The meaning of R-*isomorphism*, etc., is obvious.

For $a$ in an R-module $M$, the order $o(a)$ is defined as the set of all annihilators of $a$ in R:

$$o(a) = \{\alpha \in R \mid \alpha a = 0\}.$$

Thus $o(a)$ is a left ideal of R. The case $o(a) = (0)$ corresponds to elements of infinite order in groups.

*Example 1.* If R is the ring Z of integers, then every abelian group $A$ can be regarded as a Z-module under the natural definition of multiplication of $n \in Z$ and $a \in A$, namely, $na$ is the $n$th multiple of $a$.

*Example 2.* If R is the ring $Q_p$ of rationals with denominators prime to $p$, then a $p$-group $A$ can be made into a $Q_p$-module in a natural way. Namely, if $(n, p) = 1$, then for every $a \in A$, the product $n^{-1}a$ is a uniquely determined element of $\langle a \rangle$. Indeed, if $r$, $s$ are integers such that $nr + o(a)s = 1$, then $a = nra + o(a)sa = nra$ shows that $n^{-1}a = ra$ [and it is easy to see that in $A$ no element $\neq ra$ gives $a$ on multiplication by $n$].

*Example 3.* In a similar way, we conclude that every $p$-group $A$ is a $Q_p^*$-module in the natural way: if $\pi = s_0 + s_1 p + \cdots + s_n p^n + \cdots \in Q_p^*$, and if $a \in A$ is of order $p^n$, then

$$\pi a = (s_0 + s_1 p + \cdots + s_{n-1}p^{n-1})a,$$

where the element on the right does not change if we use a larger partial sum of $\pi$.

Let us notice that in all of our examples, submodules, R-homomorphisms are simply subgroups, group homomorphisms.

Modules over $Q_p^*$ are also called *p-adic modules*.

EXERCISES

1.  The cyclic R-module generated by $a$ is R-isomorphic to the R-module $R/o(a)$.

2.  If R is a principal ideal domain, if $M$ is an R-module, and $a, b \in M$ such that $o(a) = (\alpha) \neq 0$ and $o(b) = (\beta) \neq 0$, then $o(a + b)$ is a divisor of $\alpha\beta(\alpha, \beta)^{-1}$ and a multiple of $\alpha\beta(\alpha, \beta)^{-2}$.

3.  Let R be a commutative domain and $M$ an R-module. Then the elements $a \in M$ such that $o(a) \neq 0$ form a submodule $N$ of $M$ such that in the quotient module $M/N$ all the elements $\neq 0$ have order 0.

4.  Let $N$ be a submodule of an R-module $M$. Prove that $o(a) \subseteq o(a + N)$ where $o(a + N)$ denotes the order of the coset $a + N$ in $M/N$.

5.  If $R = Z/(m)$, $m$ an integer $> 0$, then for every element $a$ in an R-module we have $ma = 0$.

6.  If $\phi : R' \to R$ is a ring-homomorphism carrying the identity of R' into that of R, then every R-module $M$ becomes an R'-module by putting $\alpha'a = \phi(\alpha')a$ for all $\alpha' \in R'$, $a \in M$.

7.  A Q-module is—as a group—torsion-free.

## 5.  CATEGORIES OF ABELIAN GROUPS

In the theory of abelian groups, it is often convenient to express situations in terms of categories. In fact, categories and functors seem to be proper unifying concepts in a number of cases. Therefore, let us introduce categories and exhibit some important concepts connected with them.

Categories are not algebraic systems in the usual sense of the word, i.e., they are not necessarily sets equipped with algebraic operations. Categories need not even be sets, they are merely "classes." It is necessary to get rid of the assumption of being sets, since, e.g., we often consider all abelian groups which do not form a set. [Therefore, it is apt to use the Gödel–Bernays axioms of set theory, where both sets and classes are admitted. However, if one wishes to avoid the use of classes, then he may restrict himself to abelian groups belonging to some "universe" in the sense of Grothendieck.]

A *category* $\mathscr{C}$ is a class of *objects* $A, B, C, \cdots$, and *morphisms* $\alpha, \beta, \gamma, \cdots$ satisfying the following axioms:

*1.* With each ordered pair $A, B$ of objects in $\mathscr{C}$ there is associated a set Map$(A, B)$ of morphisms in $\mathscr{C}$ such that every morphism in $\mathscr{C}$ belongs to exactly one Map $(A, B)$. If $\alpha \in \mathscr{C}$ belongs to Map$(A, B)$ then we write $\alpha : A \to B$ and may call $\alpha$ a *map* of $A$ into $B$, while $A$ is the *domain*, $B$ the *range* of $\alpha$.

*2.* With $\alpha \in$ Map$(A, B)$ and $\beta \in$ Map$(B, C)$, there is associated a unique element of Map$(A, C)$, called their product $\beta\alpha$.

*3.* Whenever the products are defined, associativity prevails:

$$\gamma(\beta\alpha) = (\gamma\beta)\alpha.$$

*4.* For each $A \in \mathscr{C}$ there exists a morphism $1_A \in$ Map$(A, A)$, called the *identity morphism* of $A$, such that $1_A \alpha = \alpha$ and $\beta 1_A = \beta$ whenever the products make sense.

One verifies at once that $1_A$ is uniquely determined by the object $A$; indeed, if $\iota_A \in \text{Map}(A, A)$ has the same property, then $1_A \iota_A$ must be equal both to $1_A$ and to $\iota_A$. Calling a morphism $\iota \in \mathscr{C}$ an *identity* if $\iota\alpha = \alpha$ and $\beta\iota = \beta$ whenever the products are defined, we conclude that there is a one-to-one correspondence between the objects $A$ and the identities $1_A$ of $\mathscr{C}$, and, therefore, categories can also be defined in terms of morphisms only.

There are numerous examples for categories: the sets with mappings as morphisms, the not necessarily commutative groups or rings with homomorphisms as morphisms, R-modules with R-homomorphisms, topological spaces with continuous mappings, etc. For our present topic, the most important example is *the category $\mathscr{A}$ of all abelian groups* where the objects are the abelian groups, and the morphisms are the homomorphisms between them. Obviously, the torsion (abelian) groups, the *p*-groups, the torsion-free groups, *et al.* form categories if the morphisms are again the homomorphisms. [In general, if the objects of some category are groups, then it is implicitly understood, unless otherwise stated, that the morphisms of this category are just the homomorphisms between the groups in the category.]

Just as we have homomorphisms between algebraic systems, correspondingly we have functors between categories. If $\mathscr{C}$ and $\mathscr{D}$ are categories, then a *covariant functor*

$$F : \mathscr{C} \to \mathscr{D}$$

[on $\mathscr{C}$ to $\mathscr{D}$] assigns to each object $A \in \mathscr{C}$ an object $F(A) \in \mathscr{D}$, and to each morphism $\alpha : A \to B$ in $\mathscr{C}$ a morphism $F(\alpha) : F(A) \to F(B)$ in $\mathscr{D}$ satisfying the following conditions:

(i) if the product $\beta\alpha$ of $\alpha, \beta \in \mathscr{C}$ is defined, then $[F(\beta)F(\alpha)$ is defined in $\mathscr{D}$ and]

$$F(\beta\alpha) = F(\beta)F(\alpha);$$

(ii) $F$ carries the identity of $A \in \mathscr{C}$ into that of $F(A) \in \mathscr{D}$, i.e., for all $A \in \mathscr{C}$,

$$F(1_A) = 1_{F(A)}.$$

Thus a covariant functor preserves domains, ranges, products, and identities. The *identity functor E*, defined by $E(A) = A$, $E(\alpha) = \alpha$ for all $A, \alpha \in \mathscr{C}$, is a covariant functor on $\mathscr{C}$ to itself.

A *contravariant functor* $G : \mathscr{C} \to \mathscr{D}$ is defined similarly by reversing arrows, i.e., $G$ assigns an object $G(A) \in \mathscr{D}$ to every object $A \in \mathscr{C}$, and a morphism $G(\alpha) : G(B) \to G(A)$ in $\mathscr{D}$ to every morphism $\alpha : A \to B$ in $\mathscr{C}$, and it is subject to the conditions

$$G(\beta\alpha) = G(\alpha)G(\beta), \qquad G(1_A) = 1_{G(A)}.$$

The unqualified term "functor" will usually mean covariant functor.

If $F$ is a functor on $\mathscr{C}$ to $\mathscr{D}$, and $G$ is a functor on $\mathscr{D}$ to a category $\mathscr{E}$, then the composite $GF$ is a functor on $\mathscr{C}$ to $\mathscr{E}$ [where $GF(A) = G(F(A))$, and $GF(\alpha) = G(F(\alpha))$, for all $A$, $\alpha \in \mathscr{C}$]. Clearly, $GF$ is covariant if $F$, $G$ are both co- or both contravariant, and is contravariant if one of $F$, $G$ is co-, while the other is contravariant.

We shall have to consider functors in several variables, covariant in some of their variables and contravariant in others. For instance, if $\mathscr{C}$, $\mathscr{D}$, $\mathscr{E}$ are categories, then a *bifunctor* $F$ on $\mathscr{C} \times \mathscr{D}$ to $\mathscr{E}$, covariant in $\mathscr{C}$ and contravariant in $\mathscr{D}$, assigns to each couple $(C, D)$ with $C \in \mathscr{C}$, $D \in \mathscr{D}$ an object $F(C, D) \in \mathscr{E}$ and to each pair $\alpha : A \to C$, $\beta : B \to D$ of morphisms ($\alpha \in \mathscr{C}$, $\beta \in \mathscr{D}$) a morphism $F(\alpha, \beta) : F(A, D) \to F(C, B)$ such that

(1) $\qquad F(\gamma\alpha, \delta\beta) = F(\gamma, \beta) F(\alpha, \delta) \qquad$ and $\qquad F(1_C, 1_D) = 1_{F(C,D)}$

whenever $\gamma\alpha$, $\delta\beta$ are defined. Letting $D \in \mathscr{D}$ be fixed, $C \mapsto F(C, D)$ and $\alpha \mapsto F(\alpha, 1_D)$ give rise to a covariant functor on $\mathscr{C}$ to $\mathscr{E}$, while for a fixed $C \in \mathscr{C}$, $D \mapsto F(C, D)$, $\beta \mapsto F(1_C, \beta)$ yield a contravariant functor on $\mathscr{D}$ to $\mathscr{E}$. Notice that (1) implies that the diagram

$$
\begin{array}{ccc}
F(A, D) & \xrightarrow{F(\alpha, 1_D)} & F(C, D) \\
{\scriptstyle F(1_A, \beta)}\downarrow & & \downarrow{\scriptstyle F(1_C, \beta)} \\
F(A, B) & \xrightarrow{F(\alpha, 1_B)} & F(C, B)
\end{array}
$$

is commutative.

Examples for functors are abundant. The most important ones in abelian groups are those which assign to a group a subgroup or a quotient group [they are discussed in the next section], and the functors Hom, Ext, $\otimes$, and Tor [defined in Chapters VIII–X]. The following example is of a different type.

Let $\mathscr{T}$ be the category whose objects are the sequences $[A]: A_1 \xrightarrow{\alpha_1} A_2 \xrightarrow{\alpha_2} A_3$ of groups and homomorphisms subject to the condition $\alpha_2 \alpha_1 = 0$, and whose morphisms are triples $[\gamma_1, \gamma_2, \gamma_3]$ of group homomorphisms $\gamma_i : A_i \to B_i$ making the diagram

$$
\begin{array}{ccc}
[A]: A_1 & \xrightarrow{\alpha_1} A_2 \xrightarrow{\alpha_2} & A_3 \\
{\scriptstyle \gamma_1}\downarrow \quad {\scriptstyle \gamma_2}\downarrow \quad {\scriptstyle \gamma_3}\downarrow & & ([A], [B] \in \mathscr{T}) \\
[B]: B_1 & \xrightarrow{\beta_1} B_2 \xrightarrow{\beta_2} & B_3
\end{array}
$$

commute. It is straightforward to check that $\mathscr{T}$ is a category.

We define the *homology functor* $H$ on $\mathscr{T}$ to $\mathscr{A}$ as follows. For $[A] \in \mathscr{T}$, let $H[A] = \operatorname{Ker} \alpha_2 / \operatorname{Im} \alpha_1$ and let $H[\gamma_1, \gamma_2, \gamma_3] : H[A] \to H[B]$ be the homomorphism

$$
\phi : a + \operatorname{Im} \alpha_1 \mapsto \gamma_2 a + \operatorname{Im} \beta_1 \qquad (a \in \operatorname{Ker} \alpha_2).
$$

It is evident that: (1) $a \in \operatorname{Ker} \alpha_2$ implies $\gamma_2 a \in \operatorname{Ker} \beta_2$; (2) $a$, $a' \in \operatorname{Ker} \alpha_2$ and $a - a' \in \operatorname{Im} \alpha_1$ imply $\gamma_2 a - \gamma_2 a' \in \operatorname{Im} \beta_1$; (3) $\phi$ preserves addition. That $H$ satisfies the covariant functor conditions (i) and (ii) is straightforward to check.

One of the basic questions concerning functors is to find out how they behave for subgroups and quotient groups. This can be investigated conveniently in terms of exact sequences. If $F$ is a covariant functor on $\mathscr{A}$ to $\mathscr{A}$ [or subcategories of $\mathscr{A}$], and if $0 \to A \xrightarrow{\alpha} B \xrightarrow{\beta} C \to 0$ is an exact sequence, then $F$ is called *left* or *right exact* according as

$$0 \to F(A) \xrightarrow{F(\alpha)} F(B) \xrightarrow{F(\beta)} F(C) \quad \text{or} \quad F(A) \xrightarrow{F(\alpha)} F(B) \xrightarrow{F(\beta)} F(C) \to 0$$

is exact; if $F$ is both left and right exact, it is called *exact*. For a contravariant $F$, the displayed sequences are replaced by

$$0 \to F(C) \xrightarrow{F(\beta)} F(B) \xrightarrow{F(\alpha)} F(A) \quad \text{and} \quad F(C) \xrightarrow{F(\beta)} F(B) \xrightarrow{F(\alpha)} F(A) \to 0,$$

respectively. The subfunctors of the identity [see next section] are always left exact, while quotient functors are right exact.

Let $F$ and $G$ be covariant functors on $\mathscr{C}$ to $\mathscr{D}$. By a *natural transformation* $\Phi : F \to G$ is meant a function assigning to each object $A \in \mathscr{C}$ a morphism $\phi_A : F(A) \to G(A)$ in $\mathscr{D}$ in such a way that for all morphisms $\alpha : A \to B$ in $\mathscr{C}$ the diagram (in $\mathscr{D}$)

$$\begin{array}{ccc} F(A) & \xrightarrow{F(\alpha)} & F(B) \\ \phi_A \downarrow & & \downarrow \phi_B \\ G(A) & \xrightarrow{G(\alpha)} & G(B) \end{array}$$

commutes. In this case $\phi_A$ is called a *natural* morphism between $F(A)$ and $G(A)$. The natural character of certain homomorphisms and isomorphisms is of utmost importance. If $\phi_A$ is an isomorphism for every $A \in \mathscr{C}$, $\Phi$ is then called a *natural equivalence*.

In the theory of abelian groups, one encounters almost exclusively *additive functors*, i.e., functors $F$ satisfying

$$F(\alpha + \beta) = F(\alpha) + F(\beta)$$

for all $\alpha, \beta \in \mathscr{A}$ whenever $\alpha + \beta$ is defined. For an additive functor $F$ on $\mathscr{A}$ to $\mathscr{A}$ one obtains $F(0) = 0$, where 0 may stand for the zero group or zero homomorphism. Also, $F(n\alpha) = nF(\alpha)$ for every $n \in \mathbf{Z}$.

Functors on one category to a second category are studied extensively in homological algebra; we refer to Cartan and Eilenberg [1] and MacLane [3].

EXERCISES

1.  Prove that for any ring R, the left R-modules [as objects] and the R-homomorphisms [as morphisms] form a category.

2. Prove that the following is a category: the objects are commutative diagrams of the form

(2)
$$\begin{array}{ccc} A_1 & \to & A_2 \\ \downarrow & & \downarrow \\ A_3 & \to & A_4 \end{array}$$

with groups $A_i$, and the morphisms are quadruples $(\gamma_1, \gamma_2, \gamma_3, \gamma_4)$ of group homomorphisms making all squares arising between (2) and another object commutative.

3. A category with one object is essentially a semigroup [of morphisms] with unit element.

4. Call a category $\mathscr{C}'$ a *subcategory* of a category $\mathscr{C}$ if: (i) all the objects of $\mathscr{C}'$ are objects of $\mathscr{C}$; (ii) for $A, B \in \mathscr{C}$, $\text{Map}_{\mathscr{C}'}(A, B)$ is a subset of $\text{Map}_{\mathscr{C}}(A, B)$; (iii) the product of two morphisms in $\mathscr{C}'$ is the same as their composition in $\mathscr{C}$; (iv) for $A \in \mathscr{C}'$, $1_A$ is the same in $\mathscr{C}'$ as in $\mathscr{C}$. Prove that $I : A \mapsto A$, $\alpha \mapsto \alpha$ (for $A, \alpha \in \mathscr{C}'$) is a functor on $\mathscr{C}'$ to $\mathscr{C}$.

5. Let $\mathscr{C}_1$, $\mathscr{C}_2$ be two categories. The *product category* $\mathscr{C}_1 \times \mathscr{C}_2$ is defined to consist of the objects $(A_1, A_2)$ with $A_i \in \mathscr{C}_i$ and morphisms $(\alpha_1, \alpha_2) : (A_1, A_2) \to (B_1, B_2)$ with $\alpha_i \in \mathscr{C}_i$, where $(\beta_1, \beta_2)(\alpha_1, \alpha_2)$ is defined if and only if $\beta_1 \alpha_1$, $\beta_2 \alpha_2$ are defined, and then it is equal to $(\beta_1 \alpha_1, \beta_2 \alpha_2)$. Prove that this is actually a category.

6. Check that the homology functor as defined above is a functor.

7. The product of natural transformations is again a natural transformation.

## 6. FUNCTORIAL SUBGROUPS AND QUOTIENT GROUPS

Some of the most important functors in abelian group theory associate with a group $A$ a subgroup or a quotient group of $A$. Let us discuss briefly this kind of functor.

To begin, we mention a few examples of such functors. The functorial properties are straightforward to verify.

*Example 1.* Let $T : \mathscr{A} \to \mathscr{B}$ be a functor on the category $\mathscr{A}$ to the category $\mathscr{B}$ of all torsion groups such that, for $A \in \mathscr{A}$, $T(A)$ is the torsion part of $A$, and for $\alpha : A \to B$ in $\mathscr{A}$, $T(\alpha)$ is the restriction map $\alpha \,|\, T(A) : T(A) \to T(B)$.

*Example 2.* If we use the socle $S(A)$ of $A$ rather than its torsion part, then we get again a functor $S : \mathscr{A} \to \mathscr{B}$ [with $S(\alpha) = \alpha \,|\, S(A)$].

*Example 3.* For a positive integer $n$, let the functor $M_n : \mathscr{A} \to \mathscr{A}$ assign to $A$ its subgroup $nA$, and to $\alpha : A \to B$ the induced homomorphism $\alpha \,|\, nA : nA \to nB$.

*Example 4.* Let $\mathscr{A}_n$ denote the category of *n-bounded* groups, i.e., groups $G$ satisfying $nG = 0$. A functor $\mathscr{A} \to \mathscr{A}_n$ is obtained by assigning $A[n]$ to $A$ and $\alpha \,|\, A[n]$ to $\alpha : A \to B$.

*Example 5.* If $\mathscr{C}$ is the category of torsion-free groups, then the function assigning to $A \in \mathscr{A}$ the quotient group $A/T(A)$ and to $\alpha : A \to B$ in $\mathscr{A}$ the induced homomorphism $\alpha^* : a + T(A) \mapsto \alpha a + T(B)$ [which map is independent of the choice of $a$ in its coset mod $T(A)$] of $A/T(A)$ into $B/T(B)$ is a functor on $\mathscr{A}$ to $\mathscr{C}$.

*Example 6.* A functor $\mathscr{A} \to \mathscr{A}_n$ arises if we set $A \mapsto A/nA$ for all $A \in \mathscr{A}$ and $\alpha \mapsto \alpha^*$ for all $\alpha : A \to B$ in $\mathscr{A}$ where $\alpha^*$ denotes the induced homomorphism $a + nA \mapsto \alpha a + nB$.

In general, assume that we are given a function $F$ that assigns to every group $A \in \mathscr{A}$ a subgroup $F(A)$ of $A$,

$$F(A) \leqq A,$$

such that if $\alpha : A \to B$ is a homomorphism of $A$ into a group $B$, then $\alpha F(A) \leqq F(B)$, i.e., the restriction map $\alpha \,|\, F(A)$ sends $F(A)$ into $F(B)$. In this case, if we agree in putting

$$F(\alpha) = \alpha \,|\, F(A),$$

then $F$ is a functor $\mathscr{A} \to \mathscr{A}$. We shall call $F(A)$ a *functorial subgroup* of $A$. [Notice that a functorial subgroup arises always *via* a functor $F$ on $\mathscr{A}$ to $\mathscr{A}$ or to a subcategory of $\mathscr{A}$; thus it has to be defined for all $A \in \mathscr{A}$, even if in a particular case we restrict our attention to a single group $A$.] Our examples 1–4 show that $T(A)$, $S(A)$, $nA$, $A[n]$ are functorial subgroups of $A$.

Next assume that $F^*$ is a function which lets a quotient group $A/A^*$ of $A$ correspond to $A$, for every $A \in \mathscr{A}$, such that if $\alpha : A \to B$ is a homomorphism, then

$$a + A^* \mapsto \alpha a + B^*$$

is a homomorphism of $A/A^*$ into $B/B^*$. In this case, it is easy to verify the functorial properties of $F^* : \mathscr{A} \to \mathscr{A}$. We call $F^*(A) = A/A^*$ a *functorial quotient group* of $A$. Examples 5 and 6 show that $A/T(A)$ and $A/nA$ are functorial quotient groups of $A$.

There is a close connection between functorial subgroups and quotient groups:

**Theorem 6.1.** *$F(A)$ is a functorial subgroup of $A$ if and only if $A/F(A) = F^*(A)$ is a functorial quotient group of $A$.*

If $\alpha : A \to B$ is a homomorphism, then $\alpha^* : a + A^* \mapsto \alpha a + B^*$ is a homomorphism $A/A^* \to B/B^*$ exactly if $\alpha a \in B^*$ for every $a \in A^*$. This is equivalent to the condition $\alpha A^* \leqq B^*$ stated for functorial subgroups $F(A) = A^*$, $F(B) = B^*$.$\square$

Let us point out two rather general methods of manufacturing functorial subgroups and functorial quotient groups. In view of the preceding theorem, there is a natural one-to-one correspondence between the classes of functorial subgroups and quotient groups; therefore, we may confine our attention to functorial subgroups only.

Let $\mathfrak{X}$ be a class of groups $X$. With every $A \in \mathscr{A}$ we associate two subgroups, namely,

$$V_{\mathfrak{X}}(A) = \bigcap_{\phi} \operatorname{Ker} \phi \qquad \text{with } \phi : A \to X \in \mathfrak{X}$$

and

$$W_{\mathfrak{X}}(A) = \sum_{\psi} \operatorname{Im} \psi \qquad \text{with } \psi : X \to A \quad (X \in \mathfrak{X}).$$

Thus we let $\phi$ range over all homomorphisms of $A$ into groups in the class $\mathfrak{X}$, and $\psi$ over all homomorphisms of groups in $\mathfrak{X}$ into $A$.

**Proposition 6.2.** *For a fixed class $\mathfrak{X}$, both $V_{\mathfrak{X}}$ and $W_{\mathfrak{X}}$ are functors on $\mathscr{A}$ to $\mathscr{A}$.*

Let $\alpha : A \to B$ and $\phi : B \to X \in \mathfrak{X}$. Then $\phi\alpha$ is a homomorphism of $A$ into $X$, and evidently, $V_{\mathfrak{X}}(A) \leq \bigcap \operatorname{Ker} \phi\alpha$ with $\phi$ running over all $B \to X \in \mathfrak{X}$. It follows that $\alpha$ maps $V_{\mathfrak{X}}(A)$ into $\bigcap_{\phi} \operatorname{Ker} \phi = V_{\mathfrak{X}}(B)$, and so $V_{\mathfrak{X}}$ is a functor $\mathscr{A} \to \mathscr{A}$. In order to prove the same for $W_{\mathfrak{X}}$, let again $\alpha : A \to B$, and $\psi : X \to A$ for some $X \in \mathfrak{X}$. Then $\alpha\psi : X \to B$, and evidently $\sum_{\psi} \operatorname{Im} \alpha\psi \leq W_{\mathfrak{X}}(B)$. This shows that $W_{\mathfrak{X}}(A) = \sum_{\psi} \operatorname{Im} \psi$ is mapped by $\alpha$ into $W_{\mathfrak{X}}(B)$.$\square$

We illustrate our functors $V_{\mathfrak{X}}$, $W_{\mathfrak{X}}$ by the following examples.

(a) Let $\mathfrak{X}$ consist of all cyclic groups of prime order $p$. Then $V_{\mathfrak{X}}(A)$ is the Frattini subgroup of $A$ [see Ex. 4 in 3], while $W_{\mathfrak{X}}(A) = S(A)$, the socle of $A$.

(b) If $\mathfrak{X}$ consists of all finite cyclic groups, then $V_{\mathfrak{X}}(A)$ is the so-called *Ulm subgroup* of $A$, which we shall denote by $U(A)$ or by $A^1$. In this case $W_{\mathfrak{X}}(A)$ is nothing else than $T(A)$.

(c) Next, let $\mathfrak{X}$ contain one group only, namely $Z(m)$. Then $V_{\mathfrak{X}}(A) = mA$ [this will follow from (17.2)], while $W_{\mathfrak{X}}(A) = A[m]$.

(d) If $\mathfrak{X}$ is again a one-element class, $\mathfrak{X} = \{Q\}$, then it will result from theorems in Chapter IV that $V_{\mathfrak{X}}(A) = T(A)$, while $W_{\mathfrak{X}}(A) = D(A)$, the maximal divisible subgroup of $A$.

If $F_1$ and $F_2$ are functors $\mathscr{A} \to \mathscr{A}$ such that $F_1(A) \leq F_2(A) \leq A$ for every $A \in \mathscr{A}$, then we write

$$F_1 \leq F_2,$$

and call $F_1$ a *subfunctor* of $F_2$. This relation $\leq$ between functors of the given type defines a partial order in the class $\mathscr{F}$ of these functors. $\mathscr{F}$ has the maximum element $E$, the identity functor, $E(A) = A$, and the minimum element 0, the zero functor : $0(A) = 0$. For obvious reasons, we shall refer to the class $\mathscr{F}$ as the class of *subfunctors of the identity*. If $F \in \mathscr{F}$, then the functor $F^*$ [as defined in (6.1)] is called a *quotient functor of the identity*.

In $\mathscr{F}$, $\leq$ is actually a lattice-order. For if $F_1, F_2 \in \mathscr{F}$, then

$$A \mapsto F_1(A) \cap F_2(A) \qquad \text{and} \qquad A \mapsto F_1(A) + F_2(A)$$

give rise to subfunctors of the identity which are $\inf(F_1, F_2)$ and $\sup(F_1, F_2)$. Therefore, we may denote them by $F_1 \wedge F_2$ and $F_1 \vee F_2$, respectively.

Moreover, if $F_i$ $(i \in I)$ is any family of functors in $\mathscr{F}$, then

$$\left(\bigwedge_i F_i\right)(A) = \bigcap_i F_i(A) \qquad \text{and} \qquad \left(\bigvee_i F_i\right)(A) = \sum_i F_i(A)$$

define their inf and sup, as is readily verified.

In addition to lattice-operations, there is a natural way of introducing a multiplication in $\mathscr{F}$: for $F_1, F_2 \in \mathscr{F}$, we set $(F_1 F_2)(A) = F_1(F_2(A))$, i.e., $F_1 F_2$ is the product of the functors $F_1, F_2$ in the usual sense. Clearly, $F_1 F_2 \in \mathscr{F}$.

Given a subfunctor $F$ of the identity, we define transfinitely the iterated functors $F^\sigma$ for ordinals $\sigma$ as follows. Let $F^0 = E$, and let

$$F^{\sigma+1} = FF^\sigma.$$

If $\sigma$ is a limit ordinal, we define

$$F^\sigma = \bigwedge_{\rho < \sigma} F^\rho.$$

[Evidently, for every $\alpha : A \to B$, $F^\sigma(\alpha)$ is the restriction of $\alpha$ to $F^\sigma(A)$.] If, for instance, $F$ is the functor $V_x$ in (a), then $F^\omega$ is just $U$ as in (b), where $\omega$ stands for the first limit ordinal.

The Ulm subgroup $U(A)$ of $A$, introduced in (b), will be of great importance later on. An alternative definition is

$$U(A) = \bigcap_n nA.$$

The equivalence with the definition given in (b) follows from results in Chapter III. More precisely, $U(A)$ is called the *first Ulm subgroup* of $A$, while the $\sigma$th iterated functor $U^\sigma(A) = A^\sigma$ yields the $\sigma$th Ulm subgroup of $A$. The quotient group

$$U^\sigma(A)/U^{\sigma+1}(A) = U_\sigma(A) = A_\sigma \qquad (\sigma = 0, 1, 2, \cdots)$$

is said to be the *$\sigma$th Ulm factor* of $A$; in particular, $A/A^1 = A_0$ is the 0th Ulm factor of $A$. For a discussion of Ulm subgroups and factors, we refer to **37**.

Notice that if $F_1$, $F_2$ are subfunctors of the identity, and $F_1 \leq F_2$, then the inclusion maps $\phi_A : F_1(A) \to F_2(A)$ yield a natural transformation $\Phi : F_1 \to F_2$, and the corresponding maps $\phi_A^* : A/F_1(A) \to A/F_2(A)$ define a natural transformation $\Phi^* : F_1^* \to F_2^*$.

### EXERCISES

1.  Prove that for subfunctors $F$, $G$, $H$ of the identity, we have $(FG)H = F(GH)$, $FG \leq F \wedge G$, and for $F \leq H$, $F \vee (G \wedge H) = (F \vee G) \wedge H$.
2.  If $T$, $S$ are as in examples 1 and 2, then $T^2 = T$, $S^2 = S$, $TS = ST = S$.

3. For a subfunctor $F$ of the identity, $C \leq A$ implies

$$F(C) \leq C \cap F(A) \qquad \text{and} \qquad (F(A) + C)/C \leq F(A/C).$$

4. Let $\mathfrak{X}$ and $\mathfrak{Y}$ be two classes of groups.
   (a) If $\mathfrak{X} \subseteq \mathfrak{Y}$, then $V_{\mathfrak{Y}} \leq V_{\mathfrak{X}}$ and $W_{\mathfrak{X}} \leq W_{\mathfrak{Y}}$.
   (b) One has always

$$V_{\mathfrak{X} \cup \mathfrak{Y}} = V_{\mathfrak{X}} \wedge V_{\mathfrak{Y}} \qquad \text{and} \qquad W_{\mathfrak{X} \cup \mathfrak{Y}} = W_{\mathfrak{X}} \vee W_{\mathfrak{Y}}.$$

5. For a class $\mathfrak{X}$ of groups, let $\mathfrak{X}_s$ and $\mathfrak{X}_q$ denote the class of all subgroups and quotient groups, respectively, of groups in $\mathfrak{X}$. Then

$$V_{\mathfrak{X}_s} = V_{\mathfrak{X}}, \qquad V_{\mathfrak{X}_q} \leq V_{\mathfrak{X}}, \qquad W_{\mathfrak{X}_q} = W_{\mathfrak{X}}, \qquad W_{\mathfrak{X}_s} \geq W_{\mathfrak{X}}.$$

6. If $F$ is a subfunctor of the identity, and $\rho, \sigma$ are ordinals, then $F^\rho F^\sigma = F^{\sigma + \rho}$.
7. For every class $\mathfrak{X}$ and for every pair $\rho, \sigma$ of ordinals, with $\rho \leq \sigma$,

$$V_{\mathfrak{X}}^\rho(A/V_{\mathfrak{X}}^\sigma(A)) = V_{\mathfrak{X}}^\rho(A)/V_{\mathfrak{X}}^\sigma(A)$$

   and

$$W_{\mathfrak{X}}^\rho(A/W_{\mathfrak{X}}^\sigma(A)) = W_{\mathfrak{X}}^\rho(A)/W_{\mathfrak{X}}^\sigma(A)$$

   hold for every $A \in \mathscr{A}$.

8. Let $F_\sigma$ be subfunctors of the identity where $\sigma$ ranges over all ordinals less than an ordinal $\tau$. Define the infinite product $\cdots F_\sigma \cdots F_1 F_0$ $(\sigma < \tau)$, and show that this is likewise a subfunctor of the identity.

9.* By making use of the notion of direct sums [see 8] verify the formula

$$F\left(\bigoplus_i A_i\right) = \bigoplus_i F(A_i)$$

   for a subfunctor $F$ of the identity. Show that the analog fails to hold for direct products.


## 7. TOPOLOGIES IN GROUPS

In abelian groups, topology can be introduced in various ways which are natural in one sense or another. These topologies are not necessarily Hausdorff; some of them do not even make the group operations continuous. The importance of topologies in abelian groups will be evident from subsequent developments.

The most important topologies to be considered here are *linear topologies*; that is to say, there is a base [fundamental system] of neighborhoods about 0 which consists of subgroups such that the cosets of these subgroups form a base of open sets. In order to formulate our definition in a reasonably general fashion, let us start with a dual ideal (i.e., filter) $\mathbf{D}$ in the lattice $\mathbf{L}(A)$ of all subgroups of $A$, and declare the subgroups $U$ of $A$ in $\mathbf{D}$ as a base of open neighborhoods about 0. Then the cosets $a + U (U \in \mathbf{D})$ will be a base of open sets about $a$. Because the intersection of two cosets $(a_1 + U_1) \cap (a_2 + U_2)$ is vacuous or a coset mod $U_1 \cap U_2$, and $U_1 \cap U_2 \in \mathbf{D}$ whenever $U_1, U_2 \in \mathbf{D}$, all the open sets will be unions of cosets $a + U$ with $U \in \mathbf{D}$. The continuity of the group operations is obvious from the simple observation that $x - y \in a + U$ implies $(x + U) - (y + U) \subseteq a + U$. We call the arising topology the $\mathbf{D}$-*topology* of $A$. Obviously, *it is Hausdorff if and only if*

$$\bigcap_{U \in \mathbf{D}} U = 0,$$

and discrete exactly if $0 \in \mathbf{D}$.

Occasionally, one has a topology which satisfies the *first axiom of countability*, i.e., there is a countable base of open neighborhoods about 0. If $U_1, U_2, \cdots, U_n, \cdots$ is such a system of neighborhoods, then $U_1, U_1 \cap U_2,$ $\cdots, U_1 \cap \cdots \cap U_n, \cdots$ is also one; thus, in this case we may without loss of generality assume the sequence decreasing.

An open subgroup $B$ is necessarily closed. Indeed, its complement in the group is the union of its cosets $a + B (a \notin B)$; these are open and so is their union. Hence the subgroups $U \in \mathbf{D}$ are both open and closed. Therefore:

**Proposition 7.1.** *If $\mathbf{D}$ is a dual ideal in the lattice $\mathbf{L}(A)$ of all subgroups of $A$, and if the $\mathbf{D}$-topology of $A$ is Hausdorff, then it makes $A$ into a 0-dimensional topological group.* $\square$

The following special cases are of significance.

1. The *Z-adic topology* where the subgroups $nA$ $(n \in \mathbf{Z}, n \neq 0)$ form a base of neighborhoods about 0. This is a $\mathbf{D}$-topology where $\mathbf{D}$ consists of all $U \leq A$ such that $A/U$ is bounded. The $Z$-adic topology is Hausdorff exactly if

$$\bigcap_n nA = 0,$$

i.e., the first Ulm subgroup $U(A)$ of $A$ vanishes. This topology is discrete if and only if $nA = 0$ for some $n$, i.e., $A$ is bounded. It is easy to see that the closure $B^-$ of a subgroup $B$ is given by the formula

$$B^- = \bigcap_n (B + nA),$$

and $B$ is a closed subgroup exactly if the first Ulm subgroup of $A/B$ vanishes.

2. The *p-adic topology* arises if only the subgroups $p^k A$ $(k = 0, 1, 2, \cdots)$ are declared to belong to the base of neighborhoods at 0. [Now the dual ideal $\mathbf{D}$ is the set of all $U \leq A$ such that $A/U$ are bounded $p$-groups.] We have the following simple result.

**Theorem 7.2.** *The following conditions on a group $A$ are equivalent:*

(a) *the p-adic topology of $A$ is Hausdorff;*
(b) *$A$ contains no elements $\neq 0$ of infinite p-height;*
(c) *$\|a\| = \exp(-h_p(a))$ is a norm on $A$ ($a \in A$);*
(d) *$\delta(a, b) = \|a - b\|$ is a metric on $A$ ($a, b \in A$) that yields the p-adic topology.*

Since the $p$-adic topology of $A$ is Hausdorff if and only if $\bigcap_k p^k A = 0$, the equivalence of (a) and (b) is evident. (b) implies that $h_p(a) = \infty$ is equivalent to $a = 0$, while referring also to $h_p(a + b) \geq \min(h_p(a), h_p(b))$, we have at once: (i) $\|a\| \geq 0$ for every $a \in A$; (ii) $\|a\| = 0$ is equivalent to $a = 0$; (iii) $\|a + b\| \leq \max(\|a\|, \|b\|)$ for all $a, b \in A$. Thus $\|a\|$ is actually a norm on $A$. In order to see that (c) implies (d), notice that the norm properties imply that $\delta(a, b) = \|a - b\|$ is a metric on $A$, in which a base for open sets consists of spheres of radius $\varepsilon$ ($\varepsilon > 0$), i.e., $\{b \in A \mid \delta(a, b) < \exp(-k + 1)\} = a + p^k A$. Thus (d) follows from (c), while (a) is an obvious consequence of (d).□

In a $p$-group $A$, the $Z$-adic and $p$-adic topologies coincide. In fact, if $A$ is a $p$-group, and $(m, p) = 1$, then $mA = A$. This is a simple consequence of what has been shown in example 2 in **4**.

3. The *Prüfer topology* is defined in terms of the dual ideal **D** consisting of all $U \leq A$ such that $A/U$ satisfies the minimum condition [see **25**]. This is always a Hausdorff topology in which all subgroups are closed.

4. In the *finite index topology*, the subgroups $U$ of finite index of $A$ constitute a base of neighborhoods. This is a Hausdorff topology exactly if the first Ulm subgroup of $A$ vanishes. This is coarser than the $Z$-adic and the Prüfer topologies.

Another general method of making a group $A$ into a topological space is to define a set $\mathfrak{S}$ of subgroups $S$ of $A$ closed and to consider all the cosets $a + S$ ($a \in A$) as a subbase of closed sets in $A$. In this case, the maps

$$x \mapsto x + a, \qquad x \mapsto -x \qquad \text{(for every } a \in A\text{)}$$

are easily seen to be continuous and open. However, addition fails in general to be continuous simultaneously in the variables. Hence $A$ is [a so-called semi-topological, but] in general, not a topological group in this topology which we shall call *closed* $\mathfrak{S}$-*topology*. In order to characterize the cases in which $A$ is topological, we first prove a lemma:

**Lemma 7.3** (B. H. Neumann). *Let $S_1, \cdots, S_n$ be subgroups of $A$ such that $A$ is the set-theoretic union of finitely many cosets*

(1) $$A = (a_1 + S_1) \cup \cdots \cup (a_n + S_n) \qquad (a_i \in A).$$

*Then one of $S_1, \cdots, S_n$ is of finite index in $A$.*

Assume (1) irredundant, in the sense that none of the cosets $a_i + S_i$ is contained in the union of the others. If all the $S_i$ are equal, then they are of index $n$ in $A$. Assume that among $S_1, \cdots, S_n$ there are $k \geq 2$ different and that the assertion has been verified for the case of $k - 1$ different $S_i$. Let $S_1, \cdots, S_m$ be distinct from $S_{m+1} = \cdots = S_n$ ($m < n$). By irredundancy, some coset $x + S_n$ is not contained in $\bigcup_{i=m+1}^{n} (a_i + S_i)$, hence $x + S_n \subseteq \bigcup_{i=1}^{m} (a_i + S_i)$. But then

$$A = \bigcup_{i=1}^{m} (a_i + S_i) \cup \bigcup_{i=1}^{m} (a_{m+1} - x + a_i + S_i) \cup \cdots \cup \bigcup_{i=1}^{m} (a_n - x + a_i + S_i)$$

has only $k - 1$ distinct subgroups among $S_1, \cdots, S_m$, so the assertion follows by induction. $\square$

The following result shows that if a topology on $A$, defined in terms of a set $\mathfrak{S}$ of closed subgroups, makes $A$ into a topological group, then it is a linear topology defined by the closed subgroups of finite index in $\mathfrak{S}$ [these are open!]. In the following theorem, the topology is not necessarily Hausdorff.

**Proposition 7.4** (A. L. S. Corner). *Let $\mathscr{S}$ be a set of subgroups $S$ of the group $A$ and let $\mathfrak{S}_f$ consist of all $S \in \mathfrak{S}$ with finite index in $A$. Then $A$ is a topological group in the closed $\mathfrak{S}$-topology if and only if this is the closed $\mathfrak{S}_f$-topology [i.e., the linear topology defined by taking $S \in \mathfrak{S}_f$ open].*

We need only prove the "only if" part. Assume $A$ topological in the closed $\mathfrak{S}$-topology. Given an open set $V = A\backslash(a + T)$ ($a \in A, T \in \mathfrak{S}$) containing 0, there is an open neighborhood $U$ of 0, such that $U - U \subseteq V$. We may write

$$U = A \backslash \bigcup_{i=1}^{n} (a_i + S_i)$$

with $a_i \in A$, $S_i \in \mathfrak{S}$, since every open set is the union of open sets of this form. $0 \in U$ implies $0 \notin \bigcup_{i=1}^{n}(a_i + S_i)$. Since $U - U \subseteq V$ implies for every $u \in U$, $u - a \notin U$, therefore $u \in \bigcup_{i=1}^{n}(a + a_i + S_i)$, and

(2) $$A = \bigcup_{i=1}^{n} (a_i + S_i) \cup \bigcup_{i=1}^{n} (a + a_i + S_i).$$

Let $S_1, \cdots, S_m$ be of finite and $S_{m+1}, \cdots, S_n$ of infinite index in $A$; by (7.3), $1 \leq m \leq n$. Let $S = S_1 \cap \cdots \cap S_m$ which is again of finite index in $A$. We claim $S \subseteq V$. For otherwise, we choose $b \in S \cap (a + T)$, and for $i = 1, \cdots, m$, we have $a_i + S_i = b + a_i + S_i$, which is disjoint from $S$ because of $0 \in U$. Intersecting (2) with $S$, we find, replacing $a$ by $b$,

$$S = \bigcup_{i=m+1}^{n} [(a_i + S_i) \cap S] \cup \bigcup_{i=m+1}^{n} [(b + a_i + S_i) \cap S].$$

Here the nonempty intersections are cosets mod $S \cap S_i$, hence by (7.3) some $S \cap S_j$ with $m + 1 \leq j \leq n$ is of finite index in $S$. Then $S \cap S_j$ and hence $S_j$ is of finite index in $A$; this contradiction proves $S \subseteq V$. Thus all open sets $V$ of the form $A \backslash (a + T)$ $(T \in \mathfrak{S})$ contain a finite intersection $S_1 \cap \cdots \cap S_m$ with $S_i \in \mathfrak{S}_f$ whence the result follows.☐

In view of (7.4), we may disregard closed $\mathfrak{S}$-topologies if we adhere to the continuity of the group operation.

Following Charles [4], we introduce the concept of functorial topologies. Assume that, for every $A \in \mathscr{A}$, there is defined a topology $t(A)$ under which $A$ is a topological group. We call $t = \{t(A) \mid A \in \mathscr{A}\}$ a *functorial topology* if every homomorphism in $\mathscr{A}$ is continuous. In this sense, the $Z$-adic, the $p$-adic, the Prüfer, and the finite index topologies are all functorial.

A more general method of obtaining a functorial topology is to choose a class $\mathfrak{X}$ of groups and to take the subgroups Ker $\phi$ with $\phi : A \to X \in \mathfrak{X}$ as a subbase of open neighborhoods about 0 in $A$. This will yield a linear topology on $A$ which will be Hausdorff whenever there are sufficiently many homomorphisms $\phi$, in the sense that for every $a \in A$, $a \neq 0$, there are $X \in \mathfrak{X}$ and $\phi : A \to X$ with $\phi a \neq 0$.

For the orientation of the reader, we now present a result showing that all infinite abelian groups can be equipped with a nondiscrete Hausdorff topology. This result is of theoretical importance, but no use will be made of it.

**Theorem 7.5** (Kertész and Szele [1]). *Every infinite abelian group can be made into a nondiscrete Hausdorff topological group.*

In the proof, we need some simple results which we shall prove later on only.

Let $A$ be an infinite group. The Prüfer topology makes $A$ into a topological group. This topology is discrete exactly if $A$ itself satisfies the minimum condition. Then by (25.1) $A$ contains a subgroup of type $p^\infty$. The embedding of $Z(p^\infty)$ in the group of complex numbers $z$ with $|z| = 1$ induces a nondiscrete Hausdorff topology on $Z(p^\infty)$, and by translations one obtains a nondiscrete topology on $A$.☐

EXERCISES

1.  $A[n]$ is closed in any topological group $A$.
2.  (a) Prove that every homomorphism between groups is continuous in the $p$-adic, Prüfer, and finite index topologies.
    (b) Every epimorphism is an open mapping in the $Z$-adic, $p$-adic, Prüfer, and finite index topologies.
    (c) The $Z$-adic topology of a subgroup of $A$ is finer than the topology induced by the $Z$-adic topology of $A$.
3.  (a) Prove that, in a group $A$, the $Z$-adic and $p$-adic topologies coincide, if $qA = A$ for every prime $q \neq p$.
    (b) Let $p$, $q$ be different primes. For which groups are the $p$-adic and

$q$-adic topologies the same?

4.  $J_p$ is compact in its $p$-adic topology.

5.  Show that every linear topology is a **D**-topology for some dual ideal **D** in the lattice of subgroups.

6.  Let $B$ be a subgroup of the topological group $A$. Prove that:
    (a) if $B$ is closed, then the natural map $A \to A/B$ is an open, continuous homomorphism [recall that in $A/B$, the neighborhoods are the images of those in $A$];
    (b) $A/B$ is discrete exactly if $B$ is open;
    (c) if $\alpha : A \to C$ is an open and continuous epimorphism, then the topological isomorphism $A/\text{Ker } \alpha \cong C$ holds.

7.  A closed subgroup $B$ of $A$ is nowhere dense [i.e., $A \backslash B$ is dense] if and only if $B$ is not open.

8.  Let $A$ have a linear topology. A subgroup $B$ is closed if and only if $B$ is the intersection of open subgroups of $A$.

9.  If $B$ and $C$ are closed subgroups of $A$, which has a linear topology, then $B + C$ is not necessarily closed.

10. (a) In the **D**-topology, $B$ is a dense subgroup of $A$ if and only if $B + U = A$ for every $U \in$ **D**.
    (b)* A subgroup $B$ of $A$ is dense in the $Z$-adic (or in the finite index) topology if and only if $A/B$ is divisible [see **20**].

11.* A group can be furnished with a nondiscrete linear Hausdorff topology if and only if it does not satisfy the minimum condition.

12.* Every infinite group admits an invariant metric. [*Hint*: equivalently, there is a nondiscrete topology satisfying the first axiom of countability; if $a$ is of infinite order, take $\langle 2^n a \rangle$ as a base; if it has infinite socle, take an infinite descending chain of type $\omega$ with 0 intersection as a base; if it contains a $Z(p^\infty)$, argue as in (7.5).]

## NOTES

Commutative groups are called abelian after the Norwegian mathematician Niels Henrik Abel (1802–1829), who studied algebraic equations with commutative Galois groups. Actually, a finite, commutative grouplike structure was considered by C. F. Gauss in 1801, in connection with quadratic forms [he proved a decomposition like (8.4)], but it was only in the last decades of the 19th century when a more or less systematic study of finite abelian groups developed. As the initial restriction of finiteness has been removed from group theory, Levi started the investigations of infinite abelian groups in his *Habilitations-schrift* [1]. Both Levi and, a little later, Prüfer (in his epochal papers [1], [2], [3]), restricted themselves to countable groups, but most of the proofs did not really make use of countability. From the 1930's on, abelian groups have received a good deal of attention, especially the contributions made by R. Baer, L. Ya. Kulikov, and T. Szele are significant. In the late 1950's the homological aspects began to play a stimulating role in abelian group theory.

Theorem (1.1) is most elementary, but fundamental. In view of it, the structure theory of abelian groups splits into the theories of torsion and torsion-free groups, and investigations of how these are glued together to form mixed groups. It is hard to trace the history of (1.1). It should be noted that it does not generalize to arbitrary modules. If a "torsion" element $a$ of an R-module $M$ is defined by $o(a) \neq 0$, then it is not true in general that the torsion elements form a submodule. It is known, however, that a ring R has the property that, in every R-module $M$, the torsion elements form a submodule $T$ such that $M/T$ is torsion-free if and only if R is a left Ore domain [i.e., a domain which satisfies the left Ore condition: $L_1 \cap L_2 \neq 0$ for any two nonzero left ideals $L_1, L_2$ of R]. For another notion of "torsion" see A. Hattori [*Nagoya Math. J.* **17**, 147–158 (1960)]. S. E. Dickson [*Trans. Amer. Math. Soc.* **121**, 223–235 (1966)] gives a systematic treatment of what he calls a "torsion theory" in abelian categories. See also J. M. Maranda [*Trans. Amer. Math. Soc.* **110**, 98–135 (1964)] where torsion preradicals are discussed.

*Problem 1.* List the functorial subgroups in important subcategories of $\mathscr{A}$.

*Problem 2.* Describe the functorial linear topologies for various categories of abelian groups [elementary, bounded, torsion, etc.].

# II

## DIRECT SUMS

The concept of direct sum is of utmost importance in the theory of abelian groups. This is due to two facts: first, if a group decomposes into a direct sum, it can be studied by investigating the components in the direct sum, and these are, in several cases, of a simpler structure; secondly, new groups can be constructed as direct sums of known groups. We shall see that almost all the structure theorems on abelian groups involve, explicitly or implicitly, some direct decomposition.

There are two ways of introducing direct sums: the internal and external direct sums. Both will be discussed here, along with their basic properties. The external definition leads to unrestricted direct sums, called direct products, which, too, are extremely useful for us. We discuss pullback and pushout diagrams as well.

Also, we are going to define direct and inverse limits, which have begun to play an increasing role in the theory of abelian groups. We shall often have occasion to use them in various context.

In the final section of this chapter, completions under a linear topology are dealt with.

### 8. DIRECT SUMS AND DIRECT PRODUCTS

Let $B$, $C$ be subgroups of $A$, and assume that they satisfy

(i) $B + C = A$;
(ii) $B \cap C = 0$.

In this case we call $A$ the [*internal*] *direct sum* of its subgroups $B$, $C$, and write

$$A = B \oplus C.$$

Condition (i) states that every $a \in A$ may be written in the form $a = b + c$ ($b \in B$, $c \in C$), and (ii) amounts to the unicity of this form. For, if $a = b + c = b' + c'$ ($b' \in B$, $c' \in C$), then $b - b' = c' - c \in B \cap C = 0$; on the other hand, the uniqueness of the form $a = b + c$ excludes the case that $b + 0 = 0 + c$ is a nonzero element common to $B$ and $C$.

If (ii) is satisfied, then we say: the subgroups $B$ and $C$ are *disjoint*. This is not consistent with the set-theoretical use of this word, but there is no danger of confusion in this abuse.

Let $B_i$ $(i \in I)$ be a family of subgroups of $A$, subject to the following two conditions:

(i) $\sum B_i = A$ [i.e., the $B_i$ together generate $A$];
(ii) for every $i \in I$,

$$B_i \cap \sum_{j \neq i} B_j = 0.$$

Then $A$ is said to be the *direct sum* of its subgroups $B_i$, in sign:

$$A = \bigoplus_{i \in I} B_i$$

or

$$A = B_1 \oplus \cdots \oplus B_n$$

if $I = \{1, \cdots, n\}$. Again, every $a \in A$ can be written in a unique form $a = b_{i_1} + \cdots + b_{i_k}$ with $b_{i_j} \neq 0$ belonging to different *components* $B_{i_j}$ $(j = 1, \cdots, k$ where $k \geq 0)$. Since every element of $\sum B_i$ is contained in a subgroup generated by a finite number of the $B_i$, condition (ii) can be replaced by the apparently weaker postulate:

$$B_i \cap (B_{i_1} + \cdots + B_{i_k}) = 0,$$

where $i_j \neq i$, and $k$ is a natural integer.

Let $a \in A = B \oplus C$, and write $a = b + c$ with $b \in B$, $c \in C$. The maps

$$\pi : a \mapsto b \qquad \text{and} \qquad \theta : a \mapsto c$$

are immediately seen to be epimorphisms of $A$ onto $B$ and $C$, respectively. Since $\pi b = b$, $\pi c = 0$, $\theta c = c$, $\theta b = 0$, and $\pi a + \theta a = a$, the endomorphisms $\pi$, $\theta$ of $A$ satisfy:

(1) $\qquad \qquad \pi^2 = \pi, \qquad \theta^2 = \theta, \qquad \theta\pi = \pi\theta = 0, \qquad \pi + \theta = 1_A.$

If we mean by a *projection* an idempotent endomorphism and by *orthogonal* endomorphisms those with 0 products [in both orders], then (1) may be expressed by saying that *a direct decomposition $A = B \oplus C$ defines a pair of orthogonal projections with sum $1_A$*. Conversely, any pair $\pi$, $\theta$ of endomorphisms of $A$ satisfying (1) yields a direct decomposition

$$A = \pi A \oplus \theta A.$$

In fact, idempotency and orthogonality imply that an element common to $\pi A$ and $\theta A$ must be both reproduced and annihilated by $\pi$ and $\theta$, so $\pi A \cap \theta A = 0$, while $\pi + \theta = 1_A$ guarantees that $\pi A + \theta A = A$.

Even if $A$ is a direct sum of several subgroups, $A = \oplus B_i$, the decomposition can be described in terms of pairwise orthogonal projections. The $i$th projection $\pi_i : A \to B_i$ assigns to $a = b_{i_1} + \cdots + b_{i_k}$ the term $b_i$ from $B_i$ [which may very well be 0]. Then we have:

(a) $\pi_i \pi_j = 0$ or $\pi_i$, according as $i \neq j$ or $i = j$;

(b) for every $a \in A$, almost all of $\pi_i a$ are 0, and their sum equals $a$.

Conversely, if $\pi_i$ $(i \in I)$ is a set of endomorphisms of $A$ satisfying (a) and (b), then $A$ is the direct sum of the $\pi_i A$, as is readily verified.

A subgroup $B$ of $A$ is called a *direct summand* of $A$, if there is a $C \leq A$ such that $A = B \oplus C$. In this case, $C$ is a *complementary direct summand*, or simply a *complement* of $B$ in $A$.

Some of the most useful properties of direct sums are listed as follows:

(a) If $A = B \oplus C$, then $C \cong A/B$. Thus the complement of $B$ in $A$ is unique up to isomorphism.

(b) If $A = B \oplus C$, and if $G$ is a subgroup of $A$ containing $B$, then

$$G = B \oplus (G \cap C).$$

(c) If $a \in A = B \oplus C$, and if $a = b + c$ $(b \in B, c \in C)$, then $o(a)$ is the least common multiple of $o(b)$ and $o(c)$.

(d) If $A = \oplus_i B_i$, and if, for every $i$, $C_i \leq B_i$, then $\sum C_i = \oplus C_i$. This is a proper subgroup of $A$ if $C_i < B_i$ for at least one $i$.

(e) If $A = \oplus_i B_i$, where each $B_i$ is a direct sum, $B_i = \oplus_j B_{ij}$, then

$$A = \oplus_i \oplus_j B_{ij}.$$

The latter is called a *refinement* of the first decomposition of $A$.

(f) If $A = \oplus_i \oplus_j B_{ij}$, then $A = \oplus_i B_i$ with $B_i = \oplus_j B_{ij}$.

Two direct decompositions of $A$, $A = \oplus_i B_i$ and $A = \oplus_j C_j$, are called *isomorphic*, if one can find a one-to-one correspondence between the two sets of components $B_i$ and $C_j$ such that corresponding components are isomorphic.

It is convenient to call an exact sequence $0 \to B \xrightarrow{\alpha} A \xrightarrow{\beta} C \to 0$ *splitting* if Im $\alpha$ is a direct summand of $A$.

Now we have come to the definition of the external direct sum. Given the groups $B$ and $C$, we wish to have a group $A$ that is the direct sum of two of its subgroups, $B'$ and $C'$, such that $B' \cong B$, $C' \cong C$. The set of all pairs $(b, c)$ with $b \in B$, $c \in C$ forms a group $A$ under the rules:

$$(b_1, c_1) = (b_2, c_2) \qquad \text{if and only if} \quad b_1 = b_2, c_1 = c_2,$$
$$(b_1, c_1) + (b_2, c_2) = (b_1 + b_2, c_1 + c_2).$$

The correspondences $b \mapsto (b, 0)$ and $c \mapsto (0, c)$ are isomorphisms of $B$, $C$ with subgroups $B'$, $C'$ of $A$. We have $A = B' \oplus C'$. If we think of $B$, $C$ being identified with $B'$, $C'$ under the mentioned isomorphisms, then we may write $A = B \oplus C$ and call $A$ the [external] direct sum of $B$ and $C$.

Let $B_i$ $(i \in I)$ be a set of groups. A vector $(\cdots, b_i, \cdots)$ over this set of $B_i$ has exactly one coordinate $b_i$ for each $i \in I$, and $b_i \in B_i$. Such a vector may also be interpreted as a function $f$ defined over $I$ such that $f(i) = b_i \in B_i$ for every $i \in I$. Equality and addition of vectors are defined coordinatewise [i.e., for functions, pointwise]. In this way, the set of all vectors becomes a group $C$ called the direct product [cartesian sum or complete direct sum] of the groups $B_i$:

$$C = \prod_{i \in I} B_i .$$

The correspondence

$$\rho_i : b_i \mapsto (\cdots, 0, b_i, 0, \cdots),$$

where $b_i$ stands on the $i$th place and 0 everywhere else, is an isomorphism of $B_i$ with a subgroup $B_i'$ of $C$. It is easy to see that the groups $B_i'$ $(i \in I)$ generate in $C$ the group $A$ of all vectors $(\cdots, b_i, \cdots)$ with $b_i = 0$ for almost all $i \in I$, and $A$ is just their direct sum. $A$ is also called the [external] direct sum of the $B_i$, $A = \oplus_i B_i$. Clearly, $A = C$ if and only if $I$ is a finite set.

We leave it to the reader to show that both the direct sum and the direct product are, up to isomorphism, uniquely determined by the set of components $B_i$.

For a group $A$ and for a cardinal $\mathfrak{m}$, $\oplus_\mathfrak{m} A$ will denote the direct sum of $\mathfrak{m}$ copies of $A$ and the symbol $\prod_\mathfrak{m} A = A^\mathfrak{m}$ the direct product of $\mathfrak{m}$ copies of $A$. In a similar fashion, $A^I$ stands for the direct product of groups isomorphic to $A$ and indexed by $I$. Clearly, $A^I$ is the group of all functions on $I$ to $A$, and we have $A^I \cong A^{|I|}$.

The external direct sums and direct products can also be described in terms of systems of homomorphisms. The functions

$$\rho_B : b \mapsto (b, 0), \qquad \rho_C : c \mapsto (0, c),$$
$$\pi_B : (b, c) \mapsto b, \qquad \pi_C : (b, c) \mapsto c,$$

are homomorphisms as indicated by the diagram

$$B \underset{\pi_B}{\overset{\rho_B}{\rightleftarrows}} B \oplus C \underset{\pi_C}{\overset{\rho_C}{\leftrightarrows}} C,$$

called [coordinate] injections and projections, respectively. They satisfy

$$\pi_B \rho_B = 1_B, \qquad \pi_C \rho_C = 1_C, \qquad \pi_B \rho_C = \pi_C \rho_B = 0, \qquad \rho_B \pi_B + \rho_C \pi_C = 1_{B \oplus C}.$$

More generally, the direct sum [product] of the $B_i$ $(i \in I)$ defines for every $i \in I$, an injection $\rho_i$ and a projection $\pi_i$

$$B_i \xrightarrow{\rho_i} \oplus B_i \xrightarrow{\pi_i} B_i \qquad [B_i \xrightarrow{\rho_i} \prod B_i \xrightarrow{\pi_i} B_i]$$

where $\rho_i b_i = (\cdots, 0, b_i, 0, \cdots)$, $\pi_i(\cdots, b_j, \cdots, b_i, \cdots) = b_i$ satisfy the following conditions:

(a) $\pi_i \rho_j = 1_{B_i}$ or 0, according as $i = j$ or $i \neq j$;
(b) if $I$ is a finite set, $I = \{1, \cdots, n\}$, and $B = B_1 \oplus \cdots \oplus B_n$, then $\rho_1 \pi_1 + \cdots + \rho_n \pi_n = 1_B$.

If, however, $I$ is infinite, then for direct sum replace (b) by:

(b′) each $b \in \oplus B_i$ can be written as a finite sum $b = \rho_{i_1} \pi_{i_1} b + \cdots + \rho_{i_n} \pi_{i_n} b$,

and for direct product by:

(b″) given a set $\{b_i\}$ with exactly one $b_i \in B_i$ for each $i \in I$, there exists a unique $b \in \prod B_i$ such that $\pi_i b = b_i$ for every $i$.

The following two results exhibit useful universal properties of direct sums and direct products.

**Theorem 8.1.** *Let $\phi_i: B_i \to A$ be homomorphisms, $i \in I$. In the diagrams*

(2)

*with $\rho_i$ the injection maps, the dotted arrow can be filled in uniquely by some $\psi$ [independently of $i$] to make all diagrams commutative.*

Write $b \in \oplus B_i$ in the form $b = \rho_1 \pi_1 b + \cdots + \rho_k \pi_k b$ [where $\pi_i$ are the corresponding projections], and define $\psi b = \phi_1 \pi_1 b + \cdots + \phi_k \pi_k b$. This $\psi$ is easily seen to be a homomorphism into $A$ satisfying $\psi \rho_i b_i = \phi_i \pi_i \rho_i b_i = \phi_i b_i$. This $\psi$ must be unique, for if $\psi': \oplus B_i \to A$ also makes (2) commutative, then $\psi - \psi'$ vanishes on all $\rho_i b_i$, hence on all $b \in \oplus B_i$, that is, $\psi - \psi' = 0$.☐

**Theorem 8.2.** *Let $\phi_i: A \to B_i$ be homomorphisms, $i \in I$. There exists a unique homomorphism $\psi$ making all the diagrams*

(3)

*commute; here $\pi_i$ denote the projections.*

*Łazaruk*

For $a \in A$, define $\psi a$ as the unique $b \in \prod B_i$ with $\pi_i b = \phi_i a$ [cf. (b'')]. This $\psi$ is evidently a homomorphism such that $\pi_i \psi = \phi_i$ for every $i$. It is unique, for if $\psi'$ is a homomorphism with the same stated property, then $\pi_i(\psi - \psi')a = 0$ for every $i$, and thus $(\psi - \psi')a \in \prod B_i$ has only 0 coordinates. That is, $(\psi - \psi')a = 0$ for every $a \in A$, $\psi - \psi' = 0$. $\square$

Let us introduce the following notations. If $A_i$ and $B_i$ ($i \in I$) are groups, and $\alpha_i : A_i \to B_i$ are homomorphisms between them, then there exist unique homomorphisms

$$\alpha : A = \bigoplus_i A_i \to B = \bigoplus_i B_i \qquad \text{and} \qquad \alpha^* : A^* = \prod A_i \to B^* = \prod B_i$$

such that

$$\pi_i' \alpha \rho_i = \alpha_i = \pi_i' \alpha^* \rho_i \qquad \text{and} \qquad \pi_j' \alpha \rho_i = 0 = \pi_j' \alpha^* \rho_t \qquad (i \neq j)$$

for the injections $\rho_i$ of the $A_i$ and projections $\pi_i'$ of the $B_i$. Namely, $\alpha$ [$\alpha^*$] sends the $i$th coordinate $a_i$ upon the $i$th coordinate $\alpha_i a_i$. We shall denote them as

$$\alpha = \bigoplus_i \alpha_i \qquad \text{and} \qquad \alpha^* = \prod_i \alpha_i.$$

For a group $G$, we introduce two maps: the *diagonal map* $\Delta_G : G \to \prod G$ [the number of components can be arbitrary] as

$$\Delta_G : g \mapsto (\cdots, g, \cdots, g, \cdots) \qquad (g \in G),$$

and the *codiagonal map* $\nabla_G : \bigoplus_i G \to G$ as

$$\nabla_G : (\cdots, g_i, \cdots) \mapsto \sum_i g_i \qquad (g_i \in G).$$

If there is no danger of confusion, we may suppress the index $G$.

**Lemma 8.3.** *If the diagram of (2.4) has exact rows and columns, and is commutative, then the following sequences are exact:*

$$0 \to A_1 \xrightarrow{\alpha_2 \lambda_1} B_2 \xrightarrow{(\mu_2 \oplus \beta_2)\Delta} B_3 \oplus C_2,$$

*and*

$$A_2 \oplus B_1 \xrightarrow{\nabla(\alpha_2 \oplus \mu_1)} B_2 \xrightarrow{\nu_2 \beta_2} C_3 \to 0.$$

Since $\lambda_1$ and $\alpha_2$ are monic, so is $\alpha_2 \lambda_1$. Furthermore, $(\mu_2 \oplus \beta_2)\Delta \alpha_2 \lambda_1 = 0$, because $\beta_2 \alpha_2 = 0$ and $\mu_2 \alpha_2 \lambda_1 = \mu_2 \mu_1 \alpha_1 = 0$. If $b_2 \in B_2$ belongs to $\mathrm{Ker}(\mu_2 \oplus \beta_2)\Delta$, then both $\mu_2 b_2 = 0$ and $\beta_2 b_2 = 0$. We can find a $b_1 \in B_1$ such that $\mu_1 b_1 = b_2$, and here $b_1 \in \mathrm{Im} \; \alpha_1$, since $\beta_1 b_1 = 0$ follows from $\nu_1 \beta_1 b_1 = \beta_2 \mu_1 b_1 = \beta_2 b_2 = 0$. Therefore $b_2 \in \mathrm{Im} \; \mu_1 \alpha_1 = \mathrm{Im} \; \alpha_2 \lambda_1$. The exactness of the second sequence follows by a similar argument. $\square$

For $a = (\cdots, b_i, \cdots) \in \prod B_i$ define the *support* of $a$ as

$$s(a) = \{i \in I \mid b_i \neq 0\}.$$

If $K$ is an ideal of the Boolean algebra $B$ of all subsets of $I$, then by the $K$-*direct sum* of the $B_i$ we mean the subset of $\prod B_i$ whose elements are the vectors $a$ with $s(a) \in K$. Since $s(a_1 - a_2) \subseteq s(a_1) \cup s(a_2)$, the $K$-direct sum is a subgroup of $\prod B_i$; it will be denoted by the symbol

$$\bigoplus_K B_i .$$

If, in particular, $K$ consists of all finite subsets of $I$, then we get the direct sum, while if $K = B$, then we arrive at the direct product.

Among the subgroups of the direct product there is an important type which frequently occurs in algebra. A subgroup $G$ of the direct product $A = \prod B_i$ is said to be a *subdirect sum* of the $B_i$ if, for every $i$, the map $\pi_i | G : G \to B_i$ is an epimorphism. This means, that for each $i \in I$ and for every $b_i \in B_i$, $G$ contains a vector $(\cdots, b_i, \cdots)$ whose $i$th coordinate is just $b_i$. In this case, $B_i$ is not necessarily a subgroup of $G$, it is an epimorphic image of $G$ under $\pi_i | G = \eta_i$. Clearly,

$$\bigcap_i \text{Ker } \eta_i = 0.$$

Conversely, if $K_i$ $(i \in I)$ is a family of subgroups of $G$ such that $\bigcap K_i = 0$, then $G$ is a subdirect sum of the quotient groups $G/K_i$ under the monomorphism

$$g \mapsto (\cdots, g + K_i, \cdots) \in \prod_i (G/K_i).$$

There are a great number of subdirect sums in a direct product of groups, and no complete survey of subdirect sums is known except for the case of subdirect sums of two groups.

Assume that the groups $B$, $C$ are mapped by epimorphisms $\beta$, $\gamma$ upon a group $F$. The elements $(b, c) \in B \oplus C$ with $\beta b = \gamma c$ form a subgroup $A$ of $B \oplus C$, and it is routine to check that $A$ is a subdirect sum of $B$ and $C$. Conversely, if $A$ is a subdirect sum of $B$ and $C$, then the elements $b \in B$ with $(b, 0) \in A$ form a subgroup $B_0 \leq B$, and the elements $c \in C$ with $(0, c) \in A$ form a subgroup $C_0 \leq C$. It is straightforward to verify that the correspondence

$$b + B_0 \mapsto c + C_0$$

whenever $(b, c) \in A$ is an isomorphism of $B/B_0$ with $C/C_0$. Now $A$ consists exactly of the pairs $(b, c)$ $(b \in B, c \in C)$ for which the canonical epimorphisms $B \to B/B_0$, $C \to C/C_0$ map $b$ and $c$ upon corresponding cosets. This shows that all subdirect sums of two groups $B$, $C$ arise in the manner described at the beginning of this paragraph. The groups $B_0$, $C_0$ are called the *kernels* of the subdirect sum; the subdirect sum is in general not determined by the kernels, it does depend also on the choice of $\beta$, $\gamma$. If $B_0$, $C_0$ are considered as subgroups of $A$, then the following isomorphisms hold:

$$A/(B_0 \oplus C_0) \cong B/B_0 \cong C/C_0, \qquad A/B_0 \cong C, \qquad A/C_0 \cong B.$$

One of the most important applications of direct sum is the next theorem.

**Theorem 8.4.** *A torsion group $A$ is the direct sum of p-groups $A_p$ belonging to different primes $p$. The $A_p$ are uniquely determined by $A$.*

Let $A_p$ consist of all $a \in A$ whose order is a power of the prime $p$. In view of $0 \in A_p$, $A_p$ is not empty. If $a, b \in A_p$, that is, if $p^m a = p^n b = 0$ for some integers $m, n \geq 0$, then $p^{\max (m,n)}(a - b) = 0$, $a - b \in A_p$, and $A_p$ is a subgroup. Every element in $A_{p_1} + \cdots + A_{p_k}$ is annihilated by a product of powers of $p_1, \cdots, p_k$, therefore

$$A_p \cap (A_{p_1} + \cdots + A_{p_k}) = 0$$

whenever $p \neq p_1, \cdots, p_k$. Thus the $A_p$ generate their direct sum $\bigoplus_p A_p$ in $A$. In order to show that every $a \in A$ lies in this direct sum, let $o(a) = m = p_1^{r_1} \cdots p_n^{r_n}$ with different primes $p_i$. The numbers $m_i = m p_i^{-r_i}$ $(i = 1, \cdots, n)$ are relatively prime, hence there are integers $s_1, \cdots, s_n$ such that $s_1 m_1 + \cdots + s_n m_n = 1$. Thus $a = s_1 m_1 a + \cdots + s_n m_n a$ where $m_i a \in A_{p_i}$ [in view of $p_i^{r_i} m_i a = ma = 0$], and so $a \in A_{p_1} + \cdots + A_{p_n} \leq \bigoplus_p A_p$.

If $A = \bigoplus_p B_p$ is any direct decomposition of $A$ into $p$-groups $B_p$ with different primes $p$, then by the definition of the $A_p$, we have $B_p \leq A_p$ for every $p$. Since the $B_p$ and the $A_p$ generate direct sums which are both equal to $A$, we must have $B_p = A_p$ for every $p$. $\square$

The subgroups $A_p$ are called the *p-components* of $A$. They are, as seen from their definition, fully invariant subgroups of $A$. If $A$ is not torsion, the $p$-component of its torsion part $T(A)$ can be called the *p-component* of $A$ [even if it fails to be a direct summand of $A$]. By virtue of (8.4), *the theory of torsion groups is essentially reduced to that of primary groups.*

*Example 1.* Let $m = p_1^{r_1} \cdots p_n^{r_n}$ be the canonical form of the integer $m > 0$. The $p$-components of $Z(m)$ [as subgroups of a cyclic group] are again cyclic, and the product of their orders is just $m$. Consequently,

$$Z(m) = Z(p_1^{r_1}) \oplus \cdots \oplus Z(p_n^{r_n}).$$

*Example 2.* The group $Q/Z$ of all complex roots of unity is a torsion group whose $p$-component consists evidently of all $p^n$th roots of unity $(n = 1, 2, \cdots)$. This means the $p$-component of $Q/Z$ is $Z(p^\infty)$. Hence

$$Q/Z = \bigoplus_p Z(p^\infty).$$

We conclude this section with:

**Theorem 8.5.** *An elementary p-group is the direct sum of cyclic groups of order $p$.*

We show that an elementary $p$-group $A$ is in the natural way a vector space over the field $F_p$ of $p$ elements. In fact, $pa = 0$ for $a \in A$, and so for $n$, $m \in Z$, one has $na = ma$ exactly if $n \equiv m$ mod $p$, that is, $n$, $m$ represent the same element of $F_p$. It is now routine to check the vector space axioms. Therefore, $A$ as a vector space over a field $F_p$ has a basis, say, $\{a_i\}_{i \in I}$. It follows that $A = \bigoplus_{i \in I} \langle a_i \rangle$. □

*Example 3.* Let $\mathfrak{m}$ be an infinite cardinal number. Then

$$Z(p)^{\mathfrak{m}} = \bigoplus_{2^{\mathfrak{m}}} Z(p).$$

In fact, $Z(p)^{\mathfrak{m}}$ is by (8.5) a direct sum of groups $Z(p)$, and its cardinality is obviously $p^{\mathfrak{m}} = 2^{\mathfrak{m}}$.

EXERCISES

1. If $m = p_1^{r_1} \cdots p_k^{r_k}$, then $A/mA$ is the direct sum of the groups $A/p_i^{r_i}A$ $(i = 1, \cdots, k)$.

2. (a) If $A$ is the direct sum [product] of the groups $B_i$ $(i \in I)$, and $C_i$ are subgroups of $B_i$, and if $C$ is the direct sum [product] of the $C_i$, then $C$ is a subgroup of $A$ such that $A/C$ is the direct sum [product] of the quotient groups $B_i/C_i$.
   (b) If $0 \to A_i \xrightarrow{\alpha_i} B_i \xrightarrow{\beta_i} C_i \to 0$ are exact sequences $(i \in I)$, then so are

   $$0 \to \bigoplus A_i \xrightarrow{\oplus \alpha_i} \bigoplus B_i \xrightarrow{\oplus \beta_i} \bigoplus C_i \to 0$$

   and

   $$0 \to \prod A_i \xrightarrow{\Pi \alpha_i} \prod B_i \xrightarrow{\Pi \beta_i} \prod C_i \to 0.$$

3. $A$ is a direct sum of its subgroups $A_i$ $(i \in I)$ if and only if

   $$\sum_i A_i = A \quad \text{and} \quad \bigcap_i A_i^* = 0 \quad \text{where} \quad A_i^* = \sum_{j \neq i} A_j.$$

4. (a) The direct sum of $p$-groups [torsion groups] is again a $p$-group [torsion group].
   (b) Determine when the direct product of torsion groups is again a torsion group.

5. (a) If $G$ is a subgroup of $A = B \oplus C$, then $G$ is a subdirect sum of $B \cap (G + C)$ and $(B + G) \cap C$.
   (b) If $G$ is a subdirect sum of $B$ and $C$, then $B + G = B \oplus C = G + C$.

6. Let $A = B \oplus C = B \oplus (\oplus A_i)$. Then $B \oplus C_i = B \oplus A_i$ and $C = \oplus_i C_i$ hold for $C_i = (B + A_i) \cap C$.

7. (a) Let $G$ be a subgroup of the direct sum $B \oplus C$. There exist subgroups $B_1$, $B_2$ of $B$ and $C_1$, $C_2$ of $C$ such that $B_2 \leq B_1$, $C_2 \leq C_1$, $B_1 \oplus C_1$ is the minimal direct sum containing $G$, and $B_2 \oplus C_2$ is the maximal direct sum contained in $G$, with components in $B$, $C$.

(b) Establish the isomorphisms

$$B_1/B_2 \cong C_1/C_2 \qquad (B_1 \oplus C_1)/G \cong G/(B_2 \oplus C_2).$$

8. (E. A. Walker [1]) (a) Let $A = B \oplus C = B^* \oplus C^*$. Then $B + B^* = B \oplus C_1 = B^* \oplus C_1^*$ where $C_1 = C \cap (B + B^*)$, $C_1^* = C^* \cap (B + B^*)$.
   (b) $A = B \oplus C = B^* \oplus C^*$ implies $B/(B \cap B^*) \cong C_1^*$ and $B^*/(B \cap B^*) \cong C_1$.

9. (F. Loonstra) Call a subdirect sum $G$ of groups $A_i$ $(i \in I)$ special, if there exist a group $F$ and epimorphisms $\alpha_i : A_i \to F$ $(i \in I)$ such that $G$ consists of all $(\cdots, a_i, \cdots, a_j, \cdots) \in \prod A_i$ with $\cdots = \alpha_i a_i = \cdots = \alpha_j a_j = \cdots$.
   (a) Show that for $|I| \geq 3$, not every subdirect sum is special.
   (b) If $I$ is finite, $G$ is special if, and only if, the maps $g + (\oplus \operatorname{Ker} \alpha_i) \mapsto \pi_j g + \operatorname{Ker} \alpha_j$ [where $\pi_j : G \to A_j$ is the coordinate projection] are all isomorphisms between $G/\oplus \operatorname{Ker} \alpha_i$ and $A_j/\operatorname{Ker} \alpha_j$, for every $j \in I$.

10. (a) A subdirect sum of $Z(p^m)$ and $Z(p^n)$ [for integers $n \geq m > 0$] with kernels $Z(p^{m-k})$ and $Z(p^{n-k})$ is isomorphic to $Z(p^n) \oplus Z(p^{m-k})$.
    (b) A subdirect sum of $Z(p^\infty)$ and $Z(p^\infty)$ with the kernels $Z(p^m)$ and $Z(p^n)$ $(n \geq m)$ is isomorphic to $Z(p^\infty) \oplus Z(p^m)$.
    (c) Every subdirect sum of a cyclic and a quasicyclic group is direct.

11. There are nonisomorphic groups among the subdirect sums of the groups $B = Z(p^2) \oplus Z(p^4)$ and $C = Z(p^3) \oplus Z(p^5)$ with kernels $B_0 = Z(p) \oplus Z(p^3)$ and $C_0 = Z(p^2) \oplus Z(p^4)$.

12. Call $A$ *subdirectly irreducible* if in any representation of $A$ as a subdirect sum of groups $A_i$, one of the coordinate projections $A \to A_i$ is an isomorphism. Show that $A$ is subdirectly irreducible if and only if it is cocyclic.

13. Let $B_i$ $(i \in I)$ be torsion subgroups of $A$. The $B_i$ generate their direct sum if and only if the socles $S(B_i)$ generate their own direct sum.

14. Let $B$, $C$ be subgroups of $A$, and let $B \oplus C$ denote their external direct sum. Prove the existence of an exact sequence

$$0 \to B \cap C \to B \oplus C \to B + C \to 0.$$

15. Assume that $A$ has two sequences of subgroups, $B_1, \cdots, B_n, \cdots$ and $C_1, \cdots, C_n, \cdots$ such that

$$C_1 = A, \qquad C_n = B_n \oplus C_{n+1} \qquad (n = 1, 2, \cdots)$$

and

$$\bigcap_n C_n = 0.$$

Prove that $A$ is a subdirect sum of the $B_n$ which contains their direct sum.

16.   Let $G$ be a subgroup of the direct product $A = \prod A_i$ such that, along with $g$, all the elements of $A$ with the same support belong to $G$. If all the $A_i$ are of order $\geq 3$, then $G$ is a **K**-direct sum.

17.   (Dlab [2]) Prove that the Frattini subgroup of a direct sum is the direct sum of the Frattini subgroups.

18.   If $F$ is the Frattini subgroup of a $p$-group $A$, then $A/F$ is the direct sum of cyclic groups of order $p$. [This does not generalize to arbitrary groups; $A = \prod_p Z(p)$ is a counterexample.]

## 9.   DIRECT SUMMANDS

We have called a subgroup $B$ of $A$ a *direct summand* of $A$ if

$$A = B \oplus C \qquad \text{for some} \quad C \leq A.$$

For the projections $\pi : A \to B$, $\theta : A \to C$ conditions (1) in **8** hold. Next we focus our attention on $B$. [It should be emphasized that $B$ alone, in general, does not define $\pi$ uniquely; but it does if $C$ is also known.] We then have the following useful lemma.

**Lemma 9.1.** *If there is a projection $\pi$ of $A$ onto its subgroup $B$, then $B$ is a direct summand of $A$.*

The map $\theta = 1_A - \pi$ is an endomorphism of $A$, satisfying conditions (1) in **8**. Hence we have $A = B \oplus \theta A$. Here $\theta A$ is nothing else than the kernel of $\pi$. $\square$

A rather trivial criterion for $B$ $(\leq A)$ to be a direct summand of $A$ is that the cosets of $A$ mod $B$ have representatives which form a subgroup $C$ $(\leq A)$. Namely, we then have $B + C = A$ and $B \cap C = 0$, thus $A = B \oplus C$. Less trivial criteria are formulated in the next result.

**Proposition 9.2.** *For a subgroup $B$ of $A$ the following conditions are equivalent $[\rho : B \to A$ denotes the injection map]:*

(a)  *$B$ is a direct summand of $A$;*

(b)  *there exists a commutative diagram*

(c)  *if*

*is a commutative diagram with exact rows, then there is a homomorphism* $\phi : V \to B$ *such that the upper triangle is commutative.*

Assuming (a), we first prove (c). If the hypotheses of (c) hold, and if $\pi : A \to B$ is a projection, then $\phi = \pi\alpha$ satisfies $\phi\gamma = \pi\alpha\gamma = \pi\rho\beta = \beta$. Next assume (c), and choose $U = B$, $V = A$, $\gamma = \rho$, $\alpha = 1_A$ and $\beta = 1_B$. Then (b) follows at once. Finally, if (b) holds, then $\pi$ is a projection of $A$ onto $B$ and hence (9.1) yields (a).□

Notice that (b) amounts to saying that *the identity automorphism of B can be extended to an endomorphism of A into B.*

For the sake of future reference we mention the following lemmas.

**Lemma 9.3.** *If* $A = B \oplus C$ *and if G is a fully invariant subgroup of A, then*

$$G = (G \cap B) \oplus (G \cap C).$$

Let $\pi, \theta$ be the projections attached to the direct decomposition $A = B \oplus C$. By the full invariance of $G$, $\pi G$ and $\theta G$ are subgroups of $G$. Now $\pi G \leq B$ and $\theta G \leq C$ imply $\pi G \cap \theta G = 0$, while $g = \pi g + \theta g$ $(g \in G)$ implies $G = \pi G + \theta G$, so that $G = \pi G \oplus \theta G$. Obviously, $\pi G \leq G \cap B$ and $\theta G \leq G \cap C$, and so necessarily $\pi G = G \cap B$ and $\theta G = G \cap C$.□

**Lemma 9.4** (Kaplansky [1]). *If the quotient group A/B is a direct sum,*

$$A/B = \bigoplus_i (A_i/B),$$

*and if B is a direct summand of every* $A_i$, *say* $A_i = B \oplus C_i$, *then B is a direct summand of A,*

$$A = B \oplus (\bigoplus_i C_i).$$

It is clear that $B$ and the $C_i$ generate $A$. Assume we have $b + c_1 + \cdots + c_n = 0$ for some $b \in B$ and $c_j \in C_j$ $(j = 1, \cdots, n)$. Passing mod $B$, we obtain $(c_1 + B) + \cdots + (c_n + B) = B$. Since $c_j + B \in A_j/B$, therefore $c_1 + B = \cdots = c_n + B = B$. Thus $c_j \in B$ for every $j$, and so $c_j \in B \cap C_j = 0$, finally, $b = 0$. Consequently, $B$ and the $C_i$ generate their direct sum.□

If $B$ is a direct summand of $A$, then the complement of $B$ in $A$ is unique up to isomorphism, but it is far from being unique as a subgroup. The following result tells us that from one complement all the others can easily be obtained.

**Lemma 9.5.** *Let* $A = B \oplus C$ *be a direct decomposition with projections* $\pi, \theta$. *If* $A = B \oplus C_1$ *with projections* $\pi_1, \theta_1$, *then*

(1) $$\pi_1 = \pi + \pi\phi\theta, \qquad \theta_1 = \theta - \pi\phi\theta$$

*for some endomorphism* $\phi$ *of A. Conversely, if* $\pi_1, \theta_1$ *are of the form* (1) *with some endomorphism* $\phi$ *of A, then* $A = B \oplus \theta_1 A$.

If $\pi_1$, $\theta_1$ are attached to $A = B \oplus C_1$, then set $\phi = \theta - \theta_1$. Clearly, $B \leq \text{Ker } \phi$, thus $\phi = \phi\pi + \phi\theta = \phi\theta$. If $a = b + c = b_1 + c_1$ with $b, b_1 \in B$, $c \in C$, $c_1 \in C_1$, then $\phi a = c - c_1 = b_1 - b \in B$ whence $\pi\phi = \phi$. Therefore $\theta_1 = \theta - \phi = \theta - \pi\phi\theta$ and $\pi_1 = 1_A - \theta_1 = \pi + \theta - \theta_1 = \pi + \pi\phi\theta$. Conversely, if $\pi_1$, $\theta_1$ are of the form (1), then $\pi_1 + \theta_1 = 1_A$, $\pi_1^2 = \pi_1$, $\theta_1^2 = \theta_1$ and $\pi_1\theta_1 = \theta_1\pi_1 = 0$, thus $A = \pi_1 A \oplus \theta_1 A$. Here $\text{Im } \pi_1 \leq \text{Im } \pi$ and $\pi_1 B = \pi B = B$, whence $\pi_1 A = B$.□

**Theorem 9.6** (Grätzer and Schmidt [1]). *If $B$ is a direct summand of $A$, then the intersection $\tilde{C}$ of all complements $C$ of $B$ in $A$ is the maximal fully invariant subgroup of $A$ that is disjoint from $B$.*

Let $A = B \oplus C$ with projections $\pi$, $\theta$, and let $\phi$ be an endomorphism of $A$. By (9.5), $C_1 = (\theta - \pi\phi\theta)A$ is again a complement of $B$, therefore for $c \in \tilde{C}$ we have $(\theta - \pi\phi\theta)c = c$ and $\theta c = c$ [for both $c \in C_1$ and $c \in C$], whence $\pi\phi c = 0$. This shows $\phi c \in C$, and since $C$ was an arbitrary complement of $B$, we obtain $\phi c \in \tilde{C}$, that is, $\tilde{C}$ is fully invariant in $A$. Evidently, $B \cap \tilde{C} = 0$. If $X$ is any fully invariant subgroup of $A$ with $B \cap X = 0$, then by virtue of (9.3), $X = (X \cap B) \oplus (X \cap C) = X \cap C$; thus $X \leq C$ and $X \leq \tilde{C}$.□

**Corollary 9.7.** *A complement of a direct summand of $A$ is unique if and only if it is fully invariant in $A$.*□

If a subgroup $B$ of a group $A$ is to be shown a direct summand of $A$, then in most cases it is not possible to find directly a projection $A \to B$. One then tries to find a complement $C$ to $B$ among the subgroups $G$ of $A$ satisfying $G \cap B = 0$.

Call a subgroup $H$ of $A$ a *B-high subgroup* (Irwin and Walker [1]) if

$$H \cap B = 0, \quad \text{and if} \quad H < H' \leq A \quad \text{implies} \quad H' \cap B \neq 0.$$

That is, $H$ is maximal with respect to the property of being disjoint from $B$. Then, in particular, $H + B = H \oplus B$. The existence of $B$-high subgroups, for every $B$, is guaranteed by Zorn's lemma. Moreover, $H$ may be chosen so as to contain a prescribed subgroup $G$ of $A$ with $G \cap B = 0$. In fact, then the set of all subgroups of $A$ that contain $G$ and are disjoint from $B$ is not empty and is inductive; thus it contains a maximal member $H$.

A complement $C$ of $B$ in $A$ is evidently a $B$-high subgroup; moreover, in view of (9.6), it must contain every fully invariant subgroup $X$ of $A$ with $X \cap B = 0$. In several cases, the construction of a complement to $B$ consists in selecting a $B$-high subgroup containing such an $X$ [cf., e.g., the proof of (27.1)].

Let us turn to the proofs of two technical lemmas.

**Lemma 9.8.** *If $B$ is a subgroup of $A$ and $C$ is a $B$-high subgroup of $A$, then $a \in A$, $pa \in C$ [$p$ a prime] implies $a \in B \oplus C \leq A$.*

If $a \in C$, there is nothing to be proved. If $a \notin C$, then $\langle C, a \rangle$ contains, owing to the choice of $C$, an element $b \in B$, $b \neq 0$, i.e., $b = c + ka$ for some $c \in C$ and integer $k$. Here $(k, p) = 1$ because of $pa \in C$ and $B \cap C = 0$. Thus, for some integers $r$, $s$, $rk + sp = 1$, and so $a = r(ka) + s(pa) = r(b - c) + s(pa) \in B \oplus C$.□

**Lemma 9.9** (G. Grätzer). *Let $A$, $B$, $C$ be as in (9.8). Then $A = B \oplus C$ if and only if $pa = b + c$ ($a \in A$, $b \in B$, $c \in C$) implies $pb' = b$ for some $b' \in B$.*

If $A = B \oplus C$ and $a = b' + c'$ ($b' \in B$, $c' \in C$), then $pa = pb' + pc' = b + c$ implies $pb' = b$. Conversely, if $pa = b + c$ implies $pb' = b$ for some $b' \in B$, then $a - b'$ satisfies the hypotheses of the preceding lemma, and so $a - b' \in B \oplus C$, $a \in B \oplus C$. This shows that the quotient group $A/(B \oplus C)$ contains no elements of prime order, and therefore it is torsion-free. But if $x \in A$ is arbitrary, not in $B \oplus C$, then $\langle C, x \rangle$ intersects $B$ in a nonzero element $b''$, $c'' + lx = b''$ for some $c'' \in C$ and integer $l$. $l \neq 0$ because of $B \cap C = 0$, thus $lx = b'' - c'' \in B \oplus C$, and $A/(B \oplus C)$ is a torsion group. We conclude $A = B \oplus C$.□

A subgroup $G$ of $A$ is called an *absolute direct summand* of $A$ if, for every $G$-high subgroup $H$ of $A$, one has $A = G \oplus H$. Absolute direct summands are rare phenomena; they are described in Ex. 8.

We conclude the discussion of direct summands with the following result.

**Proposition 9.10** (Kaplansky [3], C. P. Walker [1]). *Let $A$ be a direct sum of groups $A_i$ ($i \in I$) such that every $A_i$ is at most of power $\mathfrak{m}$ where $\mathfrak{m}$ is a fixed infinite cardinal. Then any direct summand of $A$ is again a direct sum of groups of power $\leq \mathfrak{m}$.*

Let $A = B \oplus C$, and consider some summand $A_j$. If $\{a_k\}_{k \in K}$ is a generating system of $A_j$, and if $a_k = b_k + c_k$ ($b_k \in B$, $c_k \in C$), then every $b_k$ and every $c_k$ has but a finite number of nonzero coordinates in the direct sum $A = \oplus_i A_i$. Therefore, if we collect all the $A_i$ containing at least one nonzero coordinate of some $b_k$ or $c_k$ ($k \in K$), and take their union in $A$, we obviously get a direct summand $A'_j$ of $A$, of cardinality $\leq \mathfrak{m}$. Selecting a generating system for $A'_j$ and repeating the same process with $A_j$ replaced by $A'_j$, we find a direct summand $A''_j$ of $A$ again of power $\leq \mathfrak{m}$, etc. The union of the ascending chain $A_j \leq A'_j \leq A''_j \leq \cdots$ is a direct summand $A^*_j$ of $A$, such that

$$|A^*_j| \leq \mathfrak{m} \qquad \text{and} \qquad A^*_j = (B \cap A^*_j) \oplus (C \cap A^*_j).$$

A well-ordered, increasing sequence of subgroups $S_\sigma$ of $A$ is now defined as follows. Put $S_0 = 0$. If $S_\sigma$ is defined for some ordinal $\sigma$ and $S_\sigma \neq A$, then pick out some $A_j$ not in $S_\sigma$, and let $S_{\sigma+1} = S_\sigma + A^*_j$. If $\sigma$ is a limit ordinal, then we set $S_\sigma = \bigcup_{\rho < \sigma} S_\rho$. It is evident that for some ordinal $\tau$, not exceeding $|A|$, the equality $S_\tau = A$ will hold. It is also clear that $S_{\sigma+1}/S_\sigma$ is of power

$\leq m$, every $S_\sigma$ is the direct sum of some of the $A_i$, and $S_\sigma = (B \cap S_\sigma) \oplus (C \cap S_\sigma)$ holds for every $\sigma \leq \tau$. Now $B \cap S_\sigma$ is a direct summand of $S_\sigma$, which is in turn a direct summand of $A$; thus $B \cap S_\sigma$ is a direct summand of $B \cap S_{\sigma+1}$; write $B \cap S_{\sigma+1} = (B \cap S_\sigma) \oplus B_\sigma$. The isomorphism

$$B_\sigma \cong (B \cap S_{\sigma+1})/(B \cap S_{\sigma+1} \cap S_\sigma) \cong [S_\sigma + (B \cap S_{\sigma+1})]/S_\sigma \leq S_{\sigma+1}/S_\sigma$$

shows $|B_\sigma| \leq m$. The $B_\sigma$ generate their direct sum, for if $b_{\sigma_1} + \cdots + b_{\sigma_r} = 0$ with $\sigma_1 < \cdots < \sigma_r$ ($b_{\sigma_s} \in B_{\sigma_s}$), then $b_{\sigma_1} + \cdots + b_{\sigma_{r-1}} \in B \cap S_{\sigma_r}$, $b_{\sigma_r} \in B_{\sigma_r}$, implies $b_{\sigma_r} = 0$. Since $B$ is the union of the $B_\sigma$, the assertion follows. $\square$

Let us notice that if $A$ is a direct sum of finite groups, then every direct summand of $A$ has the same property. This is a simple consequence of (15.2) and (18.1).

EXERCISES

1. If $B$, $C$ are subgroups of $A$ such that $B \cap C = 0$ and $(B + C)/C$ is a direct summand of $A/C$, then $A = B \oplus C'$ for some $C' \geq C$.
2. Extend (9.3) to infinite direct sums.
3. Let $B$ be a direct summand of $A$, and let $\pi_i$ run over all projections of $A$ onto $B$. Show that these $\pi_i$ form a semigroup under multiplication where $\pi_i \pi_j = \pi_j$. Furthermore, the projections $1 - \pi_i$ form a semigroup where $(1 - \pi_i)(1 - \pi_j) = 1 - \pi_i$.
4. Call a subgroup $G$ of $A$ *projection-invariant* if every projection $\pi$ of $A$ onto a direct summand maps $G$ into itself.
   (a) Show that $G$ is projection-invariant if and only if $\pi G = G \cap \pi A$ for every projection $\pi$.
   (b) A subgroup is projection-invariant if it is an intersection of, or generated by, projection-invariant subgroups.
   (c) (9.3) prevails for projection-invariant subgroups $G$.
   (d) A projection-invariant direct summand is fully invariant.
5. (Kulikov [4]) A direct decomposition of a group $A$, $A = \oplus A_i$, has a common refinement with any direct decomposition of $A$ if and only if every $A_i$ is projection-invariant.
6. (Fuchs [5]) Let $B \leq A$ and let $C$ be $B$-high. Then $A^* = B \oplus C \leq A$ satisfies:
   (a) $A/A^*$ is a torsion group;
   (b)* $(A/A^*)[p] \cong [(pA + C) \cap B]/pB$ for every prime $p$.
7. If $B$ is a subgroup of $A$ and $B \cong J_p$, then for any $B$-high subgroup $C$ of $A$, $A/(B \oplus C)$ is isomorphic to a subgroup of $Z(p^\infty)$.
8*. (Fuchs [5]) A direct summand $B$ of $A$ is an absolute direct summand if and only if either $B$ is divisible [see 20] or $A/B$ is a torsion group whose $p$-component is annihilated by $p^k$ whenever $B \backslash pB$ contains an element of order $p^k$.

9.  If $A = B \oplus C = B_1 \oplus C$ and $B$ is an absolute direct summand of $A$, then the same holds for $B_1$.

10* Find all absolute direct summands of a bounded group.

11. (Irwin and Walker [1]) Let $A = \oplus_i A_i$ and $B_i \leq A_i$. If $C_i$ is $B_i$-high in $A_i$, for every $i$, then $\oplus_i C_i$ is $\oplus_i B_i$-high.

12. (Enochs [1]) Let $A$ be a $p$-group and $A = B \oplus C = B_1 \oplus C_1$, two direct decompositions of $A$ such that $B[p] = B_1[p]$. Then $A = B \oplus C_1 = B_1 \oplus C$. [*Hint:* use induction on the order of $a \in C$ to show $a \in B \oplus C_1$.]

## 10.  PULLBACK AND PUSHOUT DIAGRAMS

By making use of direct sums we can prove two existence theorems for certain types of diagrams. It is useful to have them at our disposal.

**Theorem 10.1.**  *Given the homomorphisms* $\alpha : A \to C$ *and* $\beta : B \to C$, *there is a group* $G$, *unique up to isomorphism, and there are homomorphisms* $\gamma : G \to A$, $\delta : G \to B$ *such that* (i) *the diagram*

(1)
$$\begin{array}{ccc} G & \xrightarrow{\gamma} & A \\ {\scriptstyle\delta}\downarrow & & \downarrow{\scriptstyle\alpha} \\ B & \xrightarrow{\beta} & C \end{array}$$

*is commutative; and* (ii) *if*

(2)
$$\begin{array}{ccc} G' & \xrightarrow{\gamma'} & A \\ {\scriptstyle\delta'}\downarrow & & \downarrow{\scriptstyle\alpha} \\ B & \xrightarrow{\beta} & C \end{array}$$

*is a commutative diagram, then there is a unique homomorphism* $\phi : G' \to G$ *such that* $\gamma\phi = \gamma'$ *and* $\delta\phi = \delta'$.

A commutative diagram (1) satisfying (ii) is called a *pullback diagram*.

Given $\alpha$, $\beta$, define $G$ as the subgroup of the direct sum $A \oplus B$ consisting of all $(a, b)$ with $\alpha a = \beta b$, and let

$$\gamma : (a, b) \mapsto a, \qquad \delta : (a, b) \mapsto b$$

be the corresponding projections. Then (1) will obviously be commutative. Assume (2) is a commutative diagram, and define $\phi : G' \to G$ as $\phi g' = (\gamma'g', \delta'g')$ for $g' \in G'$; here $(\gamma'g', \delta'g') \in G$, because $\alpha\gamma' = \beta\delta'$. Evidently, $\gamma\phi g' = \gamma'g'$ and $\delta\phi g' = \delta'g'$ for every $g' \in G'$, and so $\phi$ is of the stated kind. It is easy to see that

(3)                Ker $\gamma = (0, \text{Ker } \beta)$      and      Ker $\delta = (\text{Ker } \alpha, 0)$.

Therefore, if $\phi' : G' \to G$ also satisfies $\gamma\phi' = \gamma'$ and $\delta\phi' = \delta'$, then $\gamma(\phi - \phi') = 0 = \delta(\phi - \phi')$, and hence Im$(\phi - \phi') \leq$ Ker $\gamma \cap$ Ker $\delta = 0$. This shows $\phi - \phi' = 0$, and $\phi$ is unique.

The uniqueness of $G$ can be proved by considering a $\bar{G}$ with the same properties; we denote the corresponding maps in (1) by $\bar{\gamma} : \bar{G} \to A$ and $\bar{\delta} : \bar{G} \to B$. Then we have unique homomorphisms $\phi : \bar{G} \to G$, $\bar{\phi} : G \to \bar{G}$, such that $\gamma\phi = \bar{\gamma}$, $\delta\phi = \bar{\delta}$, and $\bar{\gamma}\bar{\phi} = \gamma$, $\bar{\delta}\bar{\phi} = \delta$. Hence $\gamma\phi\bar{\phi} = \gamma$ and $\delta\phi\bar{\phi} = \delta$, that is, $\phi\bar{\phi} : G \to G$ preserves both the $A$- and the $B$-coordinates, and therefore $\phi\bar{\phi} = 1_G$. By (9.2), $\bar{\phi}G = \bar{\phi}\phi(\bar{\phi}G)$ is a direct summand of $\bar{G}$, $\bar{G} = \bar{\phi}G \oplus X$. Since in (ii), a unique map $G' \to \bar{G}$ was required for every $G'$, $X$ must vanish. Thus $\bar{\phi}$ is an isomorphism.□

The following is the dual of the preceding theorem.

**Theorem 10.2.** *Assume we are given homomorphisms* $\alpha : C \to A$ *and* $\beta : C \to B$. *There exist a group $G$, unique within isomorphism, and two homomorphisms* $\gamma : A \to G$, $\delta : B \to G$ *such that:* (i) *the diagram*

(4)
$$
\begin{array}{ccc}
C & \xrightarrow{\alpha} & A \\
\beta \downarrow & & \downarrow \gamma \\
B & \xrightarrow{\delta} & G
\end{array}
$$

*is commutative; and* (ii) *to every commutative diagram*

(5)
$$
\begin{array}{ccc}
C & \xrightarrow{\alpha} & A \\
\beta \downarrow & & \downarrow \gamma' \\
B & \xrightarrow{\delta'} & G'
\end{array}
$$

*there is a unique homomorphism* $\phi : G \to G'$ *satisfying* $\phi\gamma = \gamma'$ *and* $\phi\delta = \delta'$.

A commutative diagram (4) with property (ii) is said to be a *pushout diagram*.

Starting with $\alpha$, $\beta$, define $G$ as the quotient group of $A \oplus B$ modulo the subgroup $H$ of the elements of the form $(\alpha c, - \beta c)$ for $c \in C$, and let

$$
\gamma : a \mapsto (a, 0) + H, \qquad \delta : b \mapsto (0, b) + H
$$

be the maps induced by the injections. Then $\gamma\alpha c = \delta\beta c$ for every $c \in C$, and (4) is commutative. If (5) is a commutative diagram, define $\phi : G \to G'$ as $\phi : (a, b) + H \mapsto \gamma'a + \delta'b$; it is readily seen that this map is independent of the representative $(a, b)$ of the coset, and it satisfies $\phi\gamma = \gamma'$, $\phi\delta = \delta'$. The uniqueness of $\phi$ follows from the simple fact that Im $\gamma$ and Im $\delta$ generate $G$, and therefore if $\phi'\gamma = \gamma'$, $\phi'\delta = \delta'$ for some $\phi' : G \to G'$, then $(\phi - \phi')\gamma = 0 = (\phi - \phi')\delta$ imply that $\phi - \phi'$ maps the whole of $G$ upon 0. An argument analogous to that at the end of the proof of the preceding theorem establishes the uniqueness of $G$.□

The following observations on pullbacks and pushouts will be useful.

(a) *If in the pullback diagram* (1), $\alpha$ *is a monomorphism, then so is* $\delta$. *If* $\alpha$ *is an epimorphism, then so is* $\delta$.

Since $G$ is unique up to isomorphism, and moreover—as is shown by the last part of the proof of (10.1)—Im $\delta$ is unique, and Ker $\delta$ is unique up to isomorphism, it suffices to prove the statement for $G$ as constructed in the proof of (10.1). That Ker $\alpha = 0$ implies Ker $\delta = 0$ is immediately seen from (3). If $\alpha$ is epic, then to every $b \in B$ there is an $a \in A$ such that $\alpha a = \beta b$, and so $\delta$ is epic too.

(b) *If in the pushout diagram* (4), $\alpha$ *is a monomorphism, then* $\delta$ *is a monomorphism, and if* $\alpha$ *is epic, then so is* $\delta$.

Again, we need only show this for $G$ and $\delta$ as defined in the proof of (10.2). Ker $\delta$ consists of all $b \in B$ for which there is a $c \in C$ with $\alpha c = 0$ and $-\beta c = b$. Hence Ker $\delta = 0$ if Ker $\alpha = 0$. If $\alpha$ is epic, then to every $a \in A$ there is a $c \in C$ with $\alpha c = a$, and so $\delta$ maps $b + \beta c$ upon $(a, b) + H = (0, b + \beta c) + H$; hence $\delta$ is epic.

### EXERCISES

1.   Find the pullbacks for $B = 0$ [for $A = B$] in (1).
2.   Find the pushouts for $B = 0$ [for $A = B$] in (4).
3.   If $C = 0$ in the pullback diagram (1), then $G \cong A \oplus B$.
4.   If $C = 0$ in the pushout diagram (4), then $G \cong A \oplus B$.
5*.  Given a set $\alpha_i : A_i \to C$ of homomorphisms into the same group $C$, prove the generalization of (10.1).
6*.  The analog of Ex. 5 for (10.2).
7.   If in the diagram

$$A_1 \to A_2 \to A_3$$
$$\downarrow \quad \downarrow \quad \downarrow$$
$$B_1 \to B_2 \to B_3$$

each square is a pullback diagram and $A_3 \to B_3$ is a monomorphism, then the outer rectangle is again pullback.
8.   Formulate and prove the dual of Ex. 7 for pushouts.

## 11.   DIRECT LIMITS

The notion of direct limit stems from topology, and it has become very useful in algebra, too.

Let $\{A_i\}$ be a system of groups indexed by a partially ordered set $I$, which is *directed* in the sense that to $i, j \in I$ there is always a $k \in I$ such that $i \leq k$ and $j \leq k$. Suppose that for every pair $i, j \in I$ with $i \leq j$, there is given a homomorphism

$$\pi_i^j : A_i \to A_j \qquad (i \leq j)$$

subject to the conditions:

  (i)  $\pi_i^i$ is the identity map of $A_i$, for all $i \in I$;
  (ii)  if $i \leq j \leq k$, then $\pi_j^k \pi_i^j = \pi_i^k$.

In this case, the system

(1) $$A = \{A_i \ (i \in I); \ \pi_i^j\}$$

is called a *direct system*.

We form the direct sum $\bigoplus_i A_i = A$ of the groups in the direct system $A$ together with the subgroup $B$ generated by all elements of $A$ of the form

$$a_i - \pi_i^j a_i \qquad (i \leq j).$$

The *direct* [or *injective*] *limit* of the direct system $A$ is defined as the quotient group $A/B$:

$$\varinjlim_I A_i = A/B = A_*.$$

Let us list some elementary properties of direct limits.

(a) *The subgroup $B$ consists of all*

$$a = a_{i_1} + \cdots + a_{i_n} \qquad (a_i \in A_i)$$

*for which there is an $i \geq i_1, \cdots, i_n$ in $I$ such that*

$$\pi_{i_1}^i a_{i_1} + \cdots + \pi_{i_n}^i a_{i_n} = 0.$$

Since the generators $a = a_i - \pi_i^j a_i$ of $B$ have this property: $\pi_i^j a_i - \pi_j^j(\pi_i^j a_i) = 0$, all the elements of $B$ share the same property. The converse follows at once from

$$a_{i_1} + \cdots + a_{i_n} - \pi_{i_1}^i a_{i_1} - \cdots - \pi_{i_n}^i a_{i_n}$$
$$= (a_{i_1} - \pi_{i_1}^i a_{i_1}) + \cdots + (a_{i_n} - \pi_{i_n}^i a_{i_n}) \in B.$$

(b) *There exist homomorphisms*

$$\pi_i : A_i \to A_* \qquad (i \in I)$$

*such that the diagrams*

(2)

$$
\begin{array}{ccc}
A_i & \xrightarrow{\ \pi_i^j\ } & A_j \\
\ \ \pi_i \searrow & \swarrow \pi_j & \\
& A_* &
\end{array}
\qquad (i \leq j)
$$

*are all commutative.*

In fact, $\pi_i : a_i \mapsto a_i + B$ satisfy this condition. $\pi_i$ will always denote this homomorphism which we shall call *canonical*. [The name will be justified by (11.1).]

(c) *Every $a_* \in A_*$ is representable by a coset $a_i + B$ with $a_i \in A_i$ for some $i \in I$.*

Let $a = a_{i_1} + \cdots + a_{i_n}$ with $a_{i_m} \in A_{i_m}$, and choose an $i \geq i_1, \cdots, i_n$ in $I$. In this case $a - \sum \pi^i_{i_m} a_{i_m} \in B$, and so $a_i + B = a + B$, where $a_i = \sum \pi^i_{i_m} a_{i_m}$.

(d) *The groups $\operatorname{Im} \pi_i$ ($i \in I$) together overlap $A_*$.*

This is a consequence of (c) and the definition of $\pi_i$.

(e) *If $\pi_i a_i = 0$ for some $a_i \in A_i$, then there is a $j \geq i$ in $I$ such that $\pi^j_i a_i = 0$.*

By the definition of $\pi_i$, $a_i \in B$. The statement follows now from (a). Hence:

(f) *If every $\pi^j_i$ is a monomorphism, then all the $\pi_i$ are monomorphisms.*

(g) *If every $\pi^j_i$ is an epimorphism, then so is every $\pi_i$.*

Choose an $a_* \in A_*$ and write it in the form $\pi_j a_j$ for some $j \in I$ [cf. (d)]. Given $\pi_i$, there is a $k \in I$ with $i, j \leq k$, and clearly $\pi_j a_j = \pi_k \pi^k_j a_j$. By hypothesis, some $a_i \in A_i$ satisfies $\pi^k_i a_i = \pi^k_j a_j$. Therefore, $a_* = \pi_k \pi^k_j a_j = \pi_k \pi^k_i a_i = \pi_i a_i$, which shows that every $a_* \in A_*$ is contained in $\operatorname{Im} \pi_i$.

(h) *If $K$ is a cofinal subsystem of $I$, then*

$$\varinjlim_K A_j \cong \varinjlim_I A_i.$$

For, if the first group is $A'/B'$, then $a_j + B' \mapsto a_j + B$ ($j \in K$) is easily seen to be an isomorphism of $A'/B'$ onto $A/B$.

The next result points out the universal property of direct limits.

**Theorem 11.1.** *The direct limit $A_*$ of the direct system* (1) *satisfies: if $G$ is a group and there are homomorphisms $\sigma_i : A_i \to G$ with commutative diagrams*

(3)

$$
\begin{array}{ccc}
A_i & \xrightarrow{\pi_i{}^j} & A_j \\
 & \sigma_i \searrow \quad \downarrow \sigma_j & \\
 & G &
\end{array}
\qquad (i \leq j),
$$

*then there exists one, and only one, homomorphism $\sigma : A_* \to G$ such that all the diagrams*

(4)

$$
\begin{array}{ccc}
& A_i & \\
\pi_i \downarrow & \searrow \sigma_i & \\
A_* & \xrightarrow{\sigma} & G
\end{array}
\qquad (i \in I)
$$

*commute. The group $A_*$, together with the $\pi_i$, is uniquely determined by this property, up to isomorphism.*

If $a_* \in A_*$ is represented by the coset $a_i + B$ ($a_i \in A_i$), then define $\sigma a_* = \sigma_i a_i$. From the commutativity of (3) it follows that this definition is independent of the choice of $i \in I$, therefore $\sigma$ is a map $A_* \to G$. Clearly, $\sigma$ preserves addition, and hence it is a homomorphism such that $\sigma \pi_i a_i = \sigma a_* = \sigma_i a_i$, that is, (4) is commutative. If $\sigma' : A_* \to G$ also makes all diagrams of the form (4) commutative, then $(\sigma - \sigma')\pi_i = 0$ for every $i \in I$, i.e., $\sigma - \sigma'$ sends every Im $\pi_i$ into 0. By (d), $\sigma - \sigma' = 0$, and $\sigma$ is unique.

In order to establish the final statement, assume that the group $A_0$ and the homomorphisms $\tau_i : A_i \to A_0$ have the property stated for $A_*$ and $\pi_i$. By what has been shown, we have a unique homomorphism $\sigma : A_* \to A_0$, and by our assumption on $A_0$, a homomorphism $\sigma_0 : A_0 \to A_*$ such that $\tau_i = \sigma\pi_i$ and $\pi_i = \sigma_0\tau_i$. We obtain $\pi_i = \sigma_0\sigma\pi_i$ [and $\tau_i = \sigma\sigma_0\tau_i$]; thus, $\sigma_0\sigma$ is the identity map on all Im $\pi_i$, and hence on $A_*$. Therefore, $\sigma A_*$ is a direct summand of $A_0$. The uniqueness of $A_0 \to G$ for every $G$ implies $\sigma A_* = A_0$, and so $\sigma$ is an isomorphism between $A_*$ and $A_0$. $\square$

*Example 1.* Let $A$ be an arbitrary group, and $A_i (i \in I)$ the system of all finitely generated subgroups of $A$ where $I$ is partially ordered in such a way that $i \le j$ if and only if $A_i$ is a subgroup of $A_j$. Let $\pi_i^j$ be the injection map $A_i \to A_j$ $(i \le j)$. Then $A = \{A_i (i \in I); \pi_i^j\}$ is a direct system such that

$$\lim_{\longrightarrow I} A_i = A_* \cong A$$

in the canonical way: this isomorphism $\rho : A \to A_*$ is given by $a \mapsto a + B$ [where $a \in A$ is regarded as an element of some $A_i$]. For, if $\sigma_i$ is the injection $A_i \to A$, and $\sigma : A_* \to A$ is the unique homomorphism with $\sigma\pi_i = \sigma_i$ [cf. (11.1)], then from the commutativity of the small triangles in

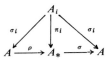

we infer that the large triangle is commutative. Since Im $\sigma_i$ exhaust $A$, and Im $\pi_i$ exhaust $A_*$, it follows that $\rho$ and $\sigma$ are inverse to each other.

*Example 2.* Let $A$ be a direct sum: $A = \bigoplus_i C_i$, and let the index set $I$ be linearly ordered in some way. Define $A_i = \bigoplus_{k < i} C_k$ and $\pi_i^j : A_i \to A_j$ as the injection map for $i \le j$. Then

$$\lim_{\longrightarrow I} A_i \cong A.$$

The proof of this runs similarly as in Example 1.

*Example 3.* Let $C_n = \langle c_n \rangle$ be a cyclic group of order $p^n$ $(n = 1, 2, \cdots)$, and define $\pi_n^{n+1} : C_n \to C_{n+1}$ to act as $\pi_n^{n+1} c_n = pc_{n+1}$. Then for all $m, n$ with $m \le n$, $\pi_m^n : C_m \to C_n$ can uniquely be determined so as to satisfy (i) and (ii). Now

$$\lim_{\longrightarrow n} C_n \cong Z(p^\infty).$$

The proof of this is left to the reader.

We are going to consider homomorphisms between direct limits which are induced by homomorphisms between the corresponding direct systems. If

$$A = \{A_i (i \in I); \pi_i^j\} \qquad \text{and} \qquad B = \{B_i (i \in I); \rho_i^j\}$$

are two direct systems with the same index set $I$, then by a *homomorphism*

$$\phi : A \to B$$

is meant a system of homomorphisms $\phi = \{\phi_i : A_i \to B_i \ (i \in I)\}$ such that the diagrams

$$
\begin{array}{ccc}
A_i & \xrightarrow{\pi_i^{\,j}} & A_j \\
\downarrow{\scriptstyle \phi_i} & & \downarrow{\scriptstyle \phi_j} \qquad (i \leq j) \\
B_i & \xrightarrow{\rho_i^{\,j}} & B_j
\end{array}
$$

are commutative.

**Theorem 11.2.** *If $A$ and $B$ are direct systems and $\phi : A \to B$ is a homomorphism, then there exists a unique homomorphism*

$$\phi_* : A_* = \varinjlim A_i \to B_* = \varinjlim B_i$$

*such that the diagrams*

(5)
$$
\begin{array}{ccc}
A_i & \xrightarrow{\pi_i} & A_* \\
\downarrow{\scriptstyle \phi_i} & & \downarrow{\scriptstyle \phi_*} \qquad (i \in I) \\
B_i & \xrightarrow{\rho_i} & B_*
\end{array}
$$

*are commutative* [$\pi_i$, $\rho_i$ *denote the canonical homomorphisms*]. $\phi_*$ *is an epimorphism* [*monomorphism*] *if all the $\phi_i$ are epimorphisms* [*monomorphisms*].

Since the maps $\rho_i \phi_i : A_i \to B_*$ satisfy the condition $\rho_j \phi_j \pi_i^{\,j} = \rho_j \rho_i^{\,j} \phi_i = \rho_i \phi_i$ for every pair $i \leq j$, (11.1) ensures the existence of a unique homomorphism $\phi_* : A_* \to B_*$ such that $\rho_i \phi_i = \phi_* \pi_i$. This proves the first assertion. If all the $\phi_i$ are epic, then $\operatorname{Im} \rho_i = \operatorname{Im} \rho_i \phi_i$ together exhaust $B_*$ and therefore $\phi_*$ must be an epimorphism. If all the $\phi_i$ are monic, then let $a \in \operatorname{Ker} \phi_*$. For some $i$, $a = \pi_i a_i$ with $a_i \in A_i$. Hence $\rho_i \phi_i a_i = \phi_* \pi_i a_i = \phi_* a = 0$, and in view of (e) we have $\rho_i^{\,j} \phi_i a_i = 0$ for some $j \geq i$. Since $\rho_i^{\,j} \phi_i = \phi_j \pi_i^{\,j}$ and $\phi_j$ is a monomorphism, $\pi_i^{\,j} a_i = 0$, whence $\pi_i a_i = 0$ and $a = 0$. This shows $\phi_*$ monic. $\square$

The following interpretation of (11.2) is of interest. Given a directed set $I$, consider the category $\mathscr{D}_I$ of all direct systems of abelian groups, indexed by $I$, with homomorphisms as defined above. Then the function $L_*$ assigning to a direct system $A$ its limit $A_*$ and to a homomorphism $\phi : A \to B$ the homomorphism $\phi_* : A_* \to B_*$ is, by virtue of (11.2), a functor on $\mathscr{D}_I$ to $\mathscr{A}$. In view of this, we are justified to consider $\phi_*$ as a natural map between direct limits.

Finally, we focus our attention on three direct systems: $A$, $B$ as above and $C = \{C_i \ (i \in I); \sigma_j^i\}$, all with the same index set $I$. If $\phi : A \to B$ and $\psi : B \to C$ are homomorphisms between them such that, for every $i$, the sequence

$$A_i \xrightarrow{\phi_i} B_i \xrightarrow{\psi_i} C_i$$

is exact, then we say, the sequence

(6)
$$A \xrightarrow{\phi} B \xrightarrow{\psi} C$$

is *exact*. We have:

**Theorem 11.3.** *If A, B, C are direct systems of groups such that (6) is an exact sequence, then the sequence*

$$A_* = \varinjlim A_i \xrightarrow{\ \phi_*\ } B_* = \varinjlim B_i \xrightarrow{\ \psi_*\ } C_* = \varinjlim C_i$$

*under the induced maps $\phi_*$, $\psi_*$ [cf. preceding theorem] is likewise exact.*

By (11.2) the diagram

$$
\begin{array}{ccccc}
A_i & \xrightarrow{\ \phi_i\ } & B_i & \xrightarrow{\ \psi_i\ } & C_i \\
\pi_i \downarrow & & \rho_i \downarrow & & \downarrow \sigma_i \\
A_* & \xrightarrow{\ \phi_*\ } & B_* & \xrightarrow{\ \psi_*\ } & C_*
\end{array}
\qquad (i \in I)
$$

is commutative. If $a \in A_*$, then $\pi_i a_i = a$ for some $i \in I$, and so $\psi_* \phi_* a = \psi_* \phi_* \pi_i a_i = \psi_* \rho_i \phi_i a_i = \sigma_i \psi_i \phi_i a_i = 0$. Let $b \in \operatorname{Ker} \psi_*$. For some $b_i \in B_i$, $\rho_i b_i = b$, whence $\sigma_i \psi_i b_i = \psi_* \rho_i b_i = \psi_* b = 0$. Therefore, for some $j \geq i$, $\sigma_i^j \psi_i b_i = 0$, and so $\psi_j \rho_i^j b_i = \sigma_i^j \psi_i b_i = 0$. The top row in the last diagram is exact, therefore there is an $a_j \in A_j$ with $\phi_j a_j = \rho_i^j b_i$. Setting $a = \pi_j a_j$, we arrive at $\phi_* a = \phi_* \pi_j a_j = \rho_j \phi_j a_j = \rho_j \rho_i^j b_i = \rho_i b_i = b$, i.e., $b \in \operatorname{Im} \phi_*$ and the bottom row is exact.☐

If we apply the second statement of (11.2) and (11.3) to the case when the $\phi_i$ are monomorphisms and the $\psi_i$ are epimorphisms, then we are led to:

**Corollary 11.4.** *If A, B, C are direct systems as in the preceding theorem and, in addition,*

$$0 \to A_i \xrightarrow{\ \phi_i\ } B_i \xrightarrow{\ \psi_i\ } C_i \to 0$$

*is exact for every $i \in I$, then the induced sequence for the direct limits*

$$0 \to A_* \xrightarrow{\ \phi_*\ } B_* \xrightarrow{\ \psi_*\ } C_* \to 0$$

*is likewise exact.*☐

This statement can be expressed less accurately, but more lively, by saying that *a direct limit of exact sequences is exact.* [In other words, the functor $L_* : \mathscr{D}_I \to \mathscr{A}$ is exact.]

EXERCISES

1.  (a) Let $A_n \cong Z$ ($n = 1, 2, \cdots$), and let $\pi_n^{n+1} : A_n \to A_{n+1}$ be multiplication by $n$. Prove that

$$\varinjlim A_n \cong Q.$$

(b) A group is locally cyclic if and only if it is a direct limit of cyclic groups.

2. If $A_*$ is the direct limit of the system $A = \{A_i \ (i \in I); \pi_i^j\}$, then for any integer $m > 0$, $mA_*$ is the direct limit of the system $\{mA_i \ (i \in I); \pi_i^j \mid mA_i\}$.

3. (a) If $A_*$ is the limit of the direct system $A = \{A_i \ (i \in I); \pi_i^j\}$, and if $a \in A_*$, then there exist an $i \in I$ and an $a_i \in A_i$ such that $\pi_i a_i = a$ and $o(a_i) = o(a)$.

   (b)* If $\alpha : G \to A_*$ where $G$ is finitely generated, then there exists an $i \in I$ and an $\alpha_i : G \to A_i$ such that $\alpha = \pi_i \alpha_i$. [*Hint:* (15.5).]

4. The direct limit of torsion [torsion-free] groups is again torsion [torsion-free].

5. Let $A = \{A_i \ (i \in I); \pi_i^j\}$ be a direct system with limit $A_*$ and with canonical maps $\pi_i : A_i \to A_*$. Define $A_\infty = A_*$ and $\pi_i^\infty = \pi_i$ with $i < \infty$ for all $i \in I$, and show that

$$A' = \{A_i \ (i \in I \cup \{\infty\}); \pi_i^j\}$$

is a direct system with limit $A_*$.

6. If $A = \{A_i \ (i \in I); \pi_i^j\}$ and $B = \{B_i \ (i \in I); \rho_i^j\}$ are direct systems of groups, then

$$\{A_i \oplus B_i \ (i \in I); \pi_i^j \oplus \rho_i^j\}$$

is again a direct system whose limit is the direct sum of the direct limits of $A$ and $B$.

7. Give an example showing that the direct limit of splitting exact sequences need not be splitting. [Cf. (29.5). Why does this not contradict Ex. 6?]

8. Show that the direct limit of longer exact sequences is likewise exact.

9. What is wrong with the following argument? Using Example 3, on page 56, the sequence $0 \to Z(p^\infty) \to Z(p^\infty) \to Z(p^\infty) \to 0$ can be represented as the direct limit of exact sequences $0 \to Z(p^n) \to Z(p^{2n}) \to Z(p^n) \to 0$, so it must be exact because of (11.4).

10. Give an example to show that a subfunctor $F$ of the identity need not commute with direct limits.

## 12. INVERSE LIMITS

Next we turn to the discussion of inverse limits; these are, in a certain sense, dual to direct limits.

Assume $A_i \ (i \in I)$ is a system of groups, indexed again by a directed set $I$, and for each pair $i, j \in I$ with $i \leq j$ there is given a homomorphism

$$\pi_i^j : A_j \to A_i \qquad (i \leq j)$$

such that

(i) $\pi_i^i$ is the identity map of $A_i$, for each $i \in I$,

(ii) for all $i \leq j \leq k$ in $I$, we have $\pi_i^j \pi_j^k = \pi_i^k$.

Then the system

$$A = \{A_i \ (i \in I); \ \pi_i^j\}$$

is called an *inverse system*. The *inverse* [or *projective*] *limit* of this system,

$$A^* = \varprojlim_I A_i,$$

is defined to consist of all vectors $a = (\cdots, a_i, \cdots)$ in the direct product $A = \prod_{i \in I} A_i$ for which

$$\pi_i^j a_j = a_i \qquad (i \leq j)$$

holds. $A^*$ is a subgroup of $A$, for if $a, a' \in A^*$, then their coordinates satisfy $\pi_i^j a_j = a_i$, $\pi_i^j a'_j = a'_i$; thus, $\pi_i^j (a_j - a'_j) = a_i - a'_i$ and $a - a' \in A^*$.

For inverse limits the following simple facts are worthwhile mentioning.

(a) *There exist homomorphisms*

$$\pi_i : A^* \to A_i \qquad (i \in I)$$

*such that the diagrams*

(1)

$$(i \leq j)$$

*are all commutative.*

In fact, $\pi_i : a \mapsto a_i$ [i.e., the restriction of the $i$th coordinate projection of $\prod A_i$ to $A^*$] satisfies this condition. These $\pi_i$ will be called *canonical*.

(b) *If every $\pi_i^j$ is a monomorphism, then all the $\pi_i$ are monomorphisms.*

Assume $\pi_i^j$ are monic, and let $\pi_i a = 0$ for some $a \in A^*$, $i \in I$. Given $j \in I$, let $k \in I$ satisfy $i, j \leq k$. Now $\pi_i^k \pi_k a = \pi_i a = 0$, whence, by our hypothesis, $\pi_k a = 0$. Therefore, $\pi_j a = \pi_j^k \pi_k a = 0$ for every $j$, and so $a = 0$.

(c) *If $K$ is a cofinal subsystem of $I$, then*

$$\varprojlim_K A_k \cong \varprojlim_I A_i.$$

To $a' \in \varprojlim_K A_k \leq \prod_{k \in K} A_k$, there exists a unique $a \in \varprojlim_I A_i$ such that, for every $k \in K$, the $k$th coordinates of $a'$ and $a$ are equal. In fact, if we define $a = (\cdots, a_i, \cdots)$ with $a_i = \pi_i^k a'_k \ (i \leq k)$, then $a' \mapsto a$ is an isomorphism between the two inverse limits.

(d) $A^*$ *is the intersection of kernels of certain endomorphisms of* $\prod_{i \in I} A_i$.

Let $(i, j)$ be a pair of elements of $I$ satisfying $j \geq i$. It gives rise to an endomorphism of the direct product $\prod A_i$:

$$\theta_{(i, j)}(\cdots, a_i, \cdots, a_j, \cdots) \mapsto (\cdots, 0, a_i - \pi_i^j a_j, 0, \cdots),$$

where at most the $i$th coordinate is not zero. If we compare this with the definition of inverse limits, it is evident that

$$A^* = \bigcap_{(i,j)} \text{Ker } \theta_{(i,j)}$$

with $(i,j)$ running over all pairs of the stated kind.

(e) *If all the groups in the inverse system* $A = \{A_i \ (i \in I); \ \pi_i^j\}$ *are (Hausdorff) topological groups and all the* $\pi_i^j$ *are continuous homomorphisms, then the inverse limit* $A^*$ *is a closed subgroup of* $\prod A_i$ *[the latter group is equipped with the product topology], and the canonical maps* $\pi_i : A^* \to A_i$ *are continuous.*

If $(\cdots, a_i, \cdots) \in \prod A_i$ is not in the inverse limit $A^*$, then there exist $i < j$ in $I$ such that $a_i \neq \pi_i^j a_j$. By the Hausdorff property, there are disjoint open subsets $U$, $V$ in $A_i$ with $a_i \in U$ and $\pi_i^j a_j \in V$. Now the set of all $(\cdots, b_i, \cdots) \in \prod A_i$ with $b_i \in U$ and $\pi_i^j b_j \in V$ is an open set in $\prod A_i$ that contains $(\cdots, a_i, \cdots)$ and is disjoint from $A^*$. The last assertion follows from the continuity of the coordinate projections.

We have the dual of (11.1):

**Theorem 12.1.** *The inverse limit* $A^*$ *of the inverse system* $A = \{A_i \ (i \in I); \ \pi_i^j\}$ *satisfies: if* $G$ *is a group and if there are homomorphisms* $\sigma_i : G \to A_i$ *with commutative diagrams*

(2)
$$
\begin{array}{ccc}
 & G & \\
\sigma_j \downarrow & & \searrow \sigma_i \\
A_j & \xrightarrow{\pi_i^j} & A_i
\end{array}
\qquad (i \leq j),
$$

*then there exists a unique homomorphism* $\sigma : G \to A^*$ *for which all the diagrams*

(3)
$$
\begin{array}{ccc}
G & \xrightarrow{\sigma} & A^* \\
 & \searrow_{\sigma_i} & \downarrow \pi_i \\
 & & A_i
\end{array}
\qquad (i \in I)
$$

*are commutative [where* $\pi_i$ *is the canonical homomorphism]. This property characterizes* $A^*$ *and* $\pi_i$ *up to isomorphism.*

For $g \in G$ set $\sigma g = (\cdots, \sigma_i g, \cdots) \in \prod A_i$. Because of the commutativity of (2), $\sigma g \in A^*$. Thus $\sigma : G \to A^*$ is a homomorphism satisfying $\sigma_i g = \pi_i \sigma g$, whence the commutativity of (3) results. If $\sigma' : G \to A^*$ also makes (3) commutative, then $\pi_i(\sigma - \sigma') = 0$ for every $i$. Hence, for $g \in G$ every coordinate of $(\sigma - \sigma')g$ vanishes and thus $\sigma - \sigma' = 0$.

In order to establish the second assertion, assume $A_0$ and maps $\tau_i : A_0 \to A_i$ have the property formulated in the first statement of the theorem. Then there exist unique homomorphisms $\sigma : A_0 \to A^*$ and $\sigma_0 : A^* \to A_0$ satisfying $\pi_i \sigma = \tau_i$ and $\tau_i \sigma_0 = \pi_i$. We infer $\pi_i = \pi_i \sigma \sigma_0$ for every $i$, and hence $\sigma \sigma_0$ is the identity map of $A^*$. This shows $\sigma_0 A^*$ a direct summand of $A_0$. From the uniqueness of $G \to A_0$ for every $G$, we conclude $\sigma_0 A^* = A_0$, and $\sigma_0$ is an isomorphism.$\square$

*Example 1.* Let

$$A = \prod_{i \in I} B_i$$

be the direct product of groups $B_i$. Define $J$ to consist of all finite subsets $\alpha$ of the index set $I$ where $\alpha \leq \beta$ means " $\alpha$ is a subset of $\beta$." For $\alpha \in J$ let $A_\alpha = \bigoplus_{i \in \alpha} B_i$ and for $\alpha \leq \beta$ let $\pi_\alpha^\beta$ denote the projection map $A_\beta \to A_\alpha$ [i.e., we drop the coordinates with indices in $\beta$ not in $\alpha$]. Then

$$A^* = \varprojlim_J A_\alpha \cong A.$$

To prove this, let $\pi_\alpha$ denote the canonical maps in (1) and $\sigma_\alpha$ the projections $A \to A_\alpha$. By (12.1) there is a unique $\sigma : A \to A^*$ such that $\pi_\alpha \sigma = \sigma_\alpha$. If $\sigma a = 0$ for some $a \in A$, then $\sigma_\alpha a = \pi_\alpha \sigma a = 0$ for all $\alpha \in J$, and so $a = 0$, $\sigma$ is a monomorphism. If $a^* = (\cdots, a_\alpha, \cdots) \in A^*$, then write $a_\alpha = b_{i_1} + \cdots + b_{i_k}$ ($b_i \in B_i$), where $\alpha = \{i_1, \cdots, i_k\}$. Because of the choice of $\pi_\alpha^\beta$, $i_1 \in \beta$ implies that the $i_1$th coordinate of $a_\beta$ must be $b_{i_1}$, hence $a^*$ defines a unique $(\cdots, b_i, \cdots) \in A$. A glance at the definition of $\sigma$ in the proof of (12.1) shows that $\sigma(\cdots, b_i, \cdots) = a^*$, so $\sigma$ is an epimorphism, and hence an isomorphism.

*Example 2.* Let $C_n = \langle c_n \rangle$ be a cyclic group of order $p^n$ ($n = 1, 2, \cdots$), and let $\pi_n^{n+1} : C_{n+1} \to C_n$ act as $\pi_n^{n+1} c_{n+1} = c_n$. The meaning of $\pi_n^m$ ($m \geq n$) is then obvious. Now $\{C_n \ (n = 1, 2, \cdots); \pi_n^m\}$ is an inverse system such that

$$C^* = \varprojlim_n C_n \cong J_p.$$

If $\pi_n$ denotes the canonical map $C^* \to C_n$, and if we define $\sigma_n : J_p \to C_n$ by $\sigma_n 1 = c_n$ ($1 \in J_p$), then there is a unique $\sigma : J_p \to C^*$ such that $\pi_n \sigma = \sigma_n$. Since no element $\neq 0$ of $J_p$ belongs to every $\mathrm{Ker}\ \sigma_n$, $\mathrm{Ker}\ \sigma = 0$. If $c = (c_1', \cdots, c_n', \cdots) \in C^*$ with $c_n' = k_n c_n$ ($k_n \in \mathbf{Z}$), then by the choice of $\pi_n^{n+1}$ we have $k_{n+1} \equiv k_n \bmod p^n$, and there is a $p$-adic integer $\tau$ such that $\tau \equiv k_n \bmod p^n$ for every $n$. Thus $\sigma_n \tau = c_n'$, and $\sigma$ must be epic.

*Example 3.* Let $A_i$ ($i \in I$) be subgroups of a group $A$ and $A_\infty = \bigcap_{i \in I} A_i$. Define $i < \infty$ for all $i \in I$, and $i \leq j$ for $i, j \in I \cup \{\infty\}$ to mean that $A_j \leq A_i$ and let $\pi_i^j$ be the injection map $A_j \to A_i$. Then

$$A = \{A_i \ (i \in I \cup \{\infty\}); \pi_i^j\}$$

is an inverse system with inverse limit $A_\infty$. In fact, $A^* = \varprojlim A$ consists of all vectors in $\prod A_i$ whose coordinates are the same element of $A_\infty$. This example shows that intersections may be regarded as inverse limits.

The notion of homomorphism for inverse systems can be defined analogously to direct systems. Let

$$A = \{A_i \ (i \in I); \pi_i^j\} \qquad \text{and} \qquad B = \{B_i \ (i \in I); \rho_i^j\}$$

be inverse systems, indexed by the same directed set $I$. A *homomorphism* $\phi : A \to B$ is a set of homomorphisms $\{\phi_i : A_i \to B_i \ (i \in I)\}$ subject to the condition that the diagrams

$$
\begin{array}{ccc}
A_j & \xrightarrow{\;\pi_i{}^j\;} & A_i \\
{\scriptstyle \phi_j}\downarrow & & \downarrow{\scriptstyle \phi_i} \\
B_j & \xrightarrow{\;\rho_i{}^j\;} & B_i
\end{array}
\qquad (i \le j)
$$

(4)

be commutative.

**Theorem 12.2.** *If* $\phi : A \to B$ *is a homomorphism between the inverse systems* $A$, $B$, *then there exists a unique homomorphism*

$$
\phi^* : A^* = \varprojlim_I A_i \to B^* = \varprojlim_I B_i
$$

*such that, for every* $i \in I$, *the diagram*

$$
\begin{array}{ccc}
A^* & \xrightarrow{\;\pi_i\;} & A_i \\
{\scriptstyle \phi^*}\downarrow & & \downarrow{\scriptstyle \phi_i} \\
B^* & \xrightarrow{\;\rho_i\;} & B_i
\end{array}
$$

(5)

*is commutative* $[\pi_i, \rho_i$ *denote the canonical maps]. If every* $\phi_i$ *is monic, then* $\phi^*$ *is a monomorphism.*

The homomorphisms $\phi_i$ $(i \in I)$ induce a homomorphism

$$
\tilde{\phi} = \prod \phi_i : \prod A_i \to \prod B_i .
$$

The commutativity of (4) implies that if $a = (\cdots, a_i, \cdots) \in A^*$, then its image under $\tilde{\phi}$ is contained in $B^*$; hence define $\phi^* : A^* \to B^*$ as the restriction of $\tilde{\phi}$ to $A^*$. With this $\phi^*$, $\phi_i \pi_i a = \phi_i a_i = \rho_i \phi^* a$, and (5) is commutative. If $\phi_0 : A^* \to B^*$ makes (5) commutative for every $i$, then $\rho_i(\phi^* - \phi_0) = 0$ for every $i$, thus all the coordinates of $(\phi^* - \phi_0)a$ vanish, for every $a \in A^*$, and so $\phi^* - \phi_0 = 0$.

If every $\phi_i$ is a monomorphism, and if $\phi^* a = 0$ for some $a \in A^*$, then $\phi_i \pi_i a = \rho_i \phi^* a = 0$ implies $\pi_i a = 0$ for every $i \in I$, whence $a = 0.\square$

For a fixed directed set $I$, consider the category $\mathscr{I}_I$ of all inverse systems indexed by $I$ with homomorphisms as defined before (12.2). The function $L^*$ that assigns to an inverse system $A$ its inverse limit $A^*$ and to a homomorphism $\phi : A \to B$ the homomorphism $\phi^* : A^* \to B^*$ is—as is readily seen from (12.2)—a functor on $\mathscr{I}_I$ to $\mathscr{A}$. Hence we may refer to $\phi^*$ as a natural homomorphism between the inverse limits.

For the inverse limit of exact sequences we have a weaker result than for direct limits.

**Theorem 12.3.** *Let*

$$
A = \{A_i \, (i \in I); \pi_i^j\}, \qquad B = \{B_i \, (i \in I); \rho_i^j\}, \qquad C = \{C_i \, (i \in I); \sigma_i^j\}
$$

*be inverse systems with the same index set* $I$, *and* $\phi : A \to B$, $\psi : B \to C$ *homomorphisms such that the sequence*

$$
0 \to A \xrightarrow{\;\phi\;} B \xrightarrow{\;\psi\;} C
$$

*is exact* [*i.e.,*

$$0 \to A_i \xrightarrow{\phi_i} B_i \xrightarrow{\psi_i} C_i$$

*is exact for every* $i$]. *Then for the inverse limits we have the exact sequence*

$$0 \to A^* \xrightarrow{\phi^*} B^* \xrightarrow{\psi^*} C^*$$

*where* $\phi^*$, $\psi^*$ *denote the maps described in* (12.2).

The exactness at $A^*$ is just the second statement of (12.2). From the definition of $\phi^*$, $\psi^*$ it is evident that $\psi^*\phi^* = 0$. If $\pi_i$, $\rho_i$, $\sigma_i$ are the canonical maps, then the diagram

$$
\begin{array}{ccccc}
0 \to A^* & \xrightarrow{\phi^*} & B^* & \xrightarrow{\psi^*} & C^* \\
\downarrow{\scriptstyle \pi_i} & & \downarrow{\scriptstyle \rho_i} & & \downarrow{\scriptstyle \sigma_i} \quad (i \in I) \\
0 \to A_i & \xrightarrow{\phi_i} & B_i & \xrightarrow{\psi_i} & C_i
\end{array}
$$

is commutative. In order to show that the top row is exact at $B^*$, let $b \in \operatorname{Ker} \psi^*$. In view of $\psi_i \rho_i b = \sigma_i \psi^* b = 0$ and the exactness of the bottom row, there is an $a_i \in A_i$, for every $i \in I$, for which $\phi_i a_i = \rho_i b$. For $j > i$, $\phi_i \pi_i^j a_j = \rho_i^j \phi_j a_j = \rho_i^j \rho_j b = \rho_i b = \phi_i a_i$, whence $\pi_i^j a_j = a_i$, $\phi_i$ being monic. We infer that $a = (\cdots, a_i, \cdots, a_j, \cdots) \in A^*$. For this $a$ we have $\rho_i \phi^* a = \phi_i \pi_i a = \phi_i a_i = \rho_i b$ for every $i$, and so $\phi^* a = b$.□

Let us note that by (12.3), the functor $L^* \colon \mathscr{I}_I \to \mathscr{A}$ is left exact.

Exercise 6 will show that (12.3) cannot be improved by putting $\to 0$ at the end of the displayed formulas.

### EXERCISES

1. Let $A_n \cong Z(n) = \langle a_n \rangle$, and for $n \mid m$, let $\pi_n^m \colon A_m \to A_n$ be the homomorphism that maps $a_m$ upon $a_n$. Then $\{A_n \ (n \in I); \pi_n^m\}$ is an inverse system [$I$ is partially ordered by the divisibility relation] whose inverse limit is $\prod_p J_p$.

2. Let $A_n \cong Z(p^\infty)$ and let $\pi_n^{n+1} \colon A_{n+1} \to A_n$ be the multiplication by $p$. Then the inverse limit is the group of all $p$-adic numbers.

3. If $A = \{A_i \ (i \in I); \pi_i^j\}$ and $B = \{B_i \ (i \in I); \rho_i^j\}$ are inverse systems, then so is

$$\{A_i \oplus B_i \ (i \in I); \pi_i^j \oplus \rho_i^j\},$$

and its inverse limit is the direct sum of the inverse limits of $A$ and $B$.

4. The inverse limit of torsion-free groups is torsion-free, but the inverse limit of torsion groups can be torsion-free $\neq 0$.

5. If in an inverse system $\{A_i \ (i \in I); \pi_i^j\}$, all the maps $\pi_i^j$ are isomorphisms, then all the $\pi_i$ are likewise isomorphisms.

6.  Let $A_n = Z$ and $\pi_n^m = 1_Z$ for all positive integers $n, m \ (\geq n)$. Let $B_n = Z(p^n)$ and let $\rho_n^m$ carry 1 of $Z(p^m)$ into 1 of $Z(p^n)$. Show that
    (a) $\{A_n; \pi_n^m\}$ and $\{B_n; \rho_n^m\}$ are inverse systems such that $\phi = \{\phi_n\}$ with $\phi_n : A_n \to B_n$ the canonical map is an epimorphism; but
    (b) the induced homomorphism $\phi^*$ between the two inverse limits is not epic.

7.  Show that the inverse limit of splitting exact sequences need not be splitting exact.

8.  Let $A = \{A_i \ (i \in I); \pi_i^j\}$ be an inverse system with limit $A^*$ and canonical maps $\pi_i : A^* \to A_i$. Define $A_\infty = A^*$ and $\pi_i^\infty = \pi_i$ with $i < \infty$ for all $i \in I$, and show that

$$A' = \{A_i \, (i \in I \cup \{\infty\}); \pi_i^j\}$$

is an inverse system with the same limit $A^*$.

## 13.   COMPLETENESS AND COMPLETIONS

Abelian groups which are complete in some topology will be of importance for us. Therefore, in this section, we wish to examine completeness and completion processes. All the topologies to be considered here are linear.

Assume a linear topology is defined on the group $A$ in terms of the dual ideal $\mathbf{D}$ of $\mathbf{L}(A)$. We label the subgroups $U$ in $\mathbf{D}$ by an index set $I$, and put $i \leq j$ for $i, j \in I$ to mean $U_i \geq U_j$. Thus $I$—as a partially ordered set—will be dual-isomorphic to $\mathbf{D}$, and hence $I$ is directed.

As usual, by a *net* in $A$ we mean a set $\{a_i\}_{i \in I}$ of elements $a_i \in A$, indexed always by $I$; in other words, a net is a function on $I$ to $A$. A net $\{a_i\}_{i \in I}$ is said to *converge* to the *limit* $a \in A$ if to every $i \in I$ there is a $j \in I$ such that $a_k - a \in U_i$ whenever $k \geq j$. If $A$ is Hausdorff in the given topology, then limits are unique; if, however, $A$ fails to be Hausdorff, then limits are determined only up to mod $\bigcap_i U_i$. The classical proof applies to show that a subgroup $B$ of $A$ is closed exactly if it contains the limits of convergent nets with members in $B$.

Evidently, if $\{a_i\}_{i \in I}$ is convergent, then so is every cofinal subnet, and it has the same limit. To facilitate discussion, we shall concentrate on nets $\{b_i\}_{i \in I}$ such that, for every $i \in I$, $b_k - a \in U_i$ for all $k \geq i$ [i.e., $j = i$ may be chosen in the definition of convergence]. We shall call such a net *neatly convergent*.

A net $\{a_i\}_{i \in I}$ is called a *Cauchy net* if to any given $i \in I$ there is a $j \in I$ such that

$$a_k - a_{k'} \in U_i \qquad \text{whenever} \quad k, k' \geq j.$$

Since in the present case $U_i$ is a subgroup, $a_k - a_j$, $a_{k'} - a_j \in U_i$ already implies $a_k - a_{k'} \in U_i$, so that, for the Cauchy character of $\{a_i\}$, it suffices to know that $a_k - a_j \in U_i$ holds for all $k \geq j$. Clearly, every cofinal subnet of a Cauchy net is again Cauchy, and if a cofinal subnet of a Cauchy net converges, then the whole net converges to the same limit. We shall consider Cauchy nets $\{b_i\}_{i \in I}$ which are *neat*, in the sense that for every $i \in I$,

$$b_k - b_i \in U_i \qquad \text{whenever} \quad k \geq i.$$

If a neat Cauchy net $\{b_i\}_{i \in I}$ converges to a limit $a$, then it neatly converges to it; for if $b_j - b \in U_i$ for some $j$ which may be assumed to be $\geq i$, then $b_j - b_i \in U_i$ implies $b_i - b \in U_i$, and hence $b_k - b_i \in U_i$ $(k \geq i)$ implies $b_k - b \in U_i$ for all $k \geq i$.

A group $A$ is called *complete* in a given topology if it is Hausdorff, and every Cauchy net in $A$ has a limit in $A$. An equivalent definition may be given by restricting ourselves to neat Cauchy nets. [Notice that, unlike many authors, we have defined completeness for Hausdorff spaces only.] We have the rather simple:

**Proposition 13.1.** *Let the group $A$ be complete in a topology. A subgroup of $A$ is closed if and only if it is complete in the induced topology.*

For the proof we refer, e.g.,to J. L. Kelley,"General Topology"(1955).□

In the next result the countability hypothesis is essential.

**Proposition 13.2.** *If $B$ is a closed subgroup of a complete group $A$ satisfying the first axiom of countability, then the quotient group $A/B$ is complete in the induced topology.*

By hypothesis, there is a base $\{U_m\}_m$ of neighborhoods of 0 in $A$ such that $U_1 \geq U_2 \geq \cdots$ and $\bigcap_m U_m = 0$. The groups $\bar{U}_m = (U_m + B)/B$ $(m = 1, 2, \cdots)$ form a base of neighborhoods of $\bar{0}$ in $A/B$. Let $a_1 + B, \cdots, a_m + B, \cdots$ be a Cauchy sequence in $A/B$ which may, without loss of generality, be assumed to be neat, i.e., $a_{m+1} - a_m + B \subseteq U_m + B$ for every $m$. Define $c_1 = a_1$, and assume we have chosen $c_1, \cdots, c_m \in A$ such that $c_i \in a_i + B$ and $c_i - c_{i-1} \in U_{i-1}$ for $i = 2, \cdots, m$. Then $a_{m+1} - c_m = u_m + b_m$ for some $u_m \in U_m$, $b_m \in B$, and we set $c_{m+1} = a_{m+1} - b_m \in a_{m+1} + B$ to have $c_{m+1} - c_m \in U_m$. Therefore, there is a sequence $\{c_m\}_m$ such that $c_m \in a_m + B$ and $c_{m+1} - c_m \in U_m$ for every $m$. In other words, the given Cauchy sequence $\{a_m + B\}$ in $A/B$ can be lifted to a Cauchy sequence $\{c_m\}$ in $A$. If $a \in A$ is the limit of $\{c_m\}$, then $\bar{a} = a + B$ is the limit of $\{a_m + B\}$. Since $B$ is closed in $A$, $A/B$ is Hausdorff.□

Let $A_t$ $(t \in J)$ be a family of groups, each equipped with a linear topology, say, defined in terms of the dual ideal $\mathbf{D}_t$ of $\mathbf{L}(A_t)$. Let $G = \prod A_t$ be the direct product of these groups and $\pi_t$ the $t$th coordinate projection, $\pi_t : G \to A_t$.

Recall that the *product topology* on $G$ is defined such that a subbase of neighborhoods of 0 is the set of all $\pi_t^{-1} U_t$ with $U_t \in \mathbf{D}_t$ and $t \in J$. This is again a linear topology, and the $\pi_t$ are continuous, open homomorphisms.

There is another topology on the direct product $\prod A_t$ which will be most important for us. Without loss of generality, one may assume that the very same index set $I$ serves to index a base of neighborhoods about 0 in each $A_t : \{U_{ti}\}_{i \in I}$ in $A_t$. [Notice that $j \geq i$ in $I$ implies $U_{ti} \geq U_{tj}$ for every $t$, but $U_{ti} = U_{tj}$ may occur for $i \neq j$.] The *box topology* on $G = \prod A_t$ is defined to have the subgroups

$$\prod_t U_{ti} = U_i^* \qquad (i \in I)$$

as a base of neighborhoods about 0. This topology is linear; it is Hausdorff exactly if all the $A_t$ are Hausdorff and satisfies the first countability hypothesis if all the $A_t$ do. The inclusion $U_i^* \leq \pi_t^{-1} U_{ti}$ (for every $t$) shows that the box topology is finer [i.e., has more open sets] than the product topology. Hence the $\pi_t$ are again continuous and open.

Notice that the box topology may depend on the way the $U_{ti}$ are indexed.

An important special case is when all the groups $A_t$ have their $Z$-adic [$p$-adic] topologies. Then the box topology in $\prod A_t$ is just the $Z$-adic [$p$-adic] topology of its own.

The box topology has the advantage that in the direct product we may use the same set $I$ to index nets. An elementary calculation shows that, in the box topology of $G = \prod A_t$, the projection $\{\pi_t g_i\}_{i \in I}$ of a (neat) Cauchy net $\{g_i\}_{i \in I}$ in $G$ is a (neat) Cauchy net in $A_t$, and a net $\{g_i\}_{i \in I}$ is a neat Cauchy net if, for every $t \in J$, $\{\pi_t g_i\}_{i \in I}$ is a neat Cauchy net in $A_t$.

**Proposition 13.3.** *A direct product is complete in the box topology if and only if every component is complete.*

Let $G = \prod A_t$ be complete in the box topology, and let $\{a_i\}_{i \in I}$ be a Cauchy net in $A_t$. Then $\{\rho_t a_i\}_{i \in I}$ is a Cauchy net in $G$ [$\rho_t$ denotes the $t$th coordinate injection], say, tending to the limit $g \in G$. Evidently, $\pi_t g$ is the limit of $\{a_i\}$. Conversely, assume all the $A_t$ complete, and let $\{g_i\}_{i \in I}$ be a neat Cauchy net in $G$. If $a_t \in A_t$ denotes the limit of the Cauchy net $\{\pi_t g_i\}_{i \in I}$, and if $g \in G$ is defined so that $\pi_t g = a_t$, then it follows at once that $g$ is the limit of $\{g_i\}_{i \in I}$. $\square$

If all the groups $A_t$ are furnished with the $Z$-adic [$p$-adic] topology, then we have:

**Corollary 13.4.** *A direct product of groups is complete in the $Z$-adic [$p$-adic] topology if and only if each component is complete in its $Z$-adic [$p$-adic] topology.* $\square$

In the rest of this section we turn our attention to the problem of embedding a group $A$, equipped with a linear topology, in a complete group $\hat{A}$. We discuss two methods, one based on Cauchy nets and another on inverse limits.

Let $A$ be a topological [not necessarily Hausdorff] group and $\{U_i\}_{i \in I}$ a base of neighborhoods of 0. A net $\{a_i\}_{i \in I}$ in $A$ can be identified with an element of the direct product $G = A^I$ under the natural correspondence

$$\{a_i\}_{i \in I} \mapsto (\cdots, a_i, \cdots).$$

In order to avoid unnecessary complications, we shall restrict ourselves to neat Cauchy nets; this can be done without loss of generality. It is readily checked that the neat Cauchy nets in $A$ form a subgroup $C$ of $G$, and the nets neatly converging to 0 form a subgroup $E$ of $C$. It follows at once that $E$ is closed in $C$ if $C$ is furnished with the product or with the box topology; hence $\hat{A} = C/E$ is a Hausdorff topological group in either induced topology.

**Lemma 13.5.** *The product and the box topologies of $A^I$ induce the same topology on $\hat{A}$.*

Since the two topologies are comparable, it suffices to show that to every neighborhood $U_i^*$ in the box topology there is a neighborhood $\pi_j^{-1} U_k$ in the product topology for which

$$(C \cap \pi_j^{-1} U_k) + E \leqq (C \cap U_i^*) + E$$

holds. By making use of the modular law, this amounts to

$$C \cap \pi_j^{-1} U_k \leqq U_i^* + E.$$

If $\{a_l\}_{l \in I}$ is a neat Cauchy net contained in $\pi_i^{-1} U_i$, then not only $a_i \in U_i$ holds, but also $a_j \in U_i$ for all $j \geqq i$, whence $\{a_l\}_{l \in I} \in U_i^* + E.\square$

An obvious consequence of this lemma is that a base of neighborhoods of 0 in $A$ is obtained by forming the subgroups $\hat{U}_i$ of $\hat{A}$, consisting of all neat Cauchy nets $\{a_l\}_{l \in I}$ such that $a_l \in U_i$ for all $l$.

The mapping

$$\mu : a \mapsto (\cdots, a, \cdots, a, \cdots) + E$$

of $A$ into $\hat{A}$ is clearly a continuous, open homomorphism. Moreover, if $A$ is Hausdorff, $\mu$ is a topological isomorphism between $A$ and $\mu A$ if $\mu A$ is taken in the induced topology of $\hat{A}$ [this topology of $A$ is described in (13.5)].

**Theorem 13.6.** *The group $\hat{A}$ is complete and $\mu A$ is a dense subgroup of $\hat{A}$.*

Given $(\cdots, a_i, \cdots) + E \in \hat{A}$ and a neighborhood $U_i^*$ of 0 in $G$, it is clear that $\mu a_i$ lies in the $U_i^*$-neighborhood of $(\cdots, a_i, \cdots) + E$. Hence $\mu A$ is dense in $\hat{A}$, and therefore, to prove completeness, we need only verify the convergence of Cauchy nets in $\mu A$ to elements of $\hat{A}$. A neat Cauchy net in $\mu A$ is of the form $\{\mu a_i\}_{i \in I}$ where $\{a_i\}_{i \in I}$ is a neat Cauchy net in $A$. It is straightforward to check that $(\cdots, a_i, \cdots) + E$ is the limit in $\hat{A}$ of the net $\{\mu a_i\}_{i \in I}$. $\square$

Another method of constructing completions is *via* inverse limits. Starting again with $\{U_i\}_{i \in I}$, let us define $C_i = A/U_i$ and, for $j \geq i$ in $I$, the homomorphism $\pi_i^j : C_j \rightarrow C_i$ as $\pi_i^j(a + U_j) = a + U_i$. Then we get an inverse system $C = \{C_i \ (i \in I); \ \pi_i^j\}$ whose inverse limit we denote by $\hat{C}$. Observe that the groups $C_i$ carry the discrete topology, and $\hat{C}$ is considered as equipped with the topology induced by the product topology on $\prod C_i$. Thus a base $\{V_i\}_{i \in I}$ of neighborhoods of 0 in $\hat{C}$ is given by the subgroups $\hat{C} \cap \pi_i^{-1}0$ [$\pi_i$ denoting the $i$th coordinate projection]. Clearly,

$$\alpha : a \mapsto (\cdots, a + U_i, \cdots) \in \hat{C}$$

is a homomorphism which is continuous and open.

**Proposition 13.7.** *There is a natural map* $\lambda : \hat{A} \rightarrow \hat{C}$ *which is a topological isomorphism such that* $\lambda\mu = \alpha$.

If $\{a_i\}_{i \in I}$ is a neat Cauchy net, representing an element of $\hat{A}$, then we define

$$\lambda : (\cdots, a_i, \cdots) + E \mapsto (\cdots, a_i + U_i, \cdots)$$

where the image is in $\hat{C}$ since $\pi_i^j(a_j + U_j) = a_j + U_i = a_i + U_i$ for $j \geq i$. If $(\cdots, a_i + U_i, \cdots) \in \hat{C}$, then $\{a_i\}_{i \in I}$ is a neat Cauchy net, and another selection of representatives yields the same coset mod $E$. That $\lambda$ maps neighborhoods upon neighborhoods is obvious, and so is $\lambda\mu = \alpha$. $\square$

Since we wish to speak of $\hat{A} \cong \hat{C}$ as *the* completion of $A$, a unicity statement is necessary. This will follow from the following universal property of $\hat{A}$.

**Theorem 13.8.** *Given a continuous homomorphism* $\phi$ *of* $A$ *into a complete group* $G$, *there exists a unique continuous homomorphism* $\hat{\phi} : \hat{A} \rightarrow G$ *such that* $\hat{\phi}\mu = \phi$.

Let $\{a_i\}_{i \in I}$ be a Cauchy net in $A$ with limit $\hat{a} \in \hat{A}$. Then continuity implies that $\{\phi a_i\}_{i \in I}$ must be Cauchy in $G$, and if $g \in G$ is its limit, then the only possible way of defining a continuous $\hat{\phi}$ is to put $\hat{\phi} : \hat{a} \mapsto g$. The rest of the assertion is self-evident. $\square$

From (13.8) it follows that $\hat{A}$ is unique up to topological isomorphism. Also, $\mu : A \rightarrow \hat{A}$ is a natural map, for if $\alpha : A \rightarrow B$ is a continuous homomorphism, then the diagram

$$\begin{array}{ccc} A & \xrightarrow{\alpha} & B \\ \mu \downarrow & & \downarrow \mu' \\ \hat{A} & \xrightarrow{\hat{\alpha}} & \hat{B} \end{array}$$

commutes where the meaning of $\mu'$ is obvious and $\hat{\alpha}$ is the homomorphism whose existence is stated in (13.8).

EXERCISES

1.  Prove that $\mu A$ [with $\mu$ as defined in the text] is topologically isomorphic to $A/U$, where $U$ is the intersection of all open subgroups of $A$.
2.  Every compact group is complete.
3.  The inverse limit of complete groups is complete.
4.  If $A$ has a base of neighborhoods about 0 consisting of subgroups of finite index, then the completion of $A$ is compact.
5.  Find the completion of $Z$ in the $p$-adic [and in the $Z$-adic] topology.
6.* (S. Lefschetz) A group $A$ is called *linearly compact* if it has a Hausdorff linear topology such that, if $C_j$ are closed subgroups of $A$ and any finite number of the cosets $a_j + C_j$ ($a_j \in A$) have a nonempty intersection, then the intersection of all $a_j + C_j$ is not vacuous either. Prove:
   (a) In this definition, "closed" can be replaced by "open."
   (b) If $A$ is linearly compact and if $\alpha : A \to B$ is a continuous epimorphism, then $B$ is linearly compact.
   (c) If $A$ is linearly compact, then every closed subgroup of $A$ has this property.
   (d) Direct products and inverse limits of linearly compact groups are linearly compact [in the product topology].
   (e) Linearly compact groups are linearly compact in any coarser topology.
7.* (a) Linearly compact groups are complete.
   (b) A group is linearly compact in the discrete topology if and only if it satisfies the minimum condition on subgroups.
   (c) Give examples for a complete group which is not linearly compact and for a linearly compact group which is not compact.

**NOTES**

Most of the results of Chapter II could have been stated more generally for certain categories; for a systematic treatment of these questions we refer to B. Mitchell [*Theory of Categories* (1965)].

Theorems (8.1), (8.2), (11.1), and (12.1) point out the universal [and co-universal] properties of the concepts under consideration. The universality of certain objects is essential in general category theory.

The primary decomposition theorem (8.4) is of central importance (in the finite case, such a decomposition characterizes the nilpotent groups). E. Matlis [ *Trans. Amer. Math. Soc.* **125,** 147–179 (1966)] considered commutative domains R such that every torsion R-module $M$ is a direct sum of submodules $M_P$ where for every $a \in M_P$, $o(a)$ is contained in exactly one maximal ideal P of R. He proved that R has this property if, and only if, every nonzero ideal of R is contained in but a finite number of maximal ideals and every prime ideal $\neq 0$ in exactly one maximal ideal.

In contrast to (8.4), (8.5) easily generalizes to arbitrary modules: if a module is the union of simple submodules, then it is a direct sum of simple modules [it is then called a semisimple module]. Semisimple modules may be characterized by the condition that every submodule is a direct summand.

*Problem 3.* (a) (J. Irwin) For which subgroups $B$ of a group $A$ are all $B$-high subgroups isomorphic?

(b) The same question with the $B$-high subgroups restricted to those containing a maximal fully invariant subgroup of $A$ disjoint from $B$.

*Problem 4.* Give an upper bound for the number of nonisomorphic $B$-high subgroups of $A$, in particular, if $B = A^1$.

*Problem 5.* Investigate the groups in which every infinite subgroup $B$ can be embedded in a direct summand of cardinality $|B|$.

See Irwin and Richards [1].

*Problem 6.* Which classes of abelian groups are closed (a) under taking subgroups and direct limits? (b) under epimorphic images and direct limits? See paper by Hill and Chromone by Chang Mo Bang.
E.g., locally cyclic groups. Naturally, analogous questions can be asked for other operations on classes; cf. **18**, Ex. 7.

*Problem 7.* (B. Charles) Describe the functorial subgroups that commute with direct products [or inverse limits].

*Problem 8.* Which groups can be complete in some metric (linear) topology?

In this connection see **39**.

# III

## DIRECT SUMS OF CYCLIC GROUPS

In this chapter we begin the study of important classes of abelian groups. First we shall be concerned with direct sums of cyclic groups. Their importance stems from the fact that they can easily be characterized by satisfactory invariants, and the analysis of other classes of abelian groups is based, to a certain extent, on our knowledge of direct sums of cyclic groups.

The first section is devoted to free groups and to a brief discussion of defining groups in terms of generators and defining relations. This is followed by a description of finitely generated groups; the so-called Fundamental Theorem is referred to in various branches of mathematics. For infinitely generated groups, one can establish criteria under which a group decomposes into a direct sum of cyclic groups; these are, however, only easy to handle in the torsion case. One of the most useful results asserts that the class of direct sums of cyclic groups is closed under passage to subgroups.

Every abelian group $A$ contains subgroups that are direct sums of cyclic groups. Those which are, in a certain sense, maximal among them, define cardinal numbers depending only on $A$. This leads to the definition of ranks which are very useful invariants for $A$.

## 14. FREE ABELIAN GROUPS—DEFINING RELATIONS

By a *free abelian group* is meant a direct sum of infinite cyclic groups. If these cyclic groups are generated by elements $x_i$ $(i \in I)$, then the free group $F$ will be

$$F = \bigoplus_{i \in I} \langle x_i \rangle.$$

Thus, $F$ consists of all finite linear combinations

$$(1) \qquad g = n_1 x_{i_1} + \cdots + n_k x_{i_k}$$

72

with different $x_{i_1}, \cdots, x_{i_k}$ where $n_j$ are integers $\neq 0$ and $k$ is a nonnegative integer. In view of the definition of direct sums, two linear combinations (1) are equal elements in $F$ if and only if they differ at most in the order of summands, and the addition of two linear combinations is to be performed formally, by adding the coefficients of the same $x_i$. Clearly, we could have equally well defined $F$ by starting with a nonempty set

$$X = \{x_i\}_{i \in I}$$

of symbols $x_i$, called *a free set of generators*, and then declaring $F$ as the collection of all formal expressions of the form (1) under the mentioned equality and addition. In view of this, $F$ is also called *the free group on the set $X$.*

It is evident that the free group $F$ is, up to isomorphism, uniquely determined by the cardinal number $\mathfrak{m}$ of the index set $I$. Thus we are justified to write $F_{\mathfrak{m}}$ for a free group with $\mathfrak{m}$ free generators. Now we have the following simple result.

**Proposition 14.1.** *The free groups $F_{\mathfrak{m}}$ and $F_{\mathfrak{n}}$ are isomorphic if and only if $\mathfrak{m} = \mathfrak{n}$ for the cardinals $\mathfrak{m}, \mathfrak{n}$.*

We need only verify the "only if" part of the assertion. Let $p$ be a prime and $F$ a free group with $\mathfrak{m}$ free generators $x_i$. Since every element $g \in F$ has a unique form (1), it is clear that $g \in pF$ is equivalent to the simultaneous fulfillment of the divisibility relations $p|n_1, \cdots, p|n_k$. Hence $F/pF$—as a vector space over the prime field of characteristic $p$—has a basis $\{x_i + pF\}$; thus, its dimension is $\mathfrak{m}$. The dimension being an invariant of vector spaces, the assertion follows.☐

Consequently, there is a one-to-one correspondence between nonisomorphic free groups and the cardinal numbers. If $\mathfrak{m}$ is the cardinal corresponding to the free group $F$, we call $F$ *of rank* $\mathfrak{m}$. The rank of a free group $F$ is uniquely determined.

One of the basic properties of free groups is contained in the following result.

**Theorem 14.2.** *A set $X = \{x_i\}_{i \in I}$ of generators of a group $F$ is a free set of generators [and hence $F$ is a free group] if and only if every mapping $\phi$ of $X$ into a group $A$ can be extended to a (unique) homomorphism $\psi : F \to A$.*

Let $X$ be a free set of generators of $F$. If $\phi : x_i \mapsto a_i$ is a mapping of $X$ into a group $A$, then define $\psi : F \to A$ as

$$\psi(n_1 x_{i_1} + \cdots + n_k x_{i_k}) = n_1 a_{i_1} + \cdots + n_k a_{i_k}.$$

The uniqueness of (1) guarantees that $\psi$ is well defined, and it is readily checked that it preserves addition. Conversely, assume that the subset $X$ in $F$ has the stated property. Then let $G$ be a free group with a free set $\{y_i\}_{i \in I}$ of generators, where $I$ is the same as for $X$. By hypothesis, $\phi : x_i \mapsto y_i \ (i \in I)$ can be lifted to a homomorphism $\psi : F \to G$, which cannot be anything else than the map $\psi : n_1 x_{i_1} + \cdots + n_k x_{i_k} \mapsto n_1 y_{i_1} + \cdots + n_k y_{i_k}$. It is evident that $\psi$ must be an isomorphism.$\square$

In particular, mapping $X$ onto a generating system of $A$, we arrive at the following result.

**Corollary 14.3.** *Every group with at most* $\mathfrak{m}$ *generators is an epimorphic image of* $F_{\mathfrak{m}}$.$\square$

For an infinite cardinal $\mathfrak{m}$, $F_{\mathfrak{m}}$ has $2^{\mathfrak{m}}$ subsets, and hence at most $2^{\mathfrak{m}}$ subgroups and quotient groups. We infer that *there exist at most* $2^{\mathfrak{m}}$ *pairwise nonisomorphic groups of cardinality* $\leq \mathfrak{m}$.

The following is an elementary, frequently used property of free groups.

**Theorem 14.4.** *If $B$ is a subgroup of $A$ such that $A/B$ is free, then $B$ is a direct summand of $A$.*

By (9.4) it suffices to verify this for the case in which $A/B$ is an infinite cyclic group, say $A/B = \langle a^* \rangle$. Select a representative $a \in a^*$ in $A$. Then the cosets $na^* \ (n = 0, \pm 1, \pm 2, \cdots)$ mod $B$ are represented by the elements $na$ of $\langle a \rangle$. Hence $A = B \oplus \langle a \rangle$.$\square$

The next theorem provides complete information about the structure of subgroups of free groups.

**Theorem 14.5.** *A subgroup of a free group is free.*

Let $F = \oplus_{i \in I} \langle a_i \rangle$ be a free group, and suppose that the index set $I$ is well ordered in some way; moreover, $I$ is the set of ordinals $< \tau$. For $\sigma \leq \tau$ we define $F_\sigma = \oplus_{\rho < \sigma} \langle a_\rho \rangle$. If $G$ is a subgroup of $F$, then set $G_\sigma = G \cap F_\sigma$. Clearly, $G_\sigma = G_{\sigma+1} \cap F_\sigma$, so $G_{\sigma+1}/G_\sigma \cong (G_{\sigma+1} + F_\sigma)/F_\sigma$. The latter quotient group is a subgroup of $F_{\sigma+1}/F_\sigma \cong \langle a_\sigma \rangle$; thus either $G_{\sigma+1} = G_\sigma$ or $G_{\sigma+1}/G_\sigma$ is an infinite cyclic group. By (14.4) we have $G_{\sigma+1} = G_\sigma \oplus \langle b_\sigma \rangle$ for some $b_\sigma \in G_{\sigma+1}$ [which is 0 if $G_{\sigma+1} = G_\sigma$]. It follows that the elements $b_\sigma$ generate the direct sum $\oplus \langle b_\sigma \rangle$. This direct sum must be $G$, because $G$ is the union of the $G_\sigma$.$\square$

There is a concept closely related to freedom which is—in the case of abelian groups—equivalent to it [but this is not so for modules in general].

Call a group $G$ *projective* if every diagram

with exact row can be completed by a suitable homomorphism $\psi : G \to B$ to a commutative diagram. Since in this case $C$ is essentially a quotient group of $B$, the projectivity of $G$ can be formulated in other words by asserting that every homomorphism of $G$ into a quotient group of any group $B$ can be lifted to a homomorphism of $G$ into $B$. Now we have:

**Theorem 14.6** (Mac Lane [1]). *A group is projective if and only if it is free.*

Let $\beta : B \to C$ be an epimorphism and $F$ a free group with $\phi : F \to C$. For each $x_i$ in a free set $\{x_i\}_{i \in I}$ of generators of $F$, we pick out some $b_i \in B$ such that $\beta b_i = \phi x_i$, which is possible, $\beta$ being epic. The correspondence $x_i \mapsto b_i$ ($i \in I$) can, owing to (14.2), be extended to a homomorphism $\psi : F \to B$. This $\psi$ satisfies $\beta\psi = \phi$; thus $F$ is projective.

Let $G$ be projective and $\beta : F \to G$ an epimorphism of a free group $F$ upon $G$. Then there exists a homomorphism $\psi : G \to F$ such that $\beta\psi = 1_G$. Hence $\psi$ is a monomorphism onto a direct summand of $F$, i.e., $G$ is isomorphic to a direct summand of $F$. By (14.5), $G$ is free. □

By (14.3), for every group $A$ we can find an exact sequence

$$0 \to H \to F \to A \to 0$$

with $F$, and hence by (14.5), $H$ free. This is called a *free* or *projective resolution* of $A$.

The results of this section have numerous consequences; in fact, we shall see that they are often referred to. They have been generalized in various directions. The case of modules over the $p$-adic integers will be of importance to us; let us therefore formulate the next theorem.

**Theorem 14.7.** *The results in* (14.1)–(14.6) *prevail if abelian groups are replaced by $p$-adic modules.*

In fact, the proofs hold *verbatim* with the obvious changes. □

The results of this section enable us to define a group in terms of generators and defining relations. Though this method is well known from general group theory, we dwell upon this subject for a little while.

Let $A$ be a group with a set $\{a_i\}_{i \in I}$ of generators, and let $\eta : F \to A$ be an epimorphism of a free group $F = \oplus_{i \in I} \langle x_i \rangle$ upon $A$, such that $\eta x_i = a_i$. Obviously, Ker $\eta$ consists of all elements $m_1 x_{i_1} + \cdots + m_k x_{i_k} \in F$ ($m_i \in \mathbf{Z}$) such that $m_1 a_{i_1} + \cdots + m_k a_{i_k} = 0$ in $A$. [Equalities of this type are called *defining relations* relative to the generating system $\{a_i\}_{i \in I}$ of $A$.] Bearing this in mind, one can define the group $A$ as

(2)     $A = \langle a_i (i \in I); m_{j1} a_{i_1} + \cdots + m_{jk} a_{i_k} = 0 \qquad (j \in J) \rangle,$

meaning thereby the group $A = F/H$ where $F$ is a free group, generated by the free generators $x_i$ ($i \in I$), and $H$ is the subgroup of $F$, generated by the elements $m_{j1} x_{i_1} + \cdots + m_{jk} x_{i_k}$ of $F$, for all $j \in J$, corresponding to the left members of the defining relations. This definition is standard. (2) is called a *presentation* of $A$. [One should, however, keep in mind that a group can be presented in various ways as a quotient group of a free group.]

Considering that $H$ is, by (14.5), again free, $H = \oplus_{k \in K} \langle y_k \rangle$, where every $y_k$ may be written in the form

$$y_k = \sum_{i \in I} n_{ki} x_i \qquad (n_{ki} \in \mathbf{Z}, \; k \in K);$$

for a fixed $k$, almost all $n_{ki}$ vanish. Thus the presentation $A = F/H$ defines a row-finite matrix

(3)                          $\|n_{ki}\|_{k \in K, \, i \in I}$        $(n_{ki} \in \mathbf{Z})$

with independent rows. Conversely, every row-finite matrix with independent rows gives rise, in the obvious way, to a group $A$. Since a group can be represented in several ways as a quotient group $F/H$ of a free group $F$, and since one can write $F$ and $H$ in various ways as direct sums of cyclic groups, different matrices of the form (3) may yield isomorphic groups. One of the major problems in this connection is to find out when two matrices define isomorphic groups. It is not hard to state conditions for this, but they do not help to decide isomorphy and have no practical application so far.

EXERCISES

1.  A generating system of $F_m$, $m$ a positive integer, is a free set of generators, if and only if it contains exactly $m$ elements.
2.  Prove the following converse of (14.4): if $F$ is a group such that $B \leq A$ and $A/B \cong F$ imply that $B$ is a direct summand of $A$, then $F$ is free.
3.  (Kertész [2]) $F$ is free if $F$ is isomorphic to a subgroup of any group $G$, whenever there is an epimorphism $G \to F$.
4.  (Kaplansky [2]) (a) Let $F$ be a free group, $G$ a subgroup, and $H$ a direct summand of $F$. Then $G \cap H$ is a direct summand of $G$. [*Hint*: $G/(G \cap H)$ is free.]
    (b) The intersection of a finite number of direct summands of a free group is again a direct summand.
5.  If $F$ is free and $\eta : F \to A$ is an epimorphism with $A$ finitely generated, then there is a direct decomposition $F = F_1 \oplus F_2$ such that $F_1$ is finitely generated and $F_2 \leq \operatorname{Ker} \eta$.
6.  If $A$ is presented by a set of generators and defining relations and $B$ by a subset of these generators and defining relations, then there is a natural homomorphism $B \to A$, by letting the generators of $B$ correspond to themselves *qua* elements of $A$.
7.  Let $A$ be presented by a set of generators and defining relations and

assume that the generators can be divided into two classes $\{b_i\}_{i \in I}$, $\{c_j\}_{j \in J}$, such that every given defining relation contains only $b_i$ or only $c_j$ terms explicitly. Then $A = B \oplus C$, where $B$ is generated by the $b_i$ and $C$ by the $c_j$.

8.  To every system of generators of a group, there is a set of defining relations relative to these generators, such that no relation can be omitted. [*Hint*: use (14.5).]

9.  Let $F$ be a free group, $G$ a subgroup of $F$, and $A$ a group with the property that every $\phi : G \to A$ can be extended to a $\psi : F \to A$. Prove that the same holds if $A$ is replaced by an epimorphic image of $A$.

10.  Let $B$ be a subgroup of $A$, and $\eta_1 : F_1 \to B$, $\eta_2 : F_2 \to A/B$ epimorphisms where $F_1$, $F_2$ are free. Then there is an epimorphism $\eta : F_1 \oplus F_2 \to A$, such that $\eta \mid F_1 = \eta_1$ and $\eta g + B = \eta_2 g$ for $g \in F_2$.

11.  Let $A_n$ ($n = 0, \pm 1, \pm 2, \cdots$) be arbitrary groups. Prove the existence of free groups $F^{(n)}$ and a sequence

$$\cdots \to F^{(n-1)} \xrightarrow{\ \alpha_{n-1}\ } F^{(n)} \xrightarrow{\ \alpha_n\ } F^{(n+1)} \to \cdots$$

such that $\alpha_n \alpha_{n-1} = 0$ and $\operatorname{Ker} \alpha_n / \operatorname{Im} \alpha_{n-1} \cong A_n$ for every $n$.

12.  Prove (14.7) in detail.

## 15.  FINITELY GENERATED GROUPS

We shall find the following lemma useful.

**Lemma 15.1.** *Let $A$ be a p-group and assume that $A$ contains an element $g$ of maximal order $p^k$. Then $\langle g \rangle$ is a direct summand of $A$.*

Let $B$ be a $\langle g \rangle$-high subgroup of $A$. In order to prove that $A = \langle g \rangle \oplus B$, we recall (9.9) and show that $pa = mg + b$ ($a \in A$, $b \in B$, and $m \in Z$) implies $p \mid m$. By the maximality of the order of $g$, $p^{k-1}mg + p^{k-1}b = p^k a = 0$. Hence $p^{k-1}mg = 0$, and $m$ must be divisible by $p$.□

Now we can turn to the first proper structure theorem in the history of group theory.

**Theorem 15.2** (Frobenius and Stickelberger [1]). *A finite group is the direct sum of a finite number of cyclic groups of prime power orders.*

Because of (8.4), we can restrict ourselves to $p$-groups. If $A \neq 0$ is a finite $p$-group, then we select an element $a \in A$ of a maximal order $p^k$. By the preceding lemma, $A = \langle a \rangle \oplus B$ for some $B < A$. Since $B$ is of smaller order than $A$, a trivial induction completes the proof.□

Having discovered the structure of finite groups, we proceed to finitely generated groups. We prove two preliminary lemmas.

**Lemma 15.3** (Rado [1]). *If $A = \langle a_1, \cdots, a_k \rangle$ and the integers $n_1, \cdots, n_k$ satisfy $(n_1, \cdots, n\text{-}) = 1$, then there exist $b_1, \cdots, b_k$ in $A$ such that*

$$A = \langle b_1, \cdots, b_k \rangle \qquad where \quad b_1 = n_1 a_1 + \cdots + n_k a_k.$$

If $n = |n_1| + \cdots + |n_k| = 1$, then $b_1 = \pm a_i$ for some $i$, and the assertion is trivial. Therefore, let $n > 1$ and apply induction on $n$. Now at least two of the $n_i$ are different from 0, say, $|n_1| \geq |n_2| > 0$. We have either $|n_1 + n_2| < |n_1|$ or $|n_1 - n_2| < |n_1|$, hence $|n_1 \pm n_2| + |n_2| + \cdots + |n_k| < n$ for one of the two signs. By $(n_1 \pm n_2, n_2, \cdots, n_k) = 1$ and by induction hypothesis,

$$A = \langle a_1, a_2, \cdots, a_k \rangle = \langle a_1, a_2 \mp a_1, a_3, \cdots, a_k \rangle = \langle b_1, b_2, \cdots, b_k \rangle$$

with

$$b_1 = (n_1 \pm n_2)a_1 + n_2(a_2 \mp a_1) + n_3 a_3 + \cdots + n_k a_k = n_1 a_1 + \cdots + n_k a_k. \square$$

It is convenient to say that $\{a_i\}_{i \in I}$ is a *basis* of $A$ if $A$ is the direct sum of the cyclic groups $\langle a_i \rangle$ $(i \in I)$.

**Lemma 15.4.** *Let $H \neq 0$ be a subgroup of a free group $F$ of rank $n$. Then $F$ has a basis $a_1, \cdots, a_n$ and $H$ has a basis $b_1, \cdots, b_n$ such that*

$$b_i = m_i a_i \qquad (i = 1, \cdots, n)$$

*where the $m_i$ are nonnegative integers satisfying*

$$m_{i-1} | m_i \qquad (i = 2, \cdots, n).$$

We start with a basis $c_1, \cdots, c_n$ of $F$ with the following minimal property: $H$ has an element $b_1 = k_1 c_1 + \cdots + k_n c_n$ with a minimal positive coefficient $k_1$. In other words, for another basis of $F$ or for another arrangement of $c_1, \cdots, c_n$, or for other elements of $H$, the first positive coefficient is never less than $k_1$. Then $k_1 | k_i$ $(i = 1, \cdots, n)$, for if $k_i = q_i k_1 + r_i$ $(0 \leq r_i < k_1)$, then we can write $b_1 = k_1 a_1 + r_2 c_2 + \cdots + r_n c_n$ where $a_1 = c_1 + q_2 c_2 + \cdots + q_n c_n$, $c_2, \cdots, c_n$ is a basis of $F$, and by the choice of $c_1, \cdots, c_n$ we must have $r_2 = \cdots = r_n = 0$. The same argument shows that if $b = s_1 c_1 + \cdots + s_n c_n \in H$, then $s_1 = qk_1$ for some integer $q$. Hence $b - qb_1 \in \langle c_2 \rangle \oplus \cdots \oplus \langle c_n \rangle$. We infer that $F = \langle a_1 \rangle \oplus \langle c_2 \rangle \oplus \cdots \oplus \langle c_n \rangle$ and $H = \langle b_1 \rangle \oplus H_1$ where $b_1 = k_1 a_1$ and $H_1 \leq \langle c_2 \rangle \oplus \cdots \oplus \langle c_n \rangle$. A simple induction on $n$ shows the existence of a basis $a_1, \cdots, a_n$ of $F$ and one $b_1, \cdots, b_n$ of $H$ satisfying $b_i = m_i a_i$.

In order to establish the last statement, we show that $m_1 | m_2$. Write $m_2 = qm_1 + r$ $(0 \leq r < m_1)$ and replace the basis element $a_1$ by $a = a_1 + qa_2$. In terms of the new basis $a, a_2, \cdots, a_n$, the element $b_1 + b_2 \in H$ has the form $b_1 + b_2 = m_1 a_1 + (qm_1 + r)a_2 = m_1 a + ra_2$. By the minimality of $m_1 = k_1$, we necessarily have $r = 0$. $\square$

The main result on finitely generated groups is the following famous theorem.

**Theorem 15.5.** *The following conditions on a group $A$ are equivalent:*

(i) *$A$ is finitely generated;*

(ii) *$A$ is the direct sum of a finite number of cyclic groups;*

(iii) *the subgroups of $A$ satisfy the maximum condition.*

Condition (i) implies (ii). Since this is the most essential portion, we give two independent proofs for it. The first is based on (15.3). Assume $A$ is finitely generated and every generating system of $A$ contains at least $k$ elements. Choose a system of $k$ generators $a_1, \cdots, a_k$ such that, say, $a_1$ is of minimal order, i.e., no other set of $k$ generators contains an element of smaller order. By the choice of $k$, $o(a_1) > 1$. As a basis of induction, we may assume that $B = \langle a_2, \ldots, a_k \rangle$ is a direct sum of cyclic groups. Consequently, it suffices to show that $A = \langle a_1 \rangle \oplus B$, which will follow if we can show that $\langle a_1 \rangle \cap B = 0$. By way of contradiction, let us assume $m_1 a_1 = m_2 a_2 + \cdots + m_k a_k \neq 0$ with $0 < m_1 < o(a_1)$. Set $(m_1, \cdots, m_k) = m$ and write $m_i = mn_i$. Then $(n_1, \cdots, n_k) = 1$, and from (15.3) we infer $\langle a_1, \cdots, a_k \rangle = \langle b_1, \cdots, b_k \rangle$ with $b_1 = -n_1 a_1 + n_2 a_2 + \cdots + n_k a_k$. Here $mb_1 = 0$, $o(b_1) < o(a_1)$ contradicts the choice of $a_1$.

The second proof makes use of (15.4). If $A$ is generated by $n$ elements, then by (14.3) we may write $A \cong F/H$, where $F$ is a free group of rank $n$. If we choose bases of $F$ and $H$ as in (15.4), then we obtain

$$A \cong \langle a_1 \rangle / \langle m_1 a_1 \rangle \oplus \cdots \oplus \langle a_n \rangle / \langle m_n a_n \rangle.$$

Thus $A$ is a direct sum of cyclic groups: if $m_i = 0$, then the $i$th summand is an infinite cyclic group, otherwise it is cyclic of order $m_i$.

Condition (ii) implies (iii). Let $A = \langle a_1 \rangle \oplus \cdots \oplus \langle a_n \rangle$. If $n = 1$, then $A$ is cyclic, and every nonzero subgroup of $A$ is of finite index; thus the subgroups satisfy the maximum condition. As a basis of induction, assume that in $B = \langle a_1 \rangle \oplus \cdots \oplus \langle a_{n-1} \rangle$, the subgroups satisfy the maximum condition. If $C_1 \leq C_2 \leq \cdots$ is an ascending chain of subgroups in $A$, then $B \cap C_1 \leq B \cap C_2 \leq \cdots$ is one in $B$, and so, from some index $k$ on, all $B \cap C_m$ are equal (to $B \cap C_k$). For $m > k$ we have

$$C_m/(B \cap C_k) = C_m/(B \cap C_m) \cong (B + C_m)/B \leq A/B,$$

where $A/B \cong \langle a_n \rangle$ is cyclic. Consequently, in the ascending chain $C_k/(B \cap C_k) \leq C_{k+1}/(B \cap C_k) \leq \cdots$, from a certain index $k + l$ on, all groups are equal, i.e., $C_{k+l} = C_{k+l+1} = \cdots$.

Finally, (iii) implies (i). Let the subgroups of $A$ satisfy the maximum condition. The set of all finitely generated subgroups of $A$ is not empty, hence $A$ has a maximal finitely generated subgroup $G$. For every $a \in A$, together with $G$, $\langle G, a \rangle$ is also finitely generated; hence $G = \langle G, a \rangle$ and $G = A$. $\square$

We notice that (15.5) implies that *subgroups of finitely generated groups are again finitely generated.*

It is natural to raise the question as to the uniqueness of the decomposition of a finitely generated group into the direct sum of cyclic groups. If the order of a finite cyclic group is divisible by at least two primes, then the group can be decomposed into a direct sum of groups of prime power orders, so—in order to get some kind of uniqueness—we may, and shall, restrict ourselves to direct decompositions into cyclic groups of infinite or prime power orders. Then uniqueness [up to isomorphism] will be a simple consequence of the subsequent theorem that gives a description of the subgroups of a finitely generated group.

**Theorem 15.6.** *Let B be a subgroup of the finitely generated group A. Then*:

(i) *the number s of infinite cyclic groups in a decomposition of B into direct sums of cyclic groups does not exceed the same number r for A;*

(ii) *if $p^{r_1} \geqq \cdots \geqq p^{r_k} > 1$ are, for a fixed p, the orders of the cyclic p-groups in a direct decomposition of A into cyclic groups of infinite and prime power orders, and if $p^{s_1} \geqq \cdots \geqq p^{s_m} > 1$ have the same meaning for B, then $k \geqq m$ and $r_i \geqq s_i$ for $i = 1, \cdots, m$;*

(iii) *if C is a direct sum of cyclic groups such that the orders in a decomposition of C satisfy (i) and (ii), then C is isomorphic to a subgroup of A.*

Because of $T(B) = B \cap T(A)$, we have $B/T(B) \cong (B + T(A))/T(A) \leqq A/T(A)$, where the quotient groups are direct sums of $s$ and $r$ infinite cyclic groups. Thus, to verify (i), we may assume $A$ torsion-free. Then $A$, $B$ are free groups of rank $r$, $s$, and, by (15.4), $B$ has a basis consisting of, at most, as many elements as some basis of $A$. Hence (14.1) yields $s \leqq r$.

In order to prove (ii), let us begin by observing that $|A[p]| = p^k$ and $|B[p]| = p^m$, whence $k \geqq m$ is obvious. By way of contradiction, assume $r_1 \geqq s_1, \cdots, r_{i-1} \geqq s_{i-1}$, but $r_i < s_i$ for some $i$. Then the $p$-component of $p^{r_i}A$ will be the direct sum of cyclic groups of orders $p^{r_1 - r_i}, \cdots, p^{r_{i-1} - r_i}$, while $p^{r_i}B$ that of cyclic groups of orders $p^{s_1 - r_i}, \cdots, p^{s_i - r_i}, \cdots$, thus $|(p^{r_i}A)[p]| = p^{i-1} < p^i \leqq |(p^{r_i}B)[p]|$ would be a contradiction.

Assertion (iii) is evident. □

If $B = A$, (15.6) yields the unicity of orders of the cyclic groups in decompositions of $A$ into direct sums of cyclic groups of orders infinity or powers of primes. [It should, however, be emphasized, that these cyclic groups are not uniquely determined subsets of $A$.] These orders are called the *invariants* of $A$; for instance, the invariants of $A = Z \oplus Z \oplus Z(p) \oplus Z(p^2) \oplus Z(q^2)$ are $\infty$, $\infty, p, p^2, q^2$. In this case, we also say: $A$ is *of type* $(\infty, \infty, p, p^2, q^2)$.

We have thus ordered to every finitely generated group $A$ a finite system consisting of symbols $\infty$ and prime powers, the invariants of $A$. Not only are these uniquely determined by the group, but they determine the group, that is to say, two finitely generated groups are necessarily isomorphic if they have the same system of invariants—this fact is usually expressed by saying that we have a *complete system of invariants*. Moreover, these invariants are *independent* in the sense that to every finite system of symbols ∞ and prime powers, there exists a finitely generated group whose system of invariants coincides with the given system.

The description of a class of groups by means of invariants is one of the major aims of group theory. In most cases, the invariants are natural integers, cardinal or ordinal numbers, but they could be matrices with integral entries, etc. The invariants must be easily describable quantities and—as their name indicates—must be uniquely determined by the groups; they are satisfactory if the isomorphy of groups can easily be recognized from their invariants. The systems of invariants for finitely generated groups are most satisfactory, and they have served as a classical model for structure theorems of algebra.

To be more precise, a structure theorem for a class of algebraic systems consists in finding a complete [and possibly independent] system of invariants along with a method of reconstructing, from the invariants, the algebraic system we started with. In the case of finitely generated groups, this construction consists in forming the direct sum of cyclic groups with the given invariants as orders.

(15.5) can be generalized in various ways to modules. For our purposes, it suffices to consider modules over the $p$-adic integers:

**Theorem 15.7.** *For modules over the $p$-adic integers, (15.3)–(15.6) hold true.* The proofs given above carry over *mutatis mutandis.*□

EXERCISES

1. (a) A group is finite if it has but a finite number of subgroups.
   (b) A group is finite if its subgroups satisfy both the maximum and the minimum condition.
2. A finite group $A$ is cyclic if and only if for every prime $p$, $|A[p]| \leq p$.
3. The number of nonisomorphic groups of order $m = p_1^{r_1} \cdots p_k^{r_k}$ is equal to $P(r_1) \cdots P(r_k)$, where $P(r)$ denotes the number of partitions of $r$ into positive integers.
4. Let $C(m)$ denote the multiplicative group of residue classes prime to $m$, modulo the integer $m = p_1^{r_1} \cdots p_k^{r_k}$.
   (a) $C(m)$ is the direct product of the $C(p_i^{r_i})$, $i = 1, \cdots, k$;
   (b) $C(p^k)$ is cyclic if $p$ is an odd prime;
   (c) $C(4)$ is cyclic, while $C(2^n)$, $n \geq 3$, is of type $(2, 2^{n-2})$;
   (d) every finite group is isomorphic to a subgroup of some $C(m)$.
5. If $m$ divides the order of $A$, then $A$ has subgroups and quotient groups of order $m$.
6. Prove assertion (i) of (15.6) with $B$ a quotient group, rather than a subgroup of $A$. Is (ii) true for quotient groups?

7.  If $A$, $B$ are finite groups such that, for every integer $m$, they contain the same number of elements of order $m$, then $A \cong B$.

8.  Given two finite groups $A$, $B$, there exists a group $C$ such that both $A$ and $B$ have direct summands isomorphic to $C$, and every group isomorphic to direct summands of both $A$ and $B$ is isomorphic to a direct summand of $C$.

9.  Let $A$ have the invariants $p^k, \cdots, p^k$ ($n$ times). A subgroup $B$ of $A$ is a direct summand of $A$ if and only if every invariant of $B$ is $p^k$.

10. (Szele [1]) Let $B$ be a subgroup of the finite $p$-group $A$. $B$ is a direct summand of $A$ exactly if there are direct decompositions $A = A_1 \oplus \cdots \oplus A_m$, $B = B_1 \oplus \cdots \oplus B_m$ such that $B_i \leq A_i$ ($i = 1, \cdots, m$) and $A_i$, $B_i$ have equal invariants $p^i$.

11. A generating system $a_1, \cdots, a_k$ of a finite group $A$ is a basis of $A$ if and only if the product $o(a_1) \cdots o(a_k)$ assumes its minimum [which must be $|A|$].

12. Establish the existence of a basis $g_1, \cdots, g_r$ of a finite group $A$ satisfying the divisibility conditions $o(g_1) | o(g_2) | \cdots | o(g_r)$, and prove that for all such bases of $A$, the orders $o(g_1), \cdots, o(g_r)$ are the same.

13. (Birkhoff [1])  Let  $A = \langle a_1 \rangle \oplus \cdots \oplus \langle a_n \rangle$, where $o(a_i) = p^{r_i}$ with $r_1 \geq \cdots \geq r_n$.  Call  $b = s_1 a_1 + \cdots + s_n a_n$  and  $c = t_1 a_1 + \cdots + t_n a_n$ orthogonal if

$$\sum_i s_i t_i \, p^{r_1 - r_i} \equiv 0 \bmod p^{r_1}.$$

Show that the elements of $A$ orthogonal to all the elements of a subgroup $B$ form a subgroup $C$ such that $|B| \cdot |C| = |A|$.

14. (a) Subgroups and quotient groups of elementary groups are elementary.
    (b) An elementary group $E$ of order $p^r$ contains

$$(p^r - 1)(p^{r-1} - 1) \cdots (p^{r-t+1} - 1)/(p - 1)(p^2 - 1) \cdots (p^t - 1)$$

    subgroups of order $p^t$ ($t \leq r$).
    (c) $E$ has $(p^r - 1)(p^{r-1} - 1) \cdots (p - 1)/(p - 1)^r$ composition series.
    (d) Determine the number of different bases of $E$.

15. (G. Frobenius) In a finite $p$-group $A$, the number of subgroups of a fixed order [dividing $|A|$] is $\equiv 1 \bmod p$.

16. (a) The sum of all the elements of a finite group $A$ is 0, unless $A$ contains a single element $a$ of order 2, in which case the sum is $a$.
    (b) Derive Wilson's congruence $(p - 1)! \equiv -1 \bmod p$ from (a).

17. Given the invariants $\infty, \cdots, p^k, \cdots, q^l$ of a finitely generated group $A$, determine the minimal number of elements in a generating system of $A$.

18. (Kaplansky [2]) If $A$ is an infinite group all of whose nonzero subgroups are isomorphic to $A$, then $A \cong Z$.

19. (Fedorov [1]) An infinite group is cyclic exactly if all of its nonzero subgroups are of finite index.
20. If the subgroups of both $A/B$ and $B$ satisfy the maximum condition, then so do the subgroups of $A$.
21. In a finitely generated group, every generating system contains a finite set of generators.
22. If $F = \langle x_1 \rangle \oplus \cdots \oplus \langle x_n \rangle$ is a free group of rank $n$, $G = \langle y_1 \rangle \oplus \cdots \oplus \langle y_n \rangle$ a subgroup of the same rank where

$$y_i = \sum_{j=1}^{n} r_{ij} x_j \qquad (i = 1, \cdots, n),$$

then the order of $F/G$ is equal to the absolute value of the determinant of the matrix $\|r_{ij}\|$.
23. The cardinal number of the set of pairwise nonisomorphic finitely generated [finite] groups is countable.
24. (Cohn [1], Honda [2], E. A. Walker [1]) Let $A = B \oplus G = C \oplus H$ where $B$ and $C$ are isomorphic finitely generated groups. Then $G \cong H$. [*Hint*: assume $G \neq H$; if $B \cong Z$, $G = Z \oplus (G \cap H)$; if $B = \langle b \rangle$ is of order $p^k$ and $p^{k-1}b \notin H$, then $A = B \oplus H$, while if $p^{k-1}b \in H$, $C = \langle c \rangle$, $p^{k-1}c \in G$, then $A = \langle b + c \rangle \oplus G = \langle b + c \rangle \oplus H$.]
25. (E. A. Walker [1]) Prove the isomorphism of direct decompositions of a finitely generated group by making use of Ex. 24.
26. (a) If $A$, $B$ are finitely generated groups each of which is isomorphic to a subgroup of the other, then $A \cong B$.
    (b) The same with "subgroup" replaced by "quotient group."

## 16. LINEAR INDEPENDENCE AND RANK

For the selection of a basis in a direct sum of cyclic groups, it is convenient to have the concept of linear independence at our disposal. This used to be defined in abelian groups in two inequivalent ways; here we shall adapt what is not trivial for torsion groups and is therefore more useful for our purposes (see Szele [2]).

A system $\{a_1, \cdots, a_k\}$ of nonzero elements of a group $A$ is called *linearly independent*, or briefly *independent*, if

(1) $$n_1 a_1 + \cdots + n_k a_k = 0 \qquad (n_i \in Z)$$

implies

$$n_1 a_1 = \cdots = n_k a_k = 0.$$

More explicitly, this means $n_i = 0$ if $o(a_i) = \infty$ and $o(a_i) \mid n_i$ if $o(a_i)$ is finite. A system of elements is *dependent* if it is not independent. [Notice that (unlike for vector spaces) it is, in general, not true that one of the elements in a dependent system can be written as a linear combination of the others.]

An infinite system $L = \{a_i\}_{i \in I}$ of elements of $A$ is called *independent*, if every finite subsystem of $L$ is independent. Thus independence is, *by definition*, a property of finite character. An independent system cannot contain equal elements, hence it is a set.

**Lemma 16.1.** *A system $L = \{a_i\}_{i \in I}$ is independent if and only if the subgroup generated by $L$ is the direct sum of the cyclic groups $\langle a_i \rangle$, $i \in I$.*

If $L$ is independent, then for each $i \in I$, the intersection of $\langle a_i \rangle$ with the subgroup generated by all $a_j$ with $a_j \in L, j \neq i$, is necessarily 0; hence $\langle L \rangle$ is the direct sum of the groups $\langle a_i \rangle$, $i \in I$. Conversely, if $\langle L \rangle = \bigoplus_{i \in I} \langle a_i \rangle$, then 0 can be written in the form $n_1 a_{i_1} + \cdots + n_k a_{i_k} = 0$ with different $i_1, \cdots, i_k$ from $I$ only in the trivial way: $n_1 a_{i_1} = \cdots = n_k a_{i_k} = 0$. This shows the independence of $L$.□

$g \in A$ is said *to depend* on a subset $L$ of $A$ if there is a *dependence relation*

(2)                              $$0 \neq ng = n_1 a_1 + \cdots + n_k a_k$$

for some $a_i \in L$ and integers $n, n_i$. A subset $K$ of $A$ *depends* on $L$ if every $g \in K$ depends on $L$. If both $K$ depends on $L$ and $L$ on $K$, then $K$ and $L$ are said to be *equivalent*.

An independent system $M$ of $A$ is *maximal* if there is no independent system in $A$ containing $M$ properly. Thus, if $g \in A, g \neq 0$, then $\{M, g\}$ is no longer independent, and $g$ depends on $M$. It is clear that any two maximal independent systems in a group $A$ are equivalent. By Zorn's lemma, *every independent system in $A$ can be extended to a maximal one.* Moreover, if the initial independent system contained only elements of infinite and prime power orders, then the same can be assumed of the maximal one. In fact, every element $a$ of finite order in an independent system can be replaced, without violating independence, by an arbitrary multiple $ma \neq 0$, and so by one of prime power order, too.

A subgroup $E$ of a group $A$ is called *essential* if $E \cap B \neq 0$ whenever $B$ is a nonzero subgroup of $A$. [Clearly, it suffices to state this for cyclic $B$.] In this case, $A$ is said to be an *essential extension* of $E$. For instance, the socle is an essential subgroup of a torsion group. We have the next simple result.

**Lemma 16.2.** *An independent system $M$ of $A$ is maximal if and only if $\langle M \rangle$ is an essential subgroup of $A$. Every maximal independent system in an essential subgroup of $A$ is maximal independent in $A$.*

Since $g \in A$ depends on $M$ if and only if $\langle M \rangle \cap \langle g \rangle \neq 0$, the first part of the lemma is evident. If $E$ is essential in $A$ and $M$ is a maximal independent system in $E$, then let $g \in A$ be arbitrary. We have a nonzero $h \in E \cap \langle g \rangle$, $h = mg$. Since $h$ depends on $M$, so does $g$, too.□

By the *rank* $r(A)$ of a group $A$ is meant the cardinal number of a maximal independent system containing only elements of infinite and prime power orders. If we restrict ourselves to elements of infinite order in $A$, i.e., we select an independent system which contains elements of order $\infty$ only and which is maximal with respect to this property, then the cardinality of this system is called the *torsion-free rank* $r_0(A)$ of $A$. The *p-rank* $r_p(A)$ of $A$ is defined analogously, by using elements whose orders are powers of a fixed prime $p$ rather than elements of infinite order. It is clear from these definitions that the following equality holds true for every group $A$:

$$(3) \qquad r(A) = r_0(A) + \sum_p r_p(A)$$

with $p$ running over all primes. Evidently, $r(A) = 0$ amounts to $A = 0$.

The main theorem on ranks reads as follows.

**Theorem 16.3.** *The ranks $r(A)$, $r_0(A)$, $r_p(A)$ of a group $A$ are invariants of $A$.*
By (3), it suffices to verify the invariance of $r_0(A)$ and $r_p(A)$.

The proof of invariance of $r_0$ is reduced to torsion-free groups if we can show that $r_0(A) = r(A/T)$, where $T$ is the torsion part of $A$. If $a_1, \cdots, a_k \in A$ are independent and of infinite order, and if the cosets $a_i^* = a_i + T$ satisfy $n_1 a_1^* + \cdots + n_k a_k^* = 0^*$, then $n_1 a_1 + \cdots + n_k a_k = b \in T$. Multiplication by $m = o(b)$ gives $mn_1 a_1 + \cdots + mn_k a_k = 0$, whence $mn_i = 0, n_i = 0$ for $i = 1, \cdots, k$, i.e., $\{a_1^*, \cdots, a_k^*\}$ is an independent system in $A/T$. Conversely, if this is an independent system in $A/T$, then $n_1 a_1 + \cdots + n_k a_k = 0$ implies $n_1 a_1^* + \cdots + n_k a_k^* = 0^*$, whence $n_i = 0$ for every $i$. This proves $r_0(A) = r(A/T)$ as claimed.

In order to verify the invariance of $r(A)$ for a torsion-free $A$, choose a maximal independent system $L = \{a_i\}_{i \in I}$ in $A$. For $g \in A, g \neq 0$, we have then $ng = n_1 a_{i_1} + \cdots + n_k a_{i_k} \neq 0$. If we associate with $g$ a system $\{i_1, \cdots, i_k; n, n_1, \cdots, n_k\}$ consisting of a finite number of indices and nonzero integers, then no other $g' \in A$ is associated with the same system, since if both $g$ and $g'$ define the same system, then $n(g' - g) = 0$ and $g' = g$. Consequently, the cardinality of $A$ does not exceed that of all systems $\{i_1, \cdots, i_k; n, n_1, \cdots, n_k\}$ which implies $|A| \leq |I| \aleph_0 = r(A) \cdot \aleph_0$. Since trivially $r(A) \leq |A|$, we see that $r(A) = |A|$ whenever $r(A)$ is infinite. If $A$ has a finite maximal independent system $\{a_1, \cdots, a_k\}$, and if $\{b_1, \cdots, b_l\}$ is any independent system in $A$, then the second system depends on the first; hence for some integer $m > 0$, $mb_i \in \langle a_1, \cdots, a_k \rangle$ for $i = 1, \cdots, l$. Thus $\langle mb_1 \rangle \oplus \cdots \oplus \langle mb_l \rangle$ is a subgroup of $\langle a_1 \rangle \oplus \cdots \oplus \langle a_k \rangle$, and so, by (15.6), the inequality $l \leq k$ holds. We infer that every independent system in $A$ contains at most $k$ elements, and therefore every maximal independent system in $A$ consists of the same number of elements. Thus $r_0(A)$ is an invariant for every $A$.

Turning to the ranks $r_p(A)$, it is clear that $r_p(A) = r(T_p)$, where $T_p$ is the $p$-component of $A$. Hence we need only prove the invariance of $r(A)$ for $p$-groups $A$.

If $A$ is a $p$-group and $S(A) = A[p]$ is its socle, then $r(A) = r(S(A))$. In fact, a system $\{a_1, \cdots, a_k\}$ in $A$ is independent if and only if $\{p^{m_1-1}a_1, \cdots, p^{m_k-1}a_k\}$ is independent where $m_i = e(a_i)$. Therefore, only the uniqueness of $r(S(A))$ is to be verified. As $S(A)$ is, in the natural way, a vector space over the field $F_p$ of $p$ elements, and independence in the sense above coincides with the vector space independence, we see that $r(S(A))$ is just the dimension of the vector space $S(A)$, and hence it is unique.☐

From the last part of the proof we get, in view of (16.1):

**Corollary 16.4.** *In an elementary p-group, every maximal independent system is a basis.*☐

It is clear how to define independence and rank for modules over the $p$-adic integers. Notice that in this case $q$-ranks for primes $q \neq p$ are trivial, so one may confine himself to torsion-free and $p$-ranks for one $p$. As in (16.3), one concludes that they are invariants of the module.

EXERCISES

1.  The subgroups $B_i$ $(i \in I)$ of $A$ generate their direct sum in $A$ if and only if every set $L = \{b_i\}_{i \in I}$ with one $b_i$ $(\neq 0)$ from each $B_i$, is independent.
2.  If $K$ is an independent and $L$ a maximal independent system in $A$ such that $L$ contains only elements of infinite and prime power orders, then $A$ has a maximal independent system containing $K$ and contained in $K \cup L$.
3.  Let $B$ be a subgroup of $A$. Show that:
    (a) $r(B) \leq r(A)$;
    (b) $r(A) \leq r(B) + r(A/B)$;
    (c) $r(A) < r(A/B)$ can happen;
    (d) $r_0(A) = r_0(B) + r_0(A/B)$.
4.  If $B_i$ $(i \in I)$ are subgroups of $A$ such that $A = \sum_{i \in I} B_i$, then $r(A) \leq \sum_i r(B_i)$ where equality holds if the sum is direct.
5.  $A$ is locally cyclic if and only if $r_0(A) + \max_p r_p(A) \leq 1$.
6.  (a) $A$ contains no subgroup which is the direct sum of two of its proper subgroups if and only if $r(A) \leq 1$.
    (b) $r(A) \leq 1$ if and only if $A$ is isomorphic to a subgroup of $Q$ or some $Z(p^\infty)$.
7.  If $B_1, \cdots, B_n$ are subgroups of $A$ such that $A/B_i$ are of finite rank, then $A/(B_1 \cap \cdots \cap B_n)$ is again of finite rank.
8.  A group of infinite rank $m$ contains exactly $2^m$ different subgroups.

9.   If $E$ is an essential subgroup of $A$ and $B \leq A$, then $E \cap B$ is essential in $B$.
10.  A subgroup $B$ of $A$ is essential if and only if $S(A) \leq B$ and $A/B$ is torsion.
11.  The essential subgroups of a group $A$ form a dual ideal in $L(A)$. This is principal if and only if $A$ is torsion.
12.  If $B$ is a subgroup of $A$, then there exists a subgroup of $A$ which is maximal with respect to the property of being an essential extension of $B$ in $A$.
13.  (a) For a $p$-group, the rank is the same if it is regarded as a Z-module or as a $Q_p^*$-module.
     (b) The group $J_p$ is as an abelian group of rank $2^{\aleph_0}$ and as a $Q_p^*$-module of rank 1.

## 17.   DIRECT SUMS OF CYCLIC $p$-GROUPS

Let $A$ be a direct sum of cyclic $p$-groups. Then, evidently, $A$ contains no elements $\neq 0$ of infinite height. We shall see by an example below that the absence of elements of infinite height does not imply that a $p$-group is a direct sum of cyclic groups. It is of importance to have criteria under which a given $p$-group is a direct sum of cyclic groups.

**Theorem 17.1** (Kulikov [1]). *A $p$-group $A$ is a direct sum of cyclic groups if and only if $A$ is the union of an ascending chain of subgroups*

$$(1) \qquad A_1 \leq A_2 \leq \cdots \leq A_n \leq \cdots, \qquad \bigcup_{n=1}^{\infty} A_n = A,$$

*such that the heights of elements $\neq 0$ of $A_n$ are less than a finite bound $k_n$.*

If $A$ is a direct sum of cyclic groups, then in a decomposition, collect the cyclic direct summands of the same order $p^n$, for every $n$, and denote their direct sum by $B_n$. Clearly, $A_n = B_1 \oplus \cdots \oplus B_n$ satisfy the conditions with $k_n = n$.

For the proof of the sufficiency, assume that the subgroups $A_n$ of $A$ satisfy the hypotheses. Since we may adjoin 0's to the beginning of (1) and repeat some terms $A_n$ [a finite number of times], it is clear that there is no loss of generality in assuming $k_n = n$, that is, $A_n \cap p^n A = 0$ for every $n$. We consider all chains

$$C_1 \leq C_2 \leq \cdots \leq C_n \leq \cdots$$

of subgroups $C_n$ of $A$, such that

$$(i) \quad A_n \leq C_n \qquad and \qquad (ii) \quad C_n \cap p^n A = 0$$

for all $n$, and define the chain of the $C_n$ less than or equal to the chain of the $B_n$ if $C_n \leq B_n$ for every $n$. Then the set of all chains with (i) and (ii) is inductive, and Zorn's lemma applies to conclude the existence of a chain $G_1 \leqq G_2 \leqq \cdots \leqq G_n \leqq \cdots$ satisfying (i) and (ii) [with $C_n$ replaced by $G_n$] which is maximal in our present sense. Evidently, $\bigcup G_n = A$.

For every $n$, we select a maximal independent set $L_n$ of elements in the subgroup $G_n[p] \cap p^{n-1}A$, and define $L$ as the union of all the $L_n$ ($n = 1, 2, \cdots$). For every $c_i \in L$ with $m_i = h(c_i)$ we choose an $a_i \in A$ satisfying $p^{m_i}a_i = c_i$. We claim that $A' = \langle \cdots, a_i, \cdots \rangle = \oplus \langle a_i \rangle$ coincides with $A$.

The first step in the proof of this is to show that $\langle L \rangle = A[p]$. Owing to (16.4), $\langle L_n \rangle = G_n[p] \cap p^{n-1}A$. Since the elements $\neq 0$ of $\langle L_n \rangle$ are all of height $n - 1$, we see that the $\langle L_n \rangle$ generate their direct sum, $\langle L \rangle = \oplus_n \langle L_n \rangle$. Assume, as a basis of induction, that every $a \in A[p]$ contained in $G_r$ belongs to $\langle L \rangle$ [which obviously holds for $r = 1$], and let $b \in G_{r+1}[p]\backslash G_r$. Then $b \notin G_r$ implies $\langle G_r, b \rangle \cap p^r A \neq 0$; let $0 \neq g + k = c \in p^r A$, where $g \in G_r$, and $k = 1$ may be assumed [for otherwise we multiply by $k'$ with $kk' \equiv 1 \mod p$]. Here $c \in G_{r+1}$ and $h(c) \geqq r$; thus, by (ii), $o(c) = p$ and $h(c) = r$. We infer $c \in \langle L_{r+1} \rangle$. Furthermore, $pg = pc - pb = 0$ and $g \in G_r$, together with the induction hypothesis, imply $g \in \langle L \rangle$. Hence $b = c - g \in \langle L \rangle$, establishing $\langle L \rangle = A[p]$.

Assume we have proved that every $a \in A$ of order $\leqq p^n$ belongs to $A'$, and let $b \in A$ be of order $p^{n+1}$ with $n \geqq 1$. As is proved in the preceding paragraph, $p^n b \in \langle L \rangle$, and so $p^n b = m_1 c_1 + \cdots + m_k c_k$ with some $c_i \in L$. Let $c_1, \cdots, c_j$ be of height $\geqq n$ and $c_{j+1}, \ldots, c_k$ of height $\leqq n - 1$. If we write $m_i c_i = p^n m_i' a_i$ for $i = 1, \cdots, j$, then we have

$$p^n(b - m_1' a_1 - \cdots - m_j' a_j) = m_{j+1}c_{j+1} + \cdots + m_k c_k \in G_{n-1}.$$

Condition (ii) implies $b - m_1' a_1 - \cdots - m_j' a_j$ is of order $\leqq p^n$, thus in $A'$, and consequently, $b \in A'$.□

As corollaries we obtain the following two important results.

**Theorem 17.2** (Prüfer [2], Baer [1]). *A bounded group is a direct sum of cyclic groups.*

If $A$ is a bounded group, then its $p$-components are again bounded. If we choose for all the members of (1) the $p$-component of $A$, then we conclude from (17.1) that the $p$-components of $A$ are direct sums of cyclic groups.□

**Theorem 17.3** (Prüfer [1]). *A countable $p$-group is a direct sum of cyclic groups, if and only if it contains no elements $\neq 0$ of infinite height.*

Only the "if" part needs a verification. Let $A$ be a countable $p$-group without elements of infinite height and $\{a_1, \cdots, a_n, \cdots\}$ a generating system for $A$. Now $A$ is the union of its finite subgroups $A_n = \{a_1, \cdots, a_n\}$ ($n = 1, 2, \cdots$), where the heights of the elements are trivially bounded. (17.1) completes the proof.□

The following example, given by Kurosh [2], shows that countability is an essential hypothesis in (17.3). Let $A$ denote the torsion part of the direct product of the cyclic groups $Z(p), Z(p^2), \cdots, Z(p^n), \cdots$. Then $A$ is a $p$-group of the power of the continuum, without elements of infinite height. Contrary to our assertion, assume $A = \oplus_n A_n$ where $A_n$ is a direct sum of cyclic groups of the same order $p^n$. The socles $S_n = \oplus_{i=n}^{\infty} A_i[p]$ form, with increasing $n$, a descending chain, where $S_n$ consists exactly of those elements of $S_1 = A[p]$ which are of height $\geq n - 1$. Clearly,

$$a = (b_1, b_2, \cdots, b_n, \cdots) \in \prod_n Z(p^n), \qquad b_n \in Z(p^n),$$

is of height $\geq n - 1$ only if $b_1 = \cdots = b_{n-1} = 0$. This shows that each quotient group $S_n/S_{n+1}$ $(n = 1, 2, \ldots)$ is of order $p$. In view of $A_n[p] \cong S_n/S_{n+1}$ we infer that the $A_n$ are finite, and hence $A$ is countable. This contradiction proves that $A$ is not a direct sum of cyclic groups.

If $A$ is a direct sum of cyclic $p$-groups, then it may have a number of such direct decompositions. But there is a unicity as far as the orders of the components are concerned:

**Theorem 17.4.** *Any two decompositions of a group into direct sums of cyclic groups of infinite and prime power orders are isomorphic.*

Let $S_n$ be the subgroup of $S = A[p]$ consisting of elements of height $\geq n - 1$. If $A$ is a direct sum of cyclic $p$-groups, then $S_n$ is the direct sum of the socles of all direct summands of orders $\geq p^n$. Consequently, $S_n/S_{n+1}$ is isomorphic to the direct sum of the socles of direct summands of order $p^n$ in a direct decomposition of $A$. We infer that the number of direct summands of order $p^n$ in a direct decomposition of $A$ with cyclic summands is equal to the rank of $S_n/S_{n+1}$. The $S_n$ were defined without any reference to direct decompositions, hence the cardinal number $\mathfrak{m}_{p^n}$ of the set of direct summands of order $p^n$ in any decomposition of $A$ into direct sums of cyclic $p$-groups is the same.

Since the number $\mathfrak{m}_0$ of direct summands $Z$ in a direct decomposition of a group $A$ into cyclic groups is equal to $r_0(A)$, we conclude from (16.3) that $\mathfrak{m}_0$, too, is uniquely determined by $A$. $\square$

The preceding theorem states that for direct sums of cyclic groups $Z$ and $Z(p^n)$, the cardinal numbers $\mathfrak{m}_0$ and $\mathfrak{m}_{p^n}$ are invariants. They constitute a complete and independent system of invariants for direct sums of cyclic groups.

EXERCISES

1. (a) A group $A$ is elementary if and only if every subgroup of $A$ is a direct summand.

(b) $A$ is elementary exactly if it is a torsion group whose Frattini subgroup vanishes.

(c) $A$ is elementary if and only if $A$ is the only essential subgroup in $A$.

2.  $A[n]$ is always a direct sum of cyclic groups.

3.  If $A$ is a direct sum of finite cyclic groups, then $A/mA \cong A[m]$ for any integer $m > 0$.

4.  Let $B$ be the direct product of $m$ copies of $Z(p^k)$ with fixed $p^k$, and $A$ the direct product of $n$ copies of $B$. Then $A$ is a direct sum of groups $Z(p^k)$; determine the cardinality of the set of the components.

5.  Describe the groups in which:
    (a) every maximal independent set is a basis;
    (b) every generating system contains a basis.

6.  Prove (17.2) for $p$-groups as follows: select an independent system of elements of maximal order $p^k$, extend it successively by elements of orders $p^{k-1}, \cdots, p$, such that in each step the independent system is as large as possible, and show that the arising set is a basis.

7.  (Szele [3]) Let $A$ be a bounded $p$-group and $p^r$ the maximal order for elements in $A$. Then $a \in (p^r/o(a))A$ for every $a \in A$ if and only if $A$ is a direct sum of cyclic groups of the same order $p^r$.

8.  (Dlab [4]) (a) If $A$ is a bounded $p$-group and $S = \{a_i\}_{i \in I}$ is a set of elements in $A$, such that $a_i + pA$ $(i \in I)$ generate $A/pA$, then $S$ generates $A$.
    (b) Every generating system of a bounded $p$-group contains a minimal generating system.
    (c)* For unbounded $p$-groups, (b) fails.

9*. (Dieudonné [2]) Generalize (17.1) as follows: if $G$ is a $p$-group containing a subgroup $A$, such that $G/A$ is a direct sum of cyclic groups and (1) holds where the heights of elements of $A_n$, taken in $G$, are bounded, then $G$ is a direct sum of cyclic groups.

10. Prove (17.3) as follows: start with a maximal independent system in the socle, $c_1, \cdots, c_n, \cdots$, and consider all $b$ with $p^k b = m_1 c_1 + \cdots + m_n c_n$ with $m_n c_n \neq 0$ for a fixed $n$. Choose a $b$, say $b_n$, with maximal order, for every $n$, and prove that $b_1, \cdots, b_n, \cdots$ is a basis.

11. Let $A$, $B$ be direct sums of cyclic groups.
    (a) $A \oplus A \cong B \oplus B$ implies $A \cong B$.
    (b) $A \oplus \cdots \oplus A \oplus \cdots \cong B \oplus \cdots \oplus B \oplus \cdots$ does not imply $A \cong B$, not even if $A$, $B$ are finitely generated.

12. Let $A$ be the direct sum of a countable set of cyclic groups of order $p^2$ and $B = A \oplus Z(p)$. Show that $A$, $B$ are not isomorphic, but both— *qua* abstract groups—have the same set of subgroups and the same set of quotient groups.

13.   Let $A = A_1 \oplus \cdots \oplus A_k$ where $A_i$ is the direct sum of $\mathfrak{m}_i$ copies of $Z(p^i)$, $i = 1, \cdots, k$. Describe the structures of subgroups and quotient groups of $A$.

14.   (a) Let $G = \oplus_{k=1}^{\infty} Z(p^k)$. Prove that every countable $p$-group is an epimorphic image of $G$.
      (b) Every $p$-group of cardinality $\mathfrak{m}$ is an epimorphic image of a direct sum of $\mathfrak{m}$ copies of $G$ in (a).

## 18.   SUBGROUPS OF DIRECT SUMS OF CYCLIC GROUPS

We have seen that subgroups of free groups are free, and from (17.3) it follows trivially that subgroups of countable direct sums of cyclic $p$-groups are again direct sums of cyclic groups. These are special cases of the following general result.

**Theorem 18.1** (Kulikov [2]). *Subgroups of direct sums of cyclic groups are again direct sums of cyclic groups.*

First, let $A$ be a direct sum of cyclic $p$-groups, and $B \leq A$. Then $A$ is the union of an ascending chain $A_1 \leq \cdots \leq A_n \leq \cdots$ of its subgroups, where the heights of elements of $A_n$ are bounded, say, $\leq k_n$. Evidently, $B$ is the union of the ascending chain

$$B_1 \leq \cdots \leq B_n \leq \cdots \qquad \text{with} \quad B_n = B \cap A_n,$$

and the heights of elements of $B_n$, relative to $B$, do not exceed $k_n$. By virtue of (17.1), $B$ is a direct sum of cyclic groups. Thus our theorem holds for torsion groups.

Let $A$ be an arbitrary direct sum of cyclic groups. If $T$ is its torsion part, then $B \cap T$ is the torsion part of the subgroup $B$ of $A$. Now $B/(B \cap T) \cong (B + T)/T \leq A/T$, where $A/T$ is a free group. By (14.5), $B/(B \cap T)$ is free, and so (14.4) implies $B = (B \cap T) \oplus C$ for some free subgroup $C$ of $B$. By what has been shown in the preceding paragraph, $B \cap T$ is a direct sum of cyclic $p$-groups; thus $B$ is a direct sum of cyclic groups.☐

**Corollary 18.2** (Kulikov [2]). *Any two direct decompositions of a group which is a direct sum of cyclic groups have isomorphic refinements.*

In view of (18.1), each direct summand is a direct sum of cyclic groups. If we replace each direct summand by a direct sum of cyclic groups of order $\infty$ or prime power, we arrive at refinements which are isomorphic, as is shown by (17.4).☐

Under certain circumstances we can conclude that $A$ is a direct sum of cyclic groups if a subgroup of $A$ is a direct sum of cyclic groups.

**Proposition 18.3** (Fuchs [1], Mostowski and Sąsiada [1]). *If $B$ is a subgroup of $A$ such that $B$ is a direct sum of cyclic groups and $A/B$ is bounded, then $A$ is a direct sum of cyclic groups.*

By hypothesis we have $nA \leqq B \leqq A$ for some integer $n$. From (18.1) we conclude that $nA$ is a direct sum of cyclic groups, and so it is enough to show that $A$ is a direct sum of cyclic groups if $pA$ is for some prime $p$.

Let $pA = \bigoplus_{i \in I} \langle b_i \rangle$, where $o(b_i)$ is infinite or a power of prime, for every $i \in I$. Choose $a_i \in A$ such that $pa_i = b_i$ for $i \in I$, with the proviso that $a_i \in \langle b_i \rangle$ if $o(b_i)$ is prime to $p$. Extend the independent system $\{a_i\}_{i \in I}$ by a set $\{c_j\}_{j \in J}$ to obtain a maximal independent system $\{a_i, c_j\}_{i,j}$ in $A$. We claim $A$ is the direct sum of the $\langle a_i \rangle$ and $\langle c_j \rangle$. We need only show that every $a \in A$ lies in this direct sum. Clearly, $pa = n_1 b_1 + \cdots + n_k b_k = p(n_1 a_1 + \cdots + n_k a_k)$, and so $a' = a - n_1 a_1 - \cdots - n_k a_k$ is in the socle of $A$. Since $a'$ depends on $\{a_i, c_j\}_{i,j}$ and therefore is expressible as a linear combination of the $a_i$, $c_j$, the same holds for $a$.□

From (18.3) we can easily derive:

**Corollary 18.4** (Kulikov [2]). *Every group $A$ is the union of an ascending chain $A_1 \leqq \cdots \leqq A_n \leqq \cdots$ of subgroups where every $A_n$ is a direct sum of cyclic groups.*

Define $A_1$ as generated by a maximal independent system of $A$. If, for some $n$, $A_{n-1}$ is defined, then let $A_n = n^{-1}A_{n-1} = \{a \in A \mid na \in A_{n-1}\}$. From the choice of $A_1$ it follows that the union $\bigcup A_n$ for all $n$ is equal to $A$, while from (18.3) we infer, successively, that $A_n$ is a direct sum of cyclic groups.□

The results of this section extend immediately to $p$-adic modules.

EXERCISES

1.  Let $A$ be a $p$-group. The torsion part of a direct product $\prod A$ with an infinite number of components is a direct sum of cyclic groups if and only if $A$ is bounded.

2.  If both $B$ and $A/B$ are direct sums of cyclic groups and if in $A/B$ the elements of finite order are of bounded order, then $A$ is a direct sum of cyclic groups.

3.  (Szele [5]) Let

$$A = \langle a_0, a_1, \cdots, a_n, \cdots; a_0 = m_1 a_1 = \cdots = m_n a_n = \cdots \rangle$$

where $m_n$ are positive integers. $A$ is a direct sum of cyclic groups if and only if the system $\{m_n\}_n$ is bounded.

4.  Let $A = B + C$ where both $B$ and $C$ are direct sums of cyclic groups. Then $A$ need not be a direct sum of cyclic groups.

5.   Relate the invariants of a direct sum of cyclic groups to those of its sub-groups.
6.   Prove (18.1) for modules over the *p*-adic integers.
7.   An *equational class* or *variety* of groups is a class of groups that is closed under formation of subgroups, quotient groups, and direct products.
   (a) Prove that the following is the list of all equational classes of abelian groups: (i) all abelian groups; (ii) for any positive integer $n$, all abelian groups annihilated by $n$.
   (b) Show that the smallest variety containing $Z$ [or $Z(n)$] is the one given in (i) [in (ii)].

## 19.   COUNTABLE FREE GROUPS

Kulikov's theorem (17.1) yields a handy necessary and sufficient condition for a torsion group to be a direct sum of cyclic groups. For arbitrary groups, however, no satisfactory condition is known so far. The known criteria refer to a selected set of elements, namely, the basis of the group, and so these can be much better described as conditions as to whether or not a given or selected subset is a basis. For criteria on the existence of a basis we may refer to Fuchs [1] or [7].

However, in the case of countable torsion-free groups there is a useful criterion.

**Theorem 19.1** (Pontryagin [1]). *A countable torsion-free group is free if and only if every subgroup of finite rank is free. Equivalently, for every integer $n$, the subgroups of rank $\leq n$ satisfy the maximum condition.*

Because of (14.5), the necessity is evident. In order to establish sufficiency, let $A = \langle a_1, \cdots, a_n, \cdots \rangle$ be a countable torsion-free group all of whose subgroups of finite rank are free. Define $A_n$ as the set of all $a \in A$ which depend on $\{a_1, \cdots, a_n\}$, with 0 adjoined. Then $A_n$ is a subgroup of rank $\leq n$. Clearly, $r(A_{n+1}) \leq r(A_n) + 1$; therefore either $A$ is of finite rank [in which case the assertion is trivial] or there is a subsequence $B_n$ of the $A_n$ such that $r(B_n) = n$ and $A$ is the union of the ascending chain $0 = B_0 < B_1 < B_2 < \cdots < B_n < \cdots$. Now $B_{n+1}/B_n$ is torsion-free [because $A/A_n$ is], of rank 1, and finitely generated; hence $B_{n+1}/B_n \cong Z$. From (14.4) we get

$$B_{n+1} = B_n \oplus \langle b_{n+1} \rangle$$

for some $b_{n+1}$. This shows that $b_1, \cdots, b_n, \cdots$ generate the direct sum $\bigoplus_n \langle b_n \rangle$, whence $A = \bigoplus_n \langle b_n \rangle$.

In view of (15.5), the second formulation is equivalent to the first one.☐

It should be noted that the preceding theorem fails to hold for groups of larger cardinalities, as will be shown by the next theorem which is of independent interest.

For a cardinal $\aleph_\sigma$, call a group $\aleph_\sigma$-*free*, if all of its subgroups of cardinality less than $\aleph_\sigma$ are free.

**Theorem 19.2** (Baer [2], Specker [1]). *A direct product of an infinite set of infinite cyclic groups is $\aleph_1$-free, but not free.*

Write $A = \prod_{i \in I} \langle a_i \rangle$ where $I$ is an infinite set and $\langle a_i \rangle \cong Z$ for each $i$. First we prove that *every finite set* $\{b_1, \cdots, b_m\} \subset A$ *can be embedded in a finitely generated direct summand of* $A$. If $m = 1$ and $b_1 \neq 0$, write $b_1 = (\cdots, n_i a_i, \cdots)$ with $n_i \in Z$. If there is a $j \in I$ with $|n_j| = 1$, then we have $A = \langle b_1 \rangle \oplus A_j$ where $A_j$ consists of all elements of $A$ with $j$th coordinates 0. If the minimum $n$ of $|n_i|$ with $n_i \neq 0$ is greater than 1, then set $n_i = q_i n + r_i$ ($0 \leq r_i < n$) and define $c_1 = (\cdots, q_i a_i, \cdots)$, $c_2 = (\cdots, r_i a_i, \cdots)$ in $A$ so that $b_1 = nc_1 + c_2$. There is some $j$ with $|q_j| = 1$ and $r_j = 0$, thus $A = \langle c_1 \rangle \oplus A_j$ where $c_2 \in A_j$ with coefficients $|r_i| < n$. By induction, $A_j$ has a finitely generated direct summand $B'$ containing $c_2$, and so $\langle c_1 \rangle \oplus B'$ is a direct summand of $A$ containing $b_1$. Assume $m > 1$ and $A$ has a finitely generated direct summand $B$ such that $b_1, \cdots, b_{m-1} \in B$. We may, in addition, suppose that $A = B \oplus C$ where $C$ is the direct product of almost all $\langle a_i \rangle$. Setting $b_m = b + c$ ($b \in B$, $c \in C$) and embedding $c$ in a finitely generated direct summand $C'$ of $C$, $B \oplus C'$ will be a finitely generated direct summand of $A$ that contains all of $b_1, \cdots, b_m$.

If $G$ is now a countable subgroup of $A$, then a maximal independent set of a subgroup $G'$ ($\leq G$) of finite rank is contained in a finitely generated direct summand $B$ of $A$, and by torsion-freeness, $G' \leq B$. Now (19.1) proves $G$ free; thus $A$ is $\aleph_1$-free.

If $A$ were free, then all the subgroups of $A$ would be free, so it suffices to exhibit a nonfree subgroup $H$ of $A$. Let $H$ be the subgroup of $A' = \prod_{i=1}^{\infty} \langle a_i \rangle$ consisting of all $b = (n_1 a_1, \cdots, n_i a_i, \cdots)$ such that for any positive integer $m$ and a fixed prime $p$, almost all coefficients $n_i$ are divisible by $p^m$. This $H$ is clearly of the power of the continuum that contains $\oplus \langle a_i \rangle$. Since each coset of $H \bmod pH$ contains some element of $\oplus \langle a_i \rangle$, $H/pH$ is countable. The comparison of the cardinalities of $H$ and $H/pH$ shows that $H$ cannot be free. $\square$

An immediate consequence is:

**Corollary 19.3** (K. Stein). *Any countable group $A$ can be written in the form*

$$A = N \oplus F$$

*where $F$ is free and $N$ has no free quotient groups. $N$ is uniquely determined by $A$.*

Define $N$ as the intersection of the kernels of all homomorphisms $\eta : A \to Z$. Then $A/N$ is isomorphic to a subgroup of a direct product of infinite cyclic groups, and so it is free by (19.2). $N$ is a direct summand owing to (14.4).

The uniqueness of $N$ follows from the observation that if $M$ is any direct summand of $A$ without free quotient groups, then its projection into $F$ must vanish by (14.5), so $M \leq N$. $\square$

EXERCISES

1. A countable group is a direct sum of cyclic groups if and only if every finite set of elements can be embedded in a finitely generated direct summand.

2. (Szele [3]) Let $L = \{a_i\}_{i \in I}$ be a generating system of a group $A$ where each $a_i$ is of infinite or prime power order. $L$ is a basis of $A$ if every finite subsystem $a_1, \cdots, a_k$ of $L$ satisfies: $\langle a_1, \cdots, a_k \rangle = \langle b_1, \cdots, b_k \rangle$ with $o(b_i) = \infty$ or $p^r$ implies

$$\min(o(a_1), \cdots, o(a_k)) \leq \min(o(b_1), \cdots, o(b_k)).$$

3. A subset $\{a_i\}_{i \in I}$ of a torsion-free group $A$ is a basis of $A$ if and only if it is a minimal generating system such that, for a finite subset $\{a_1, \cdots, a_k\}$, $a \in A$ depends on $\{a_1, \cdots, a_k\}$ implies $a \in \langle a_1, \cdots, a_k \rangle$.

4. The same with "minimal generating system" replaced by "maximal independent system."

5. (a) The intersection of direct summands of a countable free group is again a direct summand. [*Hint*: cf. Ex. 4 in **14**; apply (19.2) to the quotient group mod intersection.]
   (b) The same fails to hold for a free group of the power of the continuum.

6. In any presentation of $Z^{\aleph_0}$ there are uncountably many defining relations.

7. Given any cardinal $\mathfrak{m} \geq 2^{\aleph_0}$, there exists an $\aleph_1$-free group of cardinality $\mathfrak{m}$ which is not free.

## NOTES

The existence of free objects in the category of all abelian groups is fundamental. Though in homological algebra, projectivity is predominant, in abelian group theory freedom seems to be prevailing. Fortunately, for abelian groups, freedom and projectivity are equivalent, while for modules, the projectives are exactly the direct summands of free modules. They are free over local rings (see Kaplansky [3]) or over polynomial rings with a finite number of noncommuting indeterminates with coefficients in a commutative field (see P. M. Cohen [*J. Algebra* **1** (1964), 47–69]).

Theorem (14.5) holds for modules over left principal ideal domains. Submodules of projectives are again projective if and only if the ring is left hereditary, i.e., all left ideals are projective. (14.1) holds over commutative rings or under the hypothesis that at least one of $\mathfrak{m}$ and $\mathfrak{n}$ is infinite. There exist, however, rings R such that all free R-modules $\neq 0$ with finite sets of generators are isomorphic.

It is perhaps worth while pointing out that every R-module is free if and only if R is a field, and every R-module is projective exactly if R is a semisimple Artin ring.

A module $M$ is *quasi-projective* if, for every epimorphism $\beta: M \to N$ and every homomorphism $\phi: M \to N$, there exists an endomorphism $\psi$ of $M$ such that $\beta\psi = \phi$. Quasi-projective modules were studied by L. E. T. Wu and J. P. Jans [*Ill. J. Math.* **11** (1967), 439–448]. An abelian group is quasi-projective exactly if it is either free or a torsion group every $p$-component of which is a direct sum of cyclic groups of the same order $p^n$; see L. Fuchs and K. M. Rangaswamy [to appear in *Bull. Soc. Math. France*].

There are numerous generalizations of the results in **15**. I. Kaplansky [*J. Indian Math. Soc.* **24** (1960), 279–281] proved that, for commutative domains R, the torsion parts of all finitely generated R-modules are direct summands exactly if R is a Prüfer domain. E. Matlis [*Trans. Amer. Math. Soc.* **125** (1966), 147–179] made a good approach toward the description of commutative domains over which finitely generated torsion modules split into direct sums of cyclic modules. Supposing R is a commutative Noetherian ring, A. I. Uzkov [*Mat. Sb.* **62** (1963), 469–475] showed that every finitely generated R-module is a direct sum of cyclic modules if, and only if, R is a principal ideal ring.

Unless the ring is left Noetherian, finitely generated modules are different from finitely presented modules (where the finiteness of a set of defining relations is also assumed); these are somewhat better manageable, and it is still true that every module is the direct limit of finitely presented ones. For finitely presented R-modules, a direct sum decomposition into cyclic modules holds if, and only if, R is an elementary divisor ring, i.e., every matrix over R can be brought to a diagonal form by left and right multiplications by unimodular matrices (I. Kaplansky [*Trans. Amer. Math. Soc.* **66** (1949), 464–491]). In this case (15.4) holds true.

Little attention has been paid so far to infinite direct sums of cyclic modules; even the uniqueness of decompositions has not been dealt with. One of the difficulties lies in the fact that maximal independent systems no longer yield invariants for the modules, except for a restricted class of rings; A. Kertész [*Acta Sci. Math. Szeged.* **21** (1960), 260–269] and L. Fuchs [*Annales Univ. Sci. Budapest* **6** (1963), 71–78].

*Problem 9.* Characterize the groups in which the intersection of two direct summands is again a summand.

Cf. **14**, Ex. 4 and **19**, Ex. 5

*Problem 10.* For which ordinals $\sigma$ do there exist $\aleph_\sigma$-free groups which are not $\aleph_{\sigma+1}$-free?

*Problem 11.* Investigate **K**-direct sums of cyclic groups.

# IV

## DIVISIBLE GROUPS

This chapter is devoted to one of the most important classes of abelian groups: the divisible groups. Divisible groups are extremely easy to recognize, and they have a number of prominent properties which are characteristic for them. One of their outstanding features is that they coincide with the injective groups which embody a concept dual to projectivity, and thus they are direct summands in any group containing them. The structure of divisible groups is completely known, so that the theory of divisible groups is as satisfactory as one could expect at the present status of algebra.

The concluding topic for this chapter is concerned with a remarkable duality between maximum and minimum conditions.

## 20. DIVISIBILITY

Since multiplication of group elements by integers makes sense, it is natural to introduce divisibility of group elements by integers.

We shall say that the element $a$ of the group $A$ is *divisible* by the positive integer $n$, in symbols: $n \mid a$, if the equation

$$(1) \qquad\qquad nx = a \qquad (a \in A)$$

is solvable in $A$; that is to say, $A$ contains an element $b$ such that $x = b$ is a solution of (1). It is clear that the solvability of (1) is equivalent to $a \in nA$.

Let us list some elementary consequences of our definition.

(a) If $x = b$ is a solution of (1), then the coset $b + A[n]$ is the set of all solutions of (1).

(b) If $A$ is torsion-free, then (1) has at most one solution.

(c) If $(n, o(a)) = 1$, then (1) is always solvable. For if $r, s$ are integers such that $nr + o(a)s = 1$, then $x = ra$ satisfies $nx = nra = nra + o(a)sa = a$.

(d) $m \mid a$ and $n \mid a$ imply $[m, n] \mid a$. Let $r, s$ be integers such that $mr + ns = d = (m, n)$, and let $b, c \in A$ satisfy $mb = a, nc = a$. Then $[m, n](rc + sb) = mn\, d^{-1}(rc + sb) = md^{-1}ra + nd^{-1}sa = a$.

(e) $n \mid a$ and $n \mid b$ imply $n \mid a \pm b$.

(f) If $A$ is a direct sum, $A = B \oplus C$, then $n \mid a = b + c$ ($b \in B, c \in C$) if, and only if, $n \mid b$ and $n \mid c$. The same holds for infinite direct sums and direct products.

(g) For every homomorphism $\alpha : A \to B$, $n \mid a$ in $A$ implies $n \mid \alpha a$ in $B$.

(h) If $p$ is a prime, $p^k \mid a$ if, and only if, $k \leqq h_p(a)$.

A group $D$ is called *divisible* if $n \mid a$ for all $a \in D$ and all positive integers $n$. Thus $D$ is divisible if and only if $nD = D$ for every positive $n$. The groups $Q, Z(p^\infty), Q/Z$ are examples for divisible groups, but a direct sum of cyclic groups is not divisible [unless 0].

A group $D$ is said to be *$p$-divisible* if $p^k D = D$ for every positive integer $k$. Since $p^k D = p \cdots pD$, it is obvious that $p$-divisibility is implied by $pD = D$. The last equality is equivalent to the fact that every element of $D$ is divisible by $p$.

(A) *A group is divisible if and only if it is $p$-divisible for every prime $p$.*

If $pD = D$ for every prime $p$ and $n = p_1 \cdots p_r$, then $nD = p_1 \cdots p_r D = D$. Thus divisibility is equivalent to the fact that every element is of infinite $p$-height at every prime $p$.

(B) *A $p$-group is divisible if and only if it is $p$-divisible.*

In view of (c), for a $p$-group $D$ we always have $qD = D$, whenever the primes $p, q$ are different.

(C) *If in a $p$-group $D$, every element of order $p$ is of infinite height, then $D$ is divisible.*

Let $a \in D$ be of order $p^n$. We prove by induction on $n$ that $p \mid a$. For $n = 1$, this being a consequence of the hypothesis, assume $n > 1$, and that we have proved this for elements of order $< p^n$. By assumption, $p^{n-1}a$ is of infinite height, thus $p^{n-1}a = p^n b$ for some $b \in D$. Since $o(a - pb) \leqq p^{n-1}$, $p \mid a - pb$, which implies $p \mid a$, indeed.

(D) *Every epimorphic image of a divisible group is divisible.*

This follows from (g).

(E) *A direct sum or a direct product of groups is divisible if and only if each component is divisible.*

This is a consequence of (f).

(F) *If $D_i$ ($i \in I$) are divisible subgroups of $A$, then so is their sum $\sum D_i$.*

This is evident in view of (e).

EXERCISES

1. A group is divisible if and only if it has no maximal subgroups [i.e., it coincides with its Frattini subgroup].
2. A group is divisible if and only if it has no finite epimorphic image $\neq 0$.
3. The additive group of any field of characteristic 0 is divisible.
4. The quotient group $J_p/Z$ is divisible.
5. If $B$ is a $p$-divisible subgroup of $A$, and if $A/B$ is $p$-divisible, then $A$ is $p$-divisible.
6. Let $A$ be the direct product and $B$ the direct sum of the groups $B_n$ ($n = 1, 2, \cdots$). $A/B$ is divisible if and only if for every prime $p$, $pB_n = B_n$ holds for almost all $n$.
7. If $L = \{a_i\}_{i \in I}$ is a maximal independent system in a group $D$, and if $n \mid a_i$ for every $i \in I$ and every integer $n > 0$, then $D$ is divisible.


## 21.  INJECTIVE GROUPS

Injective groups are dual to projective groups; they are defined by dualizing the definition of projectivity.

A group $D$ is said to be *injective* if to every diagram

(1)

$$0 \to A \xrightarrow{\alpha} B$$

with exact row, there exists a homomorphism $\eta : B \to D$ making the diagram commute. If we identify $A$ with $\alpha A$, then injectivity of $D$ can be interpreted as the extensibility of any homomorphism $\xi : A \to D$ to a homomorphism of a group $B$ containing $A$ into $D$.

Our main purpose is to show that injective groups are exactly the divisible groups. (21.1) and (24.5) yield this result.

**Theorem 21.1.** (Baer [3]). *Divisible groups are injective.*
Let $D$ be divisible, and let (1) be given, where $A$ will be regarded as a subgroup of $B$. We consider all groups $G$ between $A$ and $B$, $A \leq G \leq B$, such that $\xi$ has an extension $\theta : G \to D$. We partially order the pairs $(G, \theta)$ such that $(G, \theta) \leq (G', \theta')$ means that $G \leq G'$ and $\theta$ is the restriction of $\theta' : G' \to D$ to $G$. The set of these pairs is not empty, since $(A, \xi)$ belongs to it,

and it is inductive, since every chain $(G_i, \theta_i)$ $(i \in I)$ has an upper bound, namely, $(G, \theta)$, where $G = \bigcup_i G_i$ and $\theta : G \to D$ coincides with $\theta_i$ on $G_i$. By Zorn's lemma, there is a maximal pair $(G_0, \theta_0)$ in the considered set. If $G_0 < B$, and $b \in B \backslash G_0$ satisfies $nb \in G_0$ for some $n > 0$, then choose the minimal $n$, and let $nb = g \in G_0$. By the divisibility of $D$, some $x \in D$ satisfies $nx = \theta_0 g$. It is easy to check that

$$(2) \qquad\qquad c + rb \mapsto \theta_0 c + rx \qquad (c \in G_0, 0 \leqq r < n)$$

is a homomorphism of $\langle G_0, b \rangle$ into $D$. If $nb \notin G_0$ unless $n = 0$, then (2) gives rise to a homomorphism of $\langle G_0, b \rangle$ into $D$, where $x \in D$ is arbitrary [and there is no restriction on $r$]. Thus $G_0 < B$ contradicts the maximality of $(G_0, \theta_0)$, and hence $G_0 = B$ and $\theta_0 = \eta$. $\square$

We are now able to show that divisible subgroups split the group.

**Theorem 21.2** (Baer [3]). *A divisible subgroup $D$ of a group $A$ is a direct summand of $A$, $A = D \oplus C$ for some subgroup $C$ of $A$. This $C$ can be chosen so as to contain a preassigned subgroup $B$ of $A$ with $D \cap B = 0$.*

To the injection $\alpha : D \to A$ and to the identity map $1_D : D \to D$, by (21.1) there exists a homomorphism $\eta : A \to D$ such that the diagram

commutes. (9.2) shows $A = D \oplus \operatorname{Ker} \eta$.

If $B \leq A$ satisfies $D \cap B = 0$, then $D + B = D \oplus B$, and there is a homomorphism $\xi : D \oplus B \to D$ which is the identity on $D$ and the zero on $B$. If in the preceding paragraph $1_D$ is replaced by this $\xi$, then $A = D \oplus \operatorname{Ker} \eta$ with $B \leq \operatorname{Ker} \eta$. $\square$

The second part of this theorem asserts, in other words, that divisible groups are absolute direct summands.

Given a group $A$, consider the subgroup $D$ generated by all divisible subgroups of $A$. From **20** (F), we know that $D$ is divisible; thus it is *the maximal divisible subgroup* of $A$. Clearly, $D$ is a fully invariant subgroup of $A$. By Baer's theorem, $A = D \oplus C$ where, evidently, $C$ is *reduced* in the sense that it has no divisible subgroups other than 0. We thus have:

**Theorem 21.3.** *Every group $A$ is the direct sum of a divisible group $D$ and a reduced group $C$,*

$$A = D \oplus C.$$

*Here D is a uniquely determined subgroup of A, while C is unique up to iso-morphism.*

The last statement follows from the fact that if $D$ is the maximal divisible subgroup of $A$ and $A = D' \oplus C'$, where $D'$ is divisible and $C'$ as reduced, then by (9.3) we have $D = (D \cap D') \oplus (D \cap C')$. Here $D \cap C' = 0$ as a direct summand of a divisible group contained in a reduced group, thus $D \cap D' = D$, $D \leq D'$, and so $D = D'$. $\square$

The last theorem reduces the structure problem of abelian groups to those of divisible and reduced groups.

EXERCISES

1. Let $E$ be an essential subgroup of a divisible group $D$. Show that every automorphism of $E$ extends to an automorphism of $D$.
2. If $A$ is torsion-free, then its maximal divisible subgroup is the first Ulm subgroup of $A$.
3. A group $C$ has the property that every nonzero group has a nonzero homomorphism into $C$ if and only if $Q/Z$ is a direct summand of $C$.
4. (Szélpál [1]) If all nonzero quotient groups of $A$ are isomorphic to $A$, then $A \cong Z(p^k)$ with $k = 1$ or $k = \infty$.
5. A direct sum or a direct product of groups is reduced if and only if every component is reduced.
6. Give an example of a descending chain of divisible subgroups where the intersection is not divisible. [*Hint*: in the direct sum of countably many isomorphic groups $Z(p^\infty)$, define a descending sequence with intersection of order $p$.]
7. Prove (21.2) by selecting a $D$-high subgroup and applying (9.9).
8. (a) (E. A. Walker [2]) Every torsion-free group of infinite rank is a sub-direct sum of copies of $Q$.
   (b) An unbounded $p$-group is a subdirect sum of quasicyclic groups.
9. The direct limit of injective groups is injective.

## 22. SYSTEMS OF EQUATIONS

According to the definition of divisible groups, all "linear" equations of the form $nx = a \in D$ with positive integers $n$ are solvable in a divisible group $D$. We are going to prove the much stronger result that all consistent systems of linear equations are also solvable in a divisible group.

By a *system of equations* over a group $A$, we mean a set of equations

(1)
$$\sum_{j \in J} n_{ij} x_j = a_i \qquad (a_i \in A, i \in I)$$

where $n_{ij}$ are integers such that, for a fixed $i$, $n_{ij} = 0$ for, at most, a finite number of exceptions. Here $\{x_j\}_{j \in J}$ is a set of unknowns, while $I$, $J$ are index sets of arbitrary cardinalities.

$$x_j = g_j \in A \qquad (j \in J)$$

is a *solution* of (1), if (1) is satisfied whenever the $x_j$ are replaced by the $g_j$. Clearly, a solution may be viewed as an element $(\cdots, g_j, \cdots)$ in the direct product $A^J$.

For the solvability of (1), a trivial necessary condition is that the system (1) be compatible in the sense that, if a linear combination of left members of equations vanishes, then it remains 0 when the corresponding right members are substituted. Following Kertész [3], we give another, more profound interpretation of compatibility and solvability of systems of equations.

The left member of an equation in (1) can be thought of as an element of the free group $X$ over the set $\{x_j\}_{j \in J}$ of unknowns. Let $Y$ be the subgroup of $X$ generated by all left members $f_i(x) = \sum_j n_{ij} x_j$ of equations in (1). The correspondence

$$(2) \qquad\qquad f_i(x) \mapsto a_i \qquad (i \in I)$$

induces a homomorphism $\eta : Y \to A$ if and only if every representation of 0 as a linear combination of the $f_i(x)$ will be mapped upon 0, i.e., the system (1) is compatible in the above sense. Accordingly, we call (1) *compatible* if (2) extends to a homomorphism $\eta : Y \to A$.

It is convenient to interpret a compatible system (1) as a pair $(Y, \eta)$ where $Y$ is a subgroup of the free group $X$ over the set of unknowns and $\eta$ is a homomorphism of $Y$ into $A$. Clearly, two systems define the same pair $(Y, \eta)$ exactly if the equations of either system are linear combinations of the equations of the other system, in which case the two systems obviously have the same solutions if any.

It is straightforward to show: $x_j = g_j$ $(j \in J)$ is a solution of (1) if and only if the correspondence

$$(3) \qquad\qquad x_j \mapsto g_j \qquad (j \in J)$$

extends to a homomorphism $\chi : X \to A$ whose restriction to $Y$ gives $\eta$. Thus the system $(Y, \eta)$ is solvable exactly if there is a homomorphism $\chi$ making the diagram commutative; here $\rho$ denotes the injection map. Moreover, these homomorphisms $\chi$ and the solutions of $(Y, \eta)$ are in a one-to-one correspondence, so we may use the notation $(X, \chi)$ for a solution of $(Y, \eta)$.

Now we are ready to verify:

**Theorem 22.1** (Gacsályi [1]). *Every compatible system of equations over a group A admits a solution in A if and only if A is divisible.*

Since a single equation $nx = a$ with $n \neq 0$ is a compatible system, the necessity is evident. Turning to the proof of sufficiency, let $(Y, \eta)$ be a compatible system over a divisible group $A$. By (21.1), $\eta$ can be extended to a homomorphism $\chi : X \to A$. Hence there exists a solution.☐

Compatibility being a property of finite character, we have at once:

**Corollary 22.2** (Gacsályi [1]). *A system of equations over a divisible group D is solvable in D if and only if every finite subsystem has a solution in D.*☐

In connection with systems of equations, it is worth while mentioning the next result.

**Proposition 22.3** (Gacsályi [2], Balcerzyk [1]). *A subgroup B of a group A is a direct summand of A if and only if every system of equations over B solvable in A can be solved in B.*

If $B$ is a direct summand of $A$, $A = B \oplus C$, then the $B$-components of a solution in $A$ obviously satisfy a system of equations over $B$.

Conversely, assume that systems of equations over $B$ are solvable in $B$ whenever they have a solution in $A$. For each coset $u$ of $A$ mod $B$, select a representative $a(u) \in A$, and consider the system

$$x_u + x_v - x_{u+v} = a(u) + a(v) - a(u + v) \in B \qquad \text{for all} \quad u, v \in A/B.$$

By hypothesis, this has a solution $x_u = b(u) \in B$. Then $a(u) - b(u)$ are representatives of the cosets $u \in A/B$ which form a subgroup in $A$; thus $B$ is a direct summand of $A$.☐

**EXERCISES**

1. If a system of equations over a divisible group $D$ is solvable in a group containing $D$, then it is solvable in $D$, too.
2. A system of equations over a group $A$ is compatible if and only if it is solvable in some group containing $A$. [*Hint*: for sufficiency, consider the pushout for $\eta : Y \to A$ and the injection $\rho : Y \to X$.]
3. Every system of equations over a divisible group contains a maximal solvable subsystem.

4.  (a) A homogeneous system $(Y, 0)$ over an arbitrary group $A$ admits a nontrivial solution exactly if there is a nonzero homomorphism $\phi : X/Y \to A$. The nontrivial solutions and these $\phi$ are in a one-to-one correspondence.

    (b) A homogeneous system of $n$ equations with $n + 1$ unknowns has a nontrivial solution in any group $\neq 0$.

5.  Prove that

$$x_1 - px_2 = 1, \quad x_2 - p^2x_3 = 1, \cdots, x_n - p^nx_{n+1} = 1, \cdots$$

[for any prime $p$] is a system over $Z$ which is not solvable in $Z$, though every finite subsystem has a solution in $Z$.

## 23.   THE STRUCTURE OF DIVISIBLE GROUPS

The groups $Z(p^\infty)$ and $Q$ were our first examples for divisible groups. The structure theorem on divisible groups shows that there are no divisible groups other than direct sums of $Z(p^\infty)$ and $Q$.

**Theorem 23.1.** *Any divisible group $D$ is a direct sum of quasicyclic and full rational groups. The cardinal numbers of the sets of components $Z(p^\infty)$ [for every $p$] and $Q$ form a complete and independent system of invariants for $D$.*

Clearly, the torsion part $T$ of $D$ is divisible, and (21.2) implies $D = T \oplus E$, where $E$ is torsion-free and, evidently, again divisible. If $T_p$ denotes the $p$-component of $T$, then

$$D = \bigoplus_p T_p \oplus E,$$

and it suffices to show that $T_p$ is a direct sum of groups $Z(p^\infty)$, and $E$ is a direct sum of groups $Q$.

Select a maximal independent system $\{a_i\}_{i \in I}$ in the socle of $T_p$. Owing to divisibility, there is for each $i$ an infinite sequence $a_{i1}, \cdots, a_{in}, \cdots$ in $T_p$ such that $a_{i1} = a_i, pa_{i,n+1} = a_{in}$ for $n = 1, 2, \cdots$. We conclude that every $a_i$ can be embedded in a subgroup $A_i \cong Z(p^\infty)$ of $T_p$, namely, $A_i = \langle a_{i1}, \cdots, a_{in}, \cdots \rangle$. Since $\langle a_i \rangle$ is the socle of $A_i$ and the $a_i$ $(i \in I)$ are independent, the $A_i$ generate their direct sum $A = \bigoplus_{i \in I} A_i$ in $T_p$. Here $A$ is divisible, hence a direct summand of $T_p$. But $A$ contains the socle of $T_p$, thus $A = T_p$, and $T_p$ is a direct sum of groups $Z(p^\infty)$.

We pick out a maximal independent system $\{b_j\}_{j \in J}$ in E. Because $E$ is divisible and torsion-free, for every positive $n \in Z$ there is exactly one $x \in E$ satisfying $nx = b_j$, which shows that every $b_j$ can be embedded in a subgroup $B_j \cong Q$ of $E$. Since $\{b_j\}$ is an independent system, the $B_j$ generate their direct sum $B = \bigoplus_{j \in J} B_j$ in $E$. This is a direct summand of $E$ containing a maximal independent system of $E$, hence $B = E$, and $E$ is a direct sum of groups $Q$.

The numbers of direct summands isomorphic to $Z(p^\infty)$ or to $Q$ in a decomposition of $D$ into groups $Z(p^\infty)$, $Q$ are obviously equal to the ranks $r_p(D)$ and $r_0(D)$. By (16.3), they are uniquely determined by $D$. They form a complete system of invariants, since if given $r_p(D)$ and $r_0(D)$, we can uniquely reconstruct $D$ as the direct sum of $r_p(D)$ copies of $Z(p^\infty)$ for every $p$ and $r_0(D)$ copies of $Q$. Their independence is trivial. $\square$

Consequently, for a divisible $D$ one has the decomposition

$$D \cong \bigoplus_{r_0(D)} Q \oplus \bigoplus_p \left[ \bigoplus_{r_p(D)} Z(p^\infty) \right].$$

In view of (21.3) and (23.1), the structure problem of abelian groups reduces to the case of reduced groups.

*Example 1.* If $G$ is the additive group of all real numbers, then $G$ is a torsion-free divisible group of the power of the continuum. Hence $G = \bigoplus_{\mathfrak{c}} Q$ where $\mathfrak{c}$ stands for continuum.

*Example 2.* Let

$$K = \prod_p Z(p^\infty)$$

with $p$ running over all different primes. Then the $p$-component of $K$ is $Z(p^\infty)$, and since $K$ is of the power of the continuum, we have

$$K = \bigoplus_p Z(p^\infty) \oplus (\bigoplus_{\mathfrak{c}} Q).$$

Therefore, *K is isomorphic to the group of reals* mod 1.

*Example 3.* Let $\mathfrak{m}$ be an infinite cardinal. Then

$$Z(p^\infty)^{\mathfrak{m}} = \bigoplus_{2^{\mathfrak{m}}} (Z(p^\infty) \oplus Q).$$

In fact, it is divisible, its socle is of power $2^{\mathfrak{m}}$, and it contains $2^{\mathfrak{m}}$ independent elements of infinite order.

EXERCISES

1. Divisible torsion groups are isomorphic if and only if their socles are isomorphic.
2. If $A$, $B$ are divisible groups, each containing a subgroup isomorphic to the other, then $A \cong B$.
3. If $A$ is divisible and $A \oplus A \cong B \oplus B$, then $A \cong B$.
4. If $A$ is an infinite group all of whose proper subgroups are finite, then $A \cong Z(p^\infty)$ for some $p$.
5. A group $A$ is reduced if and only if $Q$ has no nontrivial homomorphism into $A$.
6. A $p$-group is reduced if and only if its cyclic subgroups satisfy the maximum condition.

7. Let $D$ be the direct product of $\mathfrak{m}$ copies of $(Q/Z)^n$. Find the structure of $D$.

8. (Szele [4]) A group contains no two distinct isomorphic subgroups if and only if it is isomorphic to a subgroup of $Q/Z$.

9. (Kertész [1]) Let $A$ be a $p$-group where the elements of finite heights are of heights $<m$ with an integer $m > 0$. Then $A$ is the direct sum of co-cyclic groups. [*Hint*: $p^m A$ is divisible.]

10. Show that any two direct decompositions of a divisible group have isomorphic refinements.

11. A group $A$ is called *quasi-injective* if every homomorphism of every subgroup of $A$ into $A$ can be extended to an endomorphism of $A$. Prove that a quasi-injective group is either injective or is a torsion group whose $p$-components are direct sums of isomorphic cocyclic groups.

## 24.  THE DIVISIBLE HULL

Free groups are universal in a certain sense, namely, every group is an epimorphic image of some free group. The next result shows that divisible groups are universal in the dual sense:

**Theorem 24.1.** *Every group can be embedded as a subgroup in a divisible group.*

The infinite cyclic group $Z$ can clearly be embedded in a divisible group, namely, in $Q$. Hence every free group $F$ is embeddable in a direct sum of copies of $Q$, i.e., in a divisible group. Given an arbitrary group $A$, we may write $A \cong F/N$ for a suitable free group $F$. If we embed $F$ in a divisible group $D$, then $A$ will be isomorphic to the subgroup $F/N$ of the divisible group $D/N$.☐

This theorem can be improved by establishing the existence of a minimal divisible group containing a given group. To prove this, we begin with the following lemmas.

**Lemma 24.2.** *A subgroup $B$ of $A$ is essential if and only if a homomorphism $\alpha : A \to G$ with an arbitrary group $G$ is necessarily monic whenever $\alpha \,|\, B : B \to G$ is a monomorphism.*

If $B$ is an essential subgroup of $A$, and $\alpha \,|\, B$ is a monomorphism, then $\operatorname{Ker} \alpha \cap B = \operatorname{Ker}(\alpha \,|\, B) = 0$ implies $\operatorname{Ker} \alpha = 0$. Conversely, if $B$ is not essential, and if $C, 0 < C \leq A$, satisfies $C \cap B = 0$, then the canonical epimorphism $\alpha : A \to A/C$ is not monic, though $\alpha \,|\, B$ is a monomorphism.☐

Given $A$, call the divisible group $E$ containing $A$ *minimal divisible* if no proper divisible subgroup of $E$ contains $A$.

**Lemma 24.3.** *A divisible group E containing A is minimal divisible exactly if A is an essential subgroup of E.*

If $E$ is not minimal and $D$ is a proper divisible subgroup of $E$ containing $A$, then we may write $E = D \oplus C$ with $C \neq 0$. Since $A \cap C \leq D \cap C = 0$, $A$ is not essential in $E$. Conversely, if $A$ is not essential in $E$, then there is a cyclic group $C = \langle c \rangle \neq 0$ in $E$ such that $A \cap C = 0$. We may assume, without loss of generality, that $o(c) = \infty$ or $p^k$. Then $C$ can be embedded in a subgroup $B$ of $E$ such that $B \cong Q$ or $Z(p^\infty)$, and by (21.2), $E = D \oplus B$ for some group $D$ containing $A$. Thus $E$ is not minimal.$\square$

**Theorem 24.4** (Kulikov [2]). *Every divisible group containing A contains a minimal divisible group containing A. Any two minimal divisible groups containing A are isomorphic over A.*

Let $D$ be divisible and contain $A$. The divisible subgroups of $D$ that are disjoint from $A$ form an inductive set, hence there exists a maximal member $M$ in this set. By (21.2), we have $D = M \oplus E$ with $A \leq E$. Clearly, $E$ is divisible, and the maximality of $M$ guarantees that $E$ cannot have any proper direct summand and, hence, any proper divisible subgroup still containing $A$. Therefore, $D$ contains a minimal divisible subgroup $E$ containing $A$.

If $E_1$, $E_2$ are two minimal divisible groups containing $A$, then because of (21.1), the identity automorphism $1_A$ of $A$ can be extended to a homomorphism $\eta : E_1 \to E_2$. Since $\eta E_1$ is divisible and contains $A$, $\eta$ is an epimorphism. By (24.3), $A$ is an essential subgroup of $E_1$, and since $\eta \mid A = 1_A$, (24.2) shows $\eta$ monic. Consequently, $\eta$ is an isomorphism of $E_1$ with $E_2$ leaving $A$ elementwise fixed.$\square$

By the last theorem, we are justified in calling a minimal divisible group $E$ containing $A$ the *divisible hull* [*injective hull*] of $A$. Clearly, (24.3) shows that

$$r_0(E) = r_0(A) \qquad \text{and} \qquad r_p(E) = r_p(A) \qquad \text{for every prime } p.$$

Thus *the structure of the divisible hull of a group A is completely determined by the ranks of A.*

The results of this section enable us to establish the converses of (21.1) and (21.2).

**Theorem 24.5.** *For a group D, the following conditions are equivalent:*

(i) *D is divisible;*
(ii) *D is injective;*
(iii) *D is a direct summand of every group containing D.*

We need only verify that (iii) implies (i). Embed $D$ in a divisible group $E$. Then (iii) implies that $D$ is a direct summand of $E$, hence (i) holds.$\square$

It is convenient at times to call an exact sequence

$$0 \to A \to D \to D' \to 0$$

with $D$ [and hence $D'$] divisible, whose existence for every $A$ is guaranteed by (24.1), a *divisible* or *injective resolution* of $A$.

We finally introduce a theorem, mainly because it can now be proved, and its proof is based on divisible groups; it will be convenient to have it at disposal for easy reference.

**Proposition 24.6.** *If $C$ is a subgroup of a group $B$ such that $B/C$ is isomorphic to a subgroup $H$ of $G$, then there exists a group $A$ containing $B$ such that $A/C \cong G$.*

Let $D$ be a divisible group containing $B$. Now $D/C$ contains $B/C$, and if $B/C$ cannot be extended in $D/C$ to a group isomorphic to $G$, then we can always find a divisible group $E$ such that in $D/C \oplus E$ this embedding is possible. If $A/C$ is isomorphic to $G$ and contains $B/C \cong H$, then $A (\leq D \oplus E)$ is a group with the desired properties.☐

The situation of (24.6) is shown by the commutative diagram

$$0 \to C \to B \to H \to 0$$
$$\| \quad \downarrow \quad \downarrow$$
$$0 \to C \to A \to G \to 0$$

where the rows are exact, and the vertical maps are injections.

EXERCISES

1.  For a torsion-free group $A$, let $E$ be the set of all pairs $(a, m)$ with $a \in A$, $m$ an integer $> 0$, such that

    $$(a, m) = (b, n) \qquad \text{if and only if} \quad mb = na,$$

    and

    $$(a, m) + (b, n) = (na + mb, mn).$$

    Show that $E$ is a minimal divisible group containing the image of the monomorphism $a \mapsto (a, 1)$ $(a \in A)$.
2.  The divisible hull of $A$ is the divisible hull of a subgroup $B$ of $A$ if and only if $B$ is essential in $A$.
3.  (a) A divisible group $D$ containing $A$ is minimal exactly if $D/A$ is torsion and $A$ contains the socle of $D$.
    (b) If $D$ is a divisible hull of a $p$-divisible group $A$, then $D/A$ has trivial $p$-component.
4.  (Kertész [2]) (a) If $A$ is an endomorphic image of every group containing it, then $A$ is divisible.
    (b) If $A$ is a direct summand of every group containing it as an endo-

morphic image, then $A$ is divisible.

5.  (Kertész [2]) Let $A$ be such that if $A$ is an endomorphic image of a group $B$, then $B$ contains a direct summand isomorphic to $A$. Show that $A$ is a direct sum of a divisible and a free group.

6.  If $A$ is a direct summand in every group $G$ which contains $A$ such that $G/A$ is quasicyclic, then $A$ is divisible.

7.  (Dlab [2]) Every group is the Frattini subgroup of some group.

8*. (Charles [2], Khabbaz [2]) A subgroup $A$ of a divisible group $D$ is the intersection of divisible subgroups of $D$ if and only if for every prime $p$, $A[p] = D[p]$ implies $pA = A$.

9.  Let $A_n$ ($n = 0, \pm 1, \pm 2, \cdots$) be a countable system of groups. Verify the existence of divisible groups $D_n$ and a sequence

$$\cdots \to D_{n-1} \xrightarrow{\alpha_{n-1}} D_n \xrightarrow{\alpha_n} D_{n+1} \to \cdots$$

such that $\alpha_n \alpha_{n-1} = 0$, and Ker $\alpha_n /$Im $\alpha_{n-1} \cong A_n$ for every $n$.

10. (Szele [2]) Call $a \in A$ *algebraic* over a subgroup $B$ of $A$ if $a = 0$ or $\langle a \rangle \cap B \neq 0$. If every $a \in A$ is algebraic over $B$, we call $A$ *algebraic* over $B$.

(a) $A$ is algebraic over $B$ if and only if $B$ is an essential subgroup of $A$.

(b) $A$ is a maximal algebraic extension of $B$ [i.e., no group properly containing $A$ is algebraic over $B$] exactly if $A$ is a divisible hull of $B$.

11. (Szele [2]) (a) If $B \leq A$, then there exists an algebraic extension $C$ of $B$ in $A$ which is maximal among the algebraic extensions of $B$ in $A$.

(b) Conclude, by making use of (a) and Ex. 10, that every divisible group containing $B$ contains a divisible hull of $B$.

12. (E. A. Walker [4]) $G$ is said to be an *n-extension* of $A$, if $G$ is an essential extension of $A$ such that $A = nG$.

(a) If $G$ is an $n$-extension of $A$, then there is a divisible hull $D$ of $A$ containing $G$.

(b) If both $G$ and $G'$ are $n$-extensions of $A$ and $G \leq G'$, then $G = G'$.

(c) Prove the existence of an $n$-extension of any given $A$ by proving that $(D/A)[n] = G/A$ defines a group $G$ with $nG = A$.

(d) Any two $n$-extensions of $A$ are isomorphic over $A$.

(e) Every group $H$ with $nH = A$ contains an $n$-extension of $A$.


## 25.  FINITELY COGENERATED GROUPS

The duality we have noticed between free and divisible groups suggests a concept dual to finitely generated groups.

A system $L$ of elements of a group $A$ is called a *system of cogenerators* if, for every group $B$, every homomorphism $\phi : A \to B$ such that $L \cap$ Ker $\phi = \varnothing$ or 0 must be a monomorphism. This is evidently equivalent to the condition that every nonzero subgroup of $A$ contains a nonzero element of $L$. Clearly, the subgroup $\langle L \rangle$ must be an essential subgroup of $A$, and an essential subgroup is always a system of cogenerators. What has been said at the introduction of cocyclic groups [in 3] indicates that cogenerators are dual to generators.

A group will be called *finitely cogenerated* if it has a finite system of cogenerators. The following result is an analog of (15.5) and points out a beautiful duality between maximum and minimum conditions.

**Theorem 25.1** (Kurosh [1], Yahya [1]). *For a group $A$, the following conditions are equivalent*:

(i) *$A$ is finitely cogenerated*;
(ii) *$A$ is an essential extension of a finite group*;
(iii) *$A$ is a direct sum of a finite number of cocyclic groups*;
(iv) *the subgroups of $A$ satisfy the minimum condition*.

Assume (i) and let $L$ be a finite system of cogenerators for $A$. No element of $A$ is of infinite order, for otherwise we could select in the cyclic group generated by an element of infinite order a subgroup $\neq 0$ excluding all the nonzero elements of $L$. Thus $L$ is a finite system of elements of finite order, whence $\langle L \rangle$ is finite. $A$ being an essential extension of $\langle L \rangle$, (ii) follows.

Next assume (ii), i.e., $A$ is an essential extension of a finite subgroup $B$. Clearly, $A$ is torsion with a finite number of $p$-components, and in order to prove (iii), we may suppose $A$, $B$ are $p$-groups. Considering that $A[p] = B[p]$ is finite, for a fixed $a \in A$, the equation $px = a$ may have but a finite number of solutions in $A$. If $h(a) = \infty$, then the heights of the solutions $x_1, \cdots, x_k$ cannot be all finite, for if $y \in A$ satisfies $p^n y = a$, then $p^{n-1} y$ is among $x_1, \cdots, x_k$. Hence, we conclude that starting with $a \in A[p]$ of infinite height, we can ascend to obtain a quasicyclic subgroup of $A$. The union $D$ of the quasicyclic subgroups of $A$ is divisible, hence $A = D \oplus C$. Since $C[p]$ is finite, there is a finite maximal height $m$ of its elements, and so $p^{m+1} C = 0$. Thus $A$ is a direct sum of cocyclic groups where the number of summands must be finite owing to the finiteness of the socle.

We next show that (iii) implies (iv). If $r(A) = 1$, i.e., $A = Z(p^k)$ with $1 \leq k \leq \infty$, then the statement is evident. For $r(A) = n > 1$ we use induction, and assume the minimum condition in groups of rank $\leq n - 1$. Let $A = Z(p^k) \oplus B$ where $r(B) = n - 1$, and let $G_1 \geq G_2 \geq \cdots$ be a descending chain in $A$. Then $B \cap G_1 \geq B \cap G_2 \geq \cdots$, and from some index $m$ on, we have $B \cap G_m = B \cap G_{m+1} = \cdots$. From

$$G_i/(B \cap G_m) = G_i/(B \cap G_i) \cong (B + G_i)/B \leq A/B = Z(p^k) \qquad (i \geq m)$$

we infer that the $G_i/(B \cap G_m)$, and hence the $G_i$, are equal from some index on. This proves (iv).

Finally, assume (iv). $A$ contains no element $a$ of infinite order, for $\langle 2^n a \rangle$ $(n = 0, 1, 2, \cdots)$ cannot be a properly descending infinite chain in $A$. Since the socle of $A$ cannot be an infinite direct sum, $S(A)$ is finite, and hence $A$ has a finite system of cogenerators.□

We observe that from the equivalence of (i) and (iv) it results that *quotient groups of finitely cogenerated groups are finitely cogenerated.* Also, notice that (ii) is equivalent to the finiteness of the socle.

To conclude, we prove the following simple result.

**Proposition 25.2.** *Let* $a_1, \cdots, a_n$ *be a finite set of nonzero elements in* $A$, *and* $M$ *a subgroup of* $A$ *that is maximal with respect to the property of excluding each of* $a_1, \cdots, a_n$. *Then* $A/M$ *satisfies the minimum condition for subgroups. If* $n = 1$, *then* $A/M$ *is cocyclic.*

Every subgroup of $A$ which contains $M$ properly must contain at least one of $a_1, \cdots, a_n$. In other words, every nonzero subgroup of $A/M$ contains one of $a_1 + M, \cdots, a_n + M$, that is $a_1 + M, \cdots, a_n + M$ is a system of cogenerators of $A/M$. The preceding theorem completes the proof.□

Notice that (25.2) implies that every group is Hausdorff in the Prüfer topology.

EXERCISES

1.  A subdirect sum of a finite number of groups with minimum condition satisfies the minimum condition.
2.  If a subgroup $B$ of $A$ and the quotient group $A/B$ both satisfy the minimum condition, then so does $A$.
3.  A $p$-group satisfies the minimum condition if and only if it is of finite rank.
4.  If $\eta$ is an endomorphism of a group with minimum condition such that Ker $\eta = 0$, then $\eta$ is an automorphism.
5.  Find the structure of groups in which there are but a finite number of elements of any fixed order.
6.  (a) Characterize the groups in which the set of finitely generated subgroups satisfies the minimum condition.
    (b) Dually, describe the groups in which the set of subgroups with minimum condition satisfies the maximum condition.
7.  The set of endomorphic images satisfies the minimum condition if and only if the group is a direct sum of a finite number of groups $Q$ and $Z(p^k)$ $(k = 1, 2, \cdots, \infty)$. [*Hint*: $nA$.]
8.  The set of fully invariant subgroups satisfies the minimum condition if and only if the group is a direct sum of groups $Q, Z(p^\infty)$ for finitely many $p$ and $Z(p^k)$ with $p^k \mid m$ for a fixed $m$.

## NOTES

Injectivity together with the existence of injective hulls was discovered by Baer [3]. He also proved that for the injectivity of an R-module $M$, it is necessary and sufficient that every homomorphism from every left ideal L of R into $M$ extends to an R-homomorphism $R \to M$. This extensibility property, with L restricted to principal left ideals of R, is perhaps the most convenient way to define the divisibility of an R-module $M$. It is then immediately clear that an injective module is necessarily divisible. For modules over commutative domains R, the equivalence of injectivity and divisibility characterizes the Dedekind domains (see Cartan and Eilenberg [1]). For torsion-free modules over Ore domains divisibility always implies injectivity.

Epimorphic images of injective R-modules are again injective if and only if R is left hereditary. It is worthwhile noting that a commutative domain is hereditary exactly if it is a Dedekind domain.

The results in **22** prevail for injective modules in general. Let us note that the semisimple Artin rings [the regular rings in the sense of von Neumann] are characterized by the property that all modules over them are injective [divisible].

It is an easy exercise to show that over left Noetherian rings every module has a maximal injective submodule. This is not necessarily a uniquely determined submodule, unless the ring is, in addition, left hereditary. E. Matlis [*Pac. J. Math.* **8** (1958), 511–528] and Z. Papp [*Publicationes Math. Debrecen* **6** (1959), 311–327] proved that every injective module over a left Noetherian ring R is a direct sum of directly indecomposable ones. If, in addition, R is commutative, then the indecomposable R-modules are in a one-to-one correspondence with the prime ideals P of R, namely, they are the injective hulls of R/P (for R = Z take P = (0) or $(p)$); cf. Matlis [*loc. cit.*]. It is a remarkable fact that direct sums (and direct limits) of injectives are again injective if and only if the ring is left Noetherian—as was pointed out by Matlis and Papp.

*Problem 12.* Which groups are (a) projective, (b) injective over their endomorphism rings? (c) Find the projective, injective, and the weak dimensions of a group over its endomorphism ring.

See A. J. Douglas and H. K. Farahat [*Monatshefte Math.* **69** (1965), 294–305] and F. Richman and E. A. Walker [*Math. Z.* **89** (1965), 77–81].

# V

## PURE SUBGROUPS

The notion of pure subgroups is due to Prüfer [2]; it has recently become one of the most useful concepts in abelian group theory. The notion of pure subgroups is intermediate between subgroups and direct summands, and it reflects a way in which a subgroup is embedded in the whole group. This is sufficiently general to guarantee that there is an adequate supply of pure subgroups, and at the same time pure subgroups share a number of properties which are easy to handle. Their importance lies also in the methodological role they play in proving the existence of direct summands; namely, the existence of pure subgroups of one kind or another is easily established, and several criteria assure the direct summand character of certain pure subgroups.

If the exact sequences are restricted to those which we are going to call pure-exact, then the new, remarkable notion of pure-injectivity arises [which will be the topic of Chapter VII].

### 26. PURITY

A subgroup $G$ of $A$ is called *pure*, if the equation $nx = g \in G$ is solvable in $G$, whenever it is solvable in the whole group $A$. This amounts to saying that $G$ is pure in $A$ if $n \mid g$ in $A$ implies $n \mid g$ in $G$. As $n \mid g$ in $G$ is equivalent to the inclusion $g \in nG$, we see that $G$ is pure in $A$ if and only if

$$(1) \qquad nG = G \cap nA \qquad \text{for every} \quad n \in \mathbf{Z}.$$

If both $A$ and $G$ are equipped with their Z-adic topologies, then (1) evidently implies that *for a pure subgroup, the relative Z-adic topology and its own Z-adic topology are the same.*

A natural generalization of purity is $p$-purity. A subgroup $G$ of $A$ is $p$-pure ($p$ a prime) if

$$(2) \qquad p^k G = G \cap p^k A \qquad \text{for} \quad k = 1, 2, \cdots,$$

or, in other words, the $p$-heights of elements of $G$ are the same in $G$ as in $A$. For $p$-pure subgroups, the relative $p$-adic topologies coincide with their own $p$-adic topologies.

If $G$ is $p$-pure in $A$ for every prime $p$, then $G$ is pure in $A$. In fact, if $n = p_1^{r_1} \cdots p_k^{r_k}$ is the canonical representation, then

$$nG = p_1^{r_1}G \cap \cdots \cap p_k^{r_k}G = (G \cap p_1^{r_1}A) \cap \cdots \cap (G \cap p_k^{r_k}A)$$

$$= G \cap (p_1^{r_1}A \cap \cdots \cap p^{r_k}A) = G \cap nA.$$

We next list some useful facts concerning pure subgroups.

(a) Every direct summand is a pure subgroup. 0 and $A$ are (the trivial) pure subgroups of $A$.

(b) The torsion part of a mixed group and its $p$-components are pure subgroups [these fail to be, in general, direct summands].

(c) The subgroups of $Q$ and $Z(p^\infty)$ have no pure subgroups except for the trivial ones.

(d) If $A/G$ is torsion-free, then $G$ is pure in $A$. In fact, $na = g \in G$ ($n \neq 0$) for $a \in A$ implies $a \in G$.

(e) *In torsion-free groups, intersections of pure subgroups are again pure.*

Indeed, an equation $nx = g$ has at most one solution; therefore, if it is solvable in $A$, then its unique solution belongs to every pure subgroup containing $g$.

In view of this, in a torsion-free group $A$, to every subset $S$ there exists a minimal pure subgroup containing $S$, namely, the intersection $\langle S \rangle_*$ of all pure subgroups of $A$ that contain $S$; it may be called *the pure subgroup generated by $S$*. It is easy to check that $\langle S \rangle_*$ is the set of all elements of $A$ which depend on $S$.

(f) *Purity is an inductive property.*

Let $G_1 \leqq \cdots \leqq G_i \leqq \cdots$ be a chain of pure subgroups $G_i$ of $A$, and let $G$ be their union. If $nx = g \in G$ is solvable in $A$, and if $j$ is an index with $g \in G_j$, then $nx = g$ is solvable in $G_j$, and hence *a fortiori* in $G$. Similarly, $p$-purity is inductive.

(g) *A $p$-pure $p$-subgroup of a group is pure.*

This follows at once from the equation $qG = G$ which holds for every $p$-group $G$ and every prime $q \neq p$.

(h) *If $A$ is $p$-group and if the elements of order $p$ of a subgroup $G$ have the same height in $G$ as in $A$, then $G$ is pure in $A$.*

We use induction on $e(g) = n$ to prove that every $g \in G$ has the same height in $G$ as in $A$. For $e(g) = 1$, this being our assumption, suppose this holds for elements of exponent $< n$ where $n \geq 2$. If $e(g) = n$ and some $a \in A$ satisfies $p^k a = g$, then, by induction hypothesis, $p^{k+1}h = pg$ for some $h \in G$. Here $p^k h - g$ is either 0 or of order $p$, such that $p^k(h - a) = p^k h - g$. By hypothesis, $p^k g' = p^k h - g$ for a certain $g' \in G$, whence $p^k(h - g') = g$. Thus $g \in G$ is not of a smaller height in $G$ than in $A$.

(j) *If $G[p] = A[p]$ holds for a p-group $A$ and a pure subgroup $G$ of $A$, then $G = A$.*

We prove again by induction on $e(a) = n$ that every $a \in A$ belongs to $G$. Assuming $e(a) = n \geq 2$, $p^{n-1} a \in A[p] = G[p]$ implies, owing to purity, $p^{n-1} g = p^{n-1} a$ for some $g \in G$. Here $e(g - a) \leq n - 1$, thus by induction $g - a \in G$, so $a \in G$.

The following lemma will frequently be made use of.

**Lemma 26.1.** (Prüfer [2]). *Let $B, C$ be subgroups of $A$ such that $C \leq B \leq A$. Then we have:*

(i) *if $C$ is pure in $B$ and $B$ is pure in $A$, then $C$ is pure in $A$;*
(ii) *if $B$ is pure in $A$, then $B/C$ is pure in $A/C$;*
(iii) *if $C$ is pure in $A$, $B/C$ pure in $A/C$, then $B$ is pure in $A$.*

Under the hypothesis of (i),

$$nC = C \cap nB = C \cap (B \cap nA) = (C \cap B) \cap nA = C \cap nA$$

for every $n > 0$, proving the purity of $C$ in $A$. The statement (ii) follows from the equalities

$$n(B/C) = nB + C = (B \cap nA) + C = B \cap (nA + C) = B \cap n(A/C).$$

If we assume the hypotheses of (iii), then let $na = b \in B$ for some $a \in A$ and integer $n > 0$. Now $n(a + C) = b + C$ and the hypothesis imply that for some $b' \in B$, $n(b' + C) = b + C$. From $nb' = b + c$ $(c \in C)$ we get $n(b' - a) = c$; thus $nc' = c$ for some $c' \in C$. Finally, $n(b' - c') = b$ with $b' - c' \in B$ completes the proof.$\square$

Owing to (ii) and (iii), *the natural correspondence between subgroups of $A/C$ and subgroups of $A$ containing the pure subgroup $C$ preserves purity.*

A result of another nature is:

**Proposition 26.2** (T. Szele). *Every infinite subgroup can be embedded in a pure subgroup of the same power and every finite subgroup in a countable pure subgroup.*

Given a subgroup $B$ of $A$, $|B| = \mathfrak{m}$, consider all equations $nx = b \in B$ which are solvable in $A$. For each such equation we adjoin a solution $a_{n,b} \in A$ to $B$ in order to get a subgroup $B_1$ of $A$ in which all such equations over $B$ are solvable [i.e., $B_1$ is generated by $B$ and all these $a_{n,b}$]. Then we repeat this process with $B$ replaced by $B_1$ to obtain $B_2$ in which all equations with right members in $B_1$ solvable in $A$ are solvable. The union $G$ of the $B_i$ ($i = 1, 2, \cdots$) is a pure subgroup of $A$, since $nx = g \in G$ is solvable in $B_{i+1}$ if $g \in B_i$ and if the equation has a solution in $A$. Clearly, $|G| \leq \mathfrak{m}\aleph_0$ whence both parts of the assertion follow. $\square$

For torsion-free groups, (26.2) can be improved: every subgroup can be embedded in a pure subgroup of the same rank, namely, in the pure subgroup generated by it.

Purity can be defined for $p$-adic modules by replacing natural integers $n$ in (1) by $p$-adic integers. As multiplication by units does not change submodules in the whole, only the powers of $p$ must be taken into consideration and therefore, for $p$-adic modules the definition of purity [and $p$-purity] is given by (2).

EXERCISES

1.  If $G \cap H$ and $G + H$ are pure subgroups of $A$, then so are $G$ and $H$.
2.  (a) Neither the intersection nor the union of direct summands need be pure.
    (b) If $G$ is a pure subgroup of each member of a chain $\cdots \leq B_i \leq \cdots$, then $G$ is pure in $\bigcup B_i$.
3.  The **K**-direct sum of groups is pure in the direct product.
4.  If $G$ is pure in $A$, then $nG$ is pure in $nA$.
5.  If $T$ is the torsion part of $A$, and if $G$ is pure in $A$, then $T + G$ is pure in $A$.
6.  If $G$ is a pure subgroup of the group $A$, then:
    (a) $G^1 = G \cap A^1$ (first Ulm subgroups);
    (b) $(G + A^1)/A^1$ is pure in $A/A^1$;
    (c) $G \leq A^1$ implies $G$ is divisible.
7.  Give an example for a nonpure subgroup in a group, such that its own Z-adic topology is the same as the relative Z-adic topology.
8.  The only pure, essential subgroup of $A$ is $A$ itself.
9.  A group is pure in every group containing it exactly if it is divisible.
10. If in a countable $p$-group $A$, the elements of infinite height form a pure subgroup, then $A$ is a direct sum of cocyclic groups. [*Hint*: Ex. 6 (c), (17.3), and (23.1).]
11. For every pure subgroup $B$ of $A$, there exists a well-ordered ascending chain of pure subgroups of $A$:

$$B = B_0 < B_1 < \cdots < B_\sigma < \cdots < B_\tau = A \qquad (\sigma < \tau)$$

such that $B_{\sigma+1}/B_\sigma$ is countable for every $\sigma < \tau$ and $B_\sigma = \bigcup_{\rho<\sigma} B_\rho$ if $\sigma$ is a limit ordinal.
12. (Irwin [1]) Let $A^1$ be the first Ulm subgroup of $A$ and $H$ an $A^1$-high subgroup of $A$. Prove that:

(a) $H$ is pure in $A$. [*Hint*: choose a minimal $n$ such that $p^n a = h \in H$ for some $a \in A$, but for no $a \in H$, and consider $A^1 \cap \langle p^{n-1}a, H\rangle$.]

(b) $A/H$ is divisible.

(c) $A/H$ is a divisible hull of $(A^1 \oplus H)/H \cong A^1$.

13. (a) (Charles [3] Irwin [1]). Let $A$ be a $p$-group and $B$ an infinite subgroup of $A$ without elements of infinite height. Then there exists a pure subgroup $C$ of $A$ such that (i) $B \leq C$; (ii) $|B| = |C|$; (iii) $C$ contains no elements of infinite height. [*Hint*: Ex. 12.]

   (b) (Irwin and Walker [1]) Extend (a) to arbitrary groups.

14. Prove the analog of (24.6) by replacing "subgroup" by "pure subgroup" throughout.

## 27. BOUNDED PURE SUBGROUPS

The results of this section are of fundamental importance: they are probably the most frequently cited theorems in the theory of $p$-groups.

**Proposition 27.1** (Szele [6]). *Assume that the subgroup $B$ of $A$ is the direct sum of cyclic groups of the same order $p^k$. Then the following statements are equivalent*:

(a) *$B$ is a pure [p-pure] subgroup of $A$*;

(b) *$B$ satisfies $B \cap p^k A = 0$*;

(c) *$B$ is a direct summand of $A$*.

If $B$ is a $p$-pure subgroup of $A$, then $B \cap p^m A = p^m B$ for every integer $m > 0$, in particular, for $m = k$ when $p^k B = 0$. Thus (b) follows from (a). Assume (b) and let $C$ be a $B$-high subgroup of $A$ such that $p^k A \leq C$. If, for $a \in A$, $pa = b + c$ ($b \in B$, $c \in C$), then $p^k a = p^{k-1}b + p^{k-1}c$ implies $p^{k-1}b = 0$ because of $p^k a \in C$. By the assumption on the structure of $B$, there is a $b' \in B$ with $pb' = b$. A simple appeal to (9.9), together with the equation $qB = B$ for all primes $q \neq p$, shows that $A = B \oplus C$, thus (c) holds. That (c) implies (a) is trivial. □

**Corollary 27.2** (Kulikov [1]). *Every element of order $p$ and of finite height can be embedded in a finite cyclic direct summand of the group.*

Let $a \in A$ be of order $p$ and of height $k < \infty$, and let $p^k b = a$. Then $\langle b \rangle$ is pure of finite order, and (27.1) is applicable. □

**Corollary 27.3** (Kulikov [1]). *If a group contains elements of finite order, then it contains a cocyclic direct summand.*

If the group contains a subgroup $Z(p^\infty)$ for some $p$, then this is a direct summand. If it does not contain any $Z(p^\infty)$, but contains elements of order $p$, then it contains one of finite height [cf. **20** (C)], and (27.2) applies. □

**Corollary 27.4** (Kulikov [1]).  *A directly indecomposable group is either torsion-free or cocyclic.*☐

We proceed to show that pure subgroups of certain types are necessarily direct summands.

**Theorem 27.5** (Kulikov [1]).  *A bounded pure subgroup is a direct summand.*

If $B$ is a bounded group, then by virtue of (17.2) we can write $B = B_1 \oplus C$, where $B_1$ is a direct sum of cyclic groups of the same order $p^k$, and for $C$ the least upper bound of the orders of elements is less than that for $B$. If $B$ is pure in $A_1$, then so is $B_1$, and (27.1) implies $A = B_1 \oplus A_1$. Thus $B = B_1 \oplus C_1$ with $C_1 = B \cap A_1 \cong C$. Here $C_1$ is pure in $A_1$, and by induction, $C_1$ is a direct summand of $A_1$, and hence $B$ is one of $A$.☐

**Corollary 27.6.**  *A finitely cogenerated pure subgroup is a direct summand.*☐

Also, the next useful result can be derived from (27.5).

**Theorem 27.7** (Khabbaz [1]).  *For a prime power $p^n$, any $p^nA$-high subgroup of $A$ is a direct summand of $A$.*

Let $B$ be $p^nA$-high in $A$. Then $p^nB \leq B \cap p^nA = 0$, and $B$ is a bounded $p$-group. We verify the purity of $B$ in $A$ by showing $B \cap p^kA \leq p^kB$ for integers $k \geq 0$. This being trivially true for $k = 0$, we may apply induction on $k$. If $b = p^{k+1}a \neq 0$ ($b \in B$, $a \in A$), then by (9.8), $p^ka \in p^nA \oplus B$, that is, $p^ka = p^nc + d$ for some $c \in A$, $d \in B$. Because of $p^na \in p^nA$, we have $k \leq n - 1$, and therefore, $d = p^ka - p^nc \in B \cap p^kA$ which equals $p^kB$ by induction hypothesis. Hence $b = p^{k+1}a = p^{n+1}c + pd$ implies $b = pd \in p^{k+1}B$. Therefore, $B$ is a bounded pure subgroup of $A$ and (27.5) concludes the proof.☐

**Corollary 27.8** (Erdélyi [1]).  *A $p$-subgroup $B$ of a group $A$ can be embedded in a bounded direct summand of $A$ if, and only if, the heights of the elements $\neq 0$ of $B$ [relative to $A$] are bounded.*

The necessity being obvious, assume that $m$ is an upper bound for the heights. Pick out a $p^{m+1}A$-high subgroup $C$ in $A$ subject to the condition $B \leq C$, and apply the preceding theorem to conclude that $C$ is a bounded direct summand containing $B$.☐

**Corollary 27.9.**  *An element $a$ of prime power order belongs to a finite direct summand of the group exactly if $\langle a \rangle$ contains no elements $\neq 0$ of infinite height.*

It is enough to verify the sufficiency. If $a$ is of the mentioned character, then with $B = \langle a \rangle$ we apply (27.8) and embed $B$ in a bounded direct summand $C$ of $A$. Here $B$ is contained in a finite direct summand of $C$, $C$ being a direct sum of cyclic groups.☐

We conclude this section with the following characterizations of purity.

**Theorem 27.10.** *For a subgroup B of a group A the following conditions are equivalent*:

  (i) *B is pure in A*;
  (ii) *$B/nB$ is a direct summand of $A/nB$, for every $n > 0$*;
  (iii) *if $C \leq B$ is such that $B/C$ is finitely cogenerated, then $B/C$ is a direct summand of $A/C$.*

If (i) holds, then from (ii) in (26.1), we infer that $B/nB$ is pure in $A/nB$, so (ii) follows at once from (27.5). Next assume (ii) and let $C$ be as stated in (iii). Clearly, we may restrict ourselves to the case when $B/C$ is reduced; this amounts to the finiteness of $B/C$ [cf. (25.1)]. Now there is an integer $n > 0$ with $nB \leq C$, and by assumption, $B/nB$ is a direct summand of $A/nB$. Here $C/nB$ is contained in $B/nB$, and passing mod $C$, the direct summand property of $B$ mod $C$ is preserved. Finally, if (iii) holds and if $nx = b \in B$ has a solution in $A$ but not in $B$, then $b \notin nB$, and let $C$ be a subgroup of $B$ that contains $nB$ and is maximal with respect to the property of excluding $b$. Then, by virtue of (25.2), $B/C$ is cocyclic. By hypothesis, $B/C$ is a direct summand of $A/C$. Since $b + C (\neq C)$ is divisible by $n$, while $n(B/C) = 0$, the arising contradiction proves (i). $\square$

Clearly, in (iii), "finitely cogenerated" can be replaced by "finite."
It is obvious how to modify the above conditions if we wish to characterize $p$-purity rather than purity.

EXERCISES

1.  If a reduced $p$-group has elements of arbitrarily high orders, then it also has cyclic direct summands of arbitrarily high orders.
2.  Direct sums of directly indecomposable $p$-groups can be characterized by cardinal invariants.
3.  Every direct summand of a direct sum of directly indecomposable $p$-groups is again a direct sum of directly indecomposable groups.
4.  (Kulikov [1]) Any two direct decompositions of a group that is a direct sum of directly indecomposable $p$-groups have isomorphic refinements.
5.  If $A$ is a bounded group, and if $B$ is a subgroup of $A[p]$, then $A$ has a direct summand $C$ such that $C[p] = B$.
6.  Describe the groups in which every subgroup is pure.
7.  Call a group *pure-simple* if it has no pure subgroups other than the trivial ones. Prove that a group is pure-simple if and only if it is isomorphic to a subgroup of $Q$ or $Z(p^\infty)$.
8.  (a) A group satisfies the maximum [minimum] condition on pure subgroups exactly if it is of finite rank.

(b) In a group $A$ of finite rank $r$, every chain (without repetition) of pure subgroups can be embedded in a maximal one, and a maximal chain of pure subgroups is of length $r$.

9.  (de Groot [1]) If $A$ and $B$ are direct sums of cocyclic groups and each of them is isomorphic to a pure subgroup of the other, then $A \cong B$. [*Hint*: for reduced $A$, $B$ argue as in (17.4).]

10. (E. A. Walker [4]) If $A_n$ is a maximal $n$-bounded pure subgroup of $A$, then $A = A_n \oplus A_n^*$ for some subgroup $A_n^*$ of $A$ which must be an $n$-extension of $nA$ (cf. 24, Ex. 12).

11. (Cutler [1], E. A. Walker [4]) If $A$ and $B$ satisfy $nA \cong nB$ for some positive integer $n$, then there exist groups $A'$ and $B'$ such that

$$nA' = nB' = 0 \qquad \text{and} \qquad A \oplus A' \cong B \oplus B'.$$

[*Hint*: Let $A'$, $B'$ be maximal $n$-bounded pure subgroups of $B$ and $A$, respectively.]

12. If $G$ is a pure subgroup of $A = B \oplus C$, such that $G \cap C$ is essential both in $C$ and in $G$, then $A = B \oplus G$.

## 28.  QUOTIENT GROUPS MODULO PURE SUBGROUPS

Up to now we dealt with conditions for the direct summand character of pure subgroups that relate to the pure subgroups themselves. Now conditions with the same effect will be discussed which refer to the way in which pure subgroups are embedded in the group.

We start with the following useful characterization of purity.

**Theorem 28.1** (Prüfer [2]). *A subgroup $B$ of $A$ is pure if and only if every coset of $A$ modulo $B$ contains an element of the same order as this coset.*

Let $B$ be pure in $A$ and $a^* \in A/B$. If $o(a^*) = \infty$, then every representative of the coset $a^*$ is of infinite order. If $o(a^*) = n < \infty$, then for any representative $g \in a^*$, we have $ng \in B$. Purity implies $nb = ng$ for some $b \in B$. Then $a = g - b \in a^*$ is of order $\leq n$, hence of order $n$. Conversely, if the stated condition holds, and if $ng = b \in B$ for some $g \in A$, then choose $a$ in the coset $g + B$ of the order of this coset. Then $na = 0$, and $g - a \in B$ satisfies $n(g - a) = b$. $\square$

The next theorem is the main result of this section.

**Theorem 28.2** (Kulikov [1]). *If $B$ is a pure subgroup of $A$ such that $A/B$ is a direct sum of cyclic groups, then $B$ is a direct summand of $A$.*

In view of (9.4) we may restrict ourselves to the case when $A/B$ is cyclic, say, generated by $a^*$. By (28.1) we can select a representative $a \in a^*$ such that $a$ has the same order as $a^*$. Then the elements of $\langle a \rangle$ form a complete set of representatives of $A$ mod $B$, thus $A = B \oplus \langle a \rangle$. $\square$

**Corollary 28.3.** *If B is pure in A and A/B is finitely generated, then B is a direct summand of A.*□

We have the following dual of (27.10):

**Theorem 28.4.** *The following are equivalent conditions for a subgroup B of A:*

  (i) *B is pure in A;*
  (ii) *B is a direct summand of $n^{-1}B$, for every $n > 0$;*
  (iii) *if C is a group between B and A such that C/B is finitely generated, then B is a direct summand of C.*

If $B$ is pure in $A$, then the boundedness of $(n^{-1}B)/B$ implies (ii) because of (17.2) and (28.2). In order to prove that (ii) implies (iii), by (14.4) we may restrict ourselves to the case when $C/B$ is finite. Then $C \leq n^{-1}B$ for some $n$, and (iii) is an immediate consequence of (ii). Since (iii) implies that every coset of $A$ mod $B$ contains an element of the same order as the coset, (i) follows at once from (28.1).□

Pure subgroups $B$ were defined in terms of solvability of equations $nx = b \in B$, solvable in the whole group. (22.3) shows that the same does not hold for arbitrary systems of equations. If, however, we restrict ourselves to systems with a finite number of unknowns only, then we have the following result.

**Theorem 28.5** (Prüfer [3]). *If the system*

$$\sum_{j=1}^{m} n_{ij} x_j = b_i \qquad (b_i \in B, \quad i \in I)$$

*over a pure subgroup B of A with a finite number m of unknowns is solvable in A, then it possesses a solution in B, too.*

Assume $x_j = a_j (j = 1, \cdots, m)$ is a solution in $A$. (28.4, iii), $B$ is a direct summand of $\langle B, a_1, \cdots, a_m \rangle = C$, $C = B \oplus B_1$. The $B$-components of $a_j$ yield a solution in $B$.□

It is evident that this theorem characterizes pure subgroups.

Also, it is straightforward to check that all the results of this section hold true for $p$-adic modules.

## EXERCISES

1.   $B$ is pure in $A$ if and only if $n^{-1}B = B + n^{-1}0$ for every $n > 0$.
2.   If $B$ is a pure subgroup of $A$, then

$$(A/B)[n] \cong A[n]/B[n] \qquad \text{for every} \quad n.$$

3.  (Kulikov [3]) If $A = B + C$ and $B \cap C$ is pure in $B$, then $A[n] = B[n] + C[n]$ for every $n$.
4.  (Kulikov [3]) If $a$ is an element of smallest order in its coset $a + pA$, and if $o(a) \neq \infty$, then $\langle a \rangle$ is a direct summand of $A$.
5.  A torsion group has a cyclic quotient group $\neq 0$ exactly if it has a cyclic direct summand $\neq 0$.
6.  Prove (28.1)–(28.5) for $p$-adic modules.

## 29.  PURE-EXACT SEQUENCES

A short exact sequence

(1)                     $$0 \to A \xrightarrow{\alpha} B \xrightarrow{\beta} C \to 0$$

is said to be *pure-exact* if Im $\alpha$ is a pure subgroup of $B$. It is *p-pure-exact* if Im $\alpha$ is $p$-pure in $B$.

To simplify notation, in the next theorem we shall use the same letter for a homomorphism and homomorphisms induced by it.

**Theorem 29.1.** *For an exact sequence* (1), *each of the following conditions is equivalent to the pure-exactness of* (1):

(a)  $0 \to nA \xrightarrow{\alpha} nB \xrightarrow{\beta} nC \to 0$ *is exact for every* $n$;

(b)  $0 \to A[n] \xrightarrow{\alpha} B[n] \xrightarrow{\beta} C[n] \to 0$ *is exact for every* $n$;

(c)  $0 \to A/nA \xrightarrow{\alpha} B/nB \xrightarrow{\beta} C/nC \to 0$ *is exact for every* $n$;

(d)  $0 \to A/A[n] \xrightarrow{\alpha} B/B[n] \xrightarrow{\beta} C/C[n] \to 0$ *is exact for every* $n$.

That the compositions of $\alpha$ and $\beta$ are 0 throughout is evident. If (1) is exact, then $\alpha$ is monic and $\beta$ is epic in (a). The kernel of $\beta$ in (a) is $\alpha A \cap nB$ which is $\alpha(nA)$ if and only if (1) is pure-exact.

In (b), $\alpha$ is always monic, while $\beta$ is epic for every $n$ exactly if every element of $C$ of order $n$ is an image of an element of order $n$ in $B$; this is, by (28.1), equivalent to purity. Ker $\beta = \alpha A \cap B[n] = \alpha A[n]$ holds, $\alpha$ being monic.

The rest follows now from what has been proved so far and from the $3 \times 3$-lemma.☐

*Remark 1.* We shall need the fact that (1) *implies* (b) *splitting*. In view of (27.5), it suffices to show that (b) is pure-exact. The proof is straightforward and left to the reader [cf. Ex. 1].

*Remark 2.* If we restrict ourselves to $n = p^k$ with a fixed prime $p$ and $k = 1, 2, \cdots$, then (29.1) yields conditions for $p$-purity.

The next two theorems characterize pure-exact sequences in terms of homological properties.

**Theorem 29.2.** *An exact sequence* (1) *is pure-exact if and only if every cocyclic group G has the injective property relative to* (1), *i.e., for every cocyclic G the diagram*

$$0 \to A \xrightarrow{\alpha} B \xrightarrow{\beta} C \to 0$$

$$\phi \downarrow \quad {}^{\psi}\!\!\nearrow$$

$$G$$

*can be embedded in a commutative diagram for a suitable choice of* $\psi : B \to G$.

It is straightforward to check that a finite direct sum $G_1 \oplus \cdots \oplus G_m$ has the injective property relative to (1) exactly if every $G_i$ has it; thus $G$ can be assumed to be finitely cogenerated. Also, it is clear that it suffices to consider the cases when $\phi$ is an epimorphism. Then the existence of $\psi$ is equivalent to the extensibility of the isomorphism $A/\mathrm{Ker}\,\phi \cong G$ to a homomorphism $B \to G$, i.e., to the fact that $\alpha(A/\mathrm{Ker}\,\phi)$ is a direct summand of $B/\alpha\,\mathrm{Ker}\,\phi$. Now (27.10) completes the proof. □

The dual proof, with a reference to (28.4), leads us to the dual result:

**Theorem 29.3.** *A necessary and sufficient condition that an exact sequence* (1) *be pure-exact is that every cyclic group G have the projective property relative to* (1), *i.e., for every cyclic G, and for every* $\phi : G \to C$, *there be a* $\psi : G \to B$ *making*

$$G$$

$$\psi \nearrow \quad \downarrow \phi$$

$$0 \to A \xrightarrow{\alpha} B \xrightarrow{\beta} C \to 0$$

*commute.* □

We turn our attention to the behavior of pure-exact sequences towards direct limits.

**Theorem 29.4.** *If*

$$A = \{A_i\,(i \in I);\ \pi_i^j\}, \qquad B = \{B_i\,(i \in I);\ \rho_i^j\}, \qquad and \qquad C = \{C_i\,(i \in I);\ \sigma_i^j\}$$

*are direct systems of groups, and if* $\phi : A \to B$, $\psi : B \to C$ *are homomorphisms between them such that, for every* $i \in I$,

$$0 \to A_i \xrightarrow{\phi_i} B_i \xrightarrow{\psi_i} C_i \to 0$$

*is pure-exact, then the induced sequence between the direct limits*

$$0 \to A_* \xrightarrow{\phi_*} B_* \xrightarrow{\psi_*} C_* \to 0$$

*is likewise pure-exact.*

By (11.4) it suffices to verify the purity of $\phi_* A_*$ in $B_*$. Let $a \in A_*$ and $b \in B_*$ satisfy $nb = \phi_* a$ for some $n$. Then there are $i \in I$ and $a_i \in A_i$, $b_i \in B_i$ such that $a = \pi_i a_i$ and $b = \rho_i b_i$ for the canonical maps $\pi_i : A_i \to A_*$, $\rho_i : B_i \to B_*$. Because of the commutativity of diagram (5) in **11**, $\rho_i nb_i = \phi_* \pi_i a_i = \rho_i \phi_i a_i$ whence $\rho_i(nb_i - \phi_i a_i) = 0$, and $\rho_i^j(nb_i - \phi_i a_i) = 0$ for some $j \geq i$. Therefore $n\rho_i^j b_i = \rho_i^j \phi_i a_i = \phi_j \pi_i^j a_i$, and since $\rho_i^j b_i \in B_j$, $\pi_i^j a_i \in A_j$ and $0 \to A_j \xrightarrow{\phi_j} B_j \to C_j \to 0$ is pure-exact, we conclude that some $g_j \in A_j$ satisfies $\phi_j ng_j = \phi_j \pi_i^j a_i$. Apply $\rho_j$ and observe that $\rho_j \phi_j = \phi_* \pi_j$ to obtain $n\phi_* g = \phi_* a$ with $g = \pi_j g_j \in A_*$. $\square$

In other words, the preceding theorem asserts that the direct limit of pure-exact sequences is pure-exact.

**Corollary 29.5** (Yahya [1]). *Every pure-exact sequence is the direct limit of splitting exact sequences.*

Let $0 \to A \xrightarrow{\phi} B \xrightarrow{\psi} C \to 0$ be a pure-exact sequence, and $\{C_i\}_{i \in I}$ the family of all finitely generated subgroups of $C$ where $i \leq j$ means $C_i \leq C_j$. If $\sigma_i^j : C_i \to C_j$ are the injection maps for $i \leq j$, then $C$ is the direct limit of the system $C = \{C_i \, (i \in I); \sigma_i^j\}$. Clearly, $B$ is the direct limit of the system $B = \{B_i = \psi^{-1} C_i \, (i \in I); \rho_i^j\}$ where $\rho_i^j$ stands again for the injection map $B_i \to B_j$. If $A = \{A_i = A; \pi_i^j = 1_A\}$, then $\phi$ and $\psi$ induce homomorphisms $A \to B$ and $B \to C$ such that $0 \to A_i \xrightarrow{\phi_i} B_i \xrightarrow{\psi_i} C_i \to 0$ is pure-exact for every $i$ [where $\phi_i$, $\psi_i$ are the restriction maps], and the direct limit of these pure-exact sequences is just the sequence we started with. These pure-exact sequences are splitting because of (28.4). $\square$

The last corollary can immediately be applied to prove a theorem of a rather general nature.

**Theorem 29.6** (Yahya [1]). *Let $F$ be a covariant additive functor on $\mathscr{A}$ to $\mathscr{A}$ which commutes with the formation of direct limits. If*

$$0 \to A \xrightarrow{\phi} B \xrightarrow{\psi} C \to 0$$

*is a pure-exact sequence, then so is*

(2) $$0 \to F(A) \xrightarrow{F(\phi)} F(B) \xrightarrow{F(\psi)} F(C) \to 0.$$

In view of (29.5), we can represent the given pure-exact sequence as the direct limit of splitting exact sequences $0 \to A_i \xrightarrow{\phi_i} B_i \xrightarrow{\psi_i} C_i \to 0$. By the covariance and additivity of $F$, the induced sequences

(3) $$0 \to F(A_i) \xrightarrow{F(\phi_i)} F(B_i) \xrightarrow{F(\psi_i)} F(C_i) \to 0$$

are splitting exact. Also, $F$ commutes with direct limits, i.e., the direct limit of (3) is (2). A simple appeal to (29.4) completes the proof. $\square$

EXERCISES

1. Prove that if (1) is pure-exact, then all of (a), (b), (c), (d) in (29.1) are again pure-exact whatever $n$ is. Moreover, except for (a), they are splitting.
2. Give a direct proof for the equivalence of (c) [and (d)] with purity.
3. Let (1) be a pure-exact sequence. Show that the sequences $0 \to A^1 \to B^1 \to C^1 \to 0$ of first Ulm subgroups and $0 \to A_0 \to B_0 \to C_0 \to 0$ of 0th Ulm factors need not be exact; but if one of them is exact, then so is the other.
4. Let $\mathscr{E}$ be a class of exact sequences and $G_i$ ($i \in I$) a family of groups where the index set $I$ is of arbitrary cardinality. Each $G_i$ has the injective property relative to every member of $\mathscr{E}$ if and only if the same holds for their direct product $\prod G_i$.
5. The dual of Ex. 4 [replace "injective" by "projective," and "direct product" by "direct sum"].
6. Keeping the notation of (29.4), let $\phi : A \to B$ be a homomorphism such that $\phi_i A_i$ is pure in $B_i$ for every $i \in I$. Show that $\phi_* A_*$ is pure in $B_*$.
7. Let $0 \to A \to B \to C \to 0$ be an exact sequence of inverse systems such that $0 \to A_i \to B_i \to C_i \to 0$ is pure-exact for every $i$ [see (12.3)]. Prove that $0 \to A^* \to B^* \to C^*$ need not be pure-exact. [Hint: 21, Ex. 6.]
8. (Yahya [1]) Let $F$ be an additive functor on $\mathscr{A} \times \cdots \times \mathscr{A}$ ($m$ times) to $\mathscr{A}$, covariant in each variable. If $F$ commutes with direct limits and the sequences

$$0 \to A_k \xrightarrow{\alpha_k} B_k \xrightarrow{\beta_k} C_k \to 0 \qquad (k = 1, \cdots, m)$$

are pure-exact, then the sequence

$$0 \to \sum_k F(B_1, \cdots, A_k, \cdots, B_m) \xrightarrow{\xi} F(B_1, \cdots, B_m) \xrightarrow{\eta} F(C_1, \cdots, C_m) \to 0$$

is pure-exact. Here

$$\xi = \nabla \left[ \bigoplus_k F(1_{B_1}, \cdots, \alpha_k, \cdots, 1_{B_m}) \right]$$

and $\eta = F(\beta_1, \cdots, \beta_m)$.

## 30. PURE-PROJECTIVITY AND PURE-INJECTIVITY

In (29.2) and (29.3) the pure-exact sequences were characterized by the properties that the cocyclic groups have the injective, while the cyclic groups the projective property relative to them. Now starting with the class of pure-exact sequences, we wish to find all groups which have the injective and the projective property, respectively, relative to the class of pure-exact sequences.

A group $X$ is said to be *pure-projective* if it has the projective property relative to the class of pure-exact sequences; in other words, if any diagram

(1)

$$X$$
$$\psi \diagup \quad \downarrow \phi$$
$$0 \to A \xrightarrow{\ \alpha\ } B \xrightarrow{\ \beta\ } C \to 0$$

with pure-exact row can be filled in by a suitable $\psi : X \to B$ to get a commutative diagram. In order to describe the pure-projective groups, we need the following lemma.

**Lemma 30.1.** *To every group $A$ there exist a direct sum of cyclic groups,* $X = \bigoplus_{i \in I} \langle x_i \rangle$, *and an epimorphism* $\eta : X \to A$ *such that* Ker $\eta$ *is pure in* $X$.

Let $\{a_i\}_{i \in I}$ be the set of all elements of $A$, and let $A_i = \langle a_i \rangle$. For each $a_i$ we take a group $\langle x_i \rangle$ isomorphic to $A_i$, and we let $X = \bigoplus_{i \in I} \langle x_i \rangle$. The injections $\eta_i : \langle x_i \rangle \to A$ where $\eta_i x_i = a_i$ give rise to an epimorphism

$$\eta = \nabla \left[ \bigoplus_{i \in I} \eta_i \right] : X \to A$$

whose kernel we denote by $K$. The purity of $K$ in $X$ has to be verified. Let $nx = b \in K$ for some $x \in X$. If $\eta x = a_i \in A$, then $\eta x_i = a_i$ implies $x - x_i \in K$. Owing to $na_i = \eta(nx) = \eta b = 0$ and $o(x_i) = o(a_i)$, we have $nx_i = 0$, thus $n(x - x_i) = b$.□

**Theorem 30.2** (Maranda [1]). *A group is pure-projective if and only if it is a direct sum of cyclic groups.*

Assume $X$ is a direct sum of cyclic groups, $X = \bigoplus_{i \in I} \langle x_i \rangle$, and let (1) have a pure-exact row. From (29.3) we conclude that there exist $b_i \in B$ such that $\beta b_i = \phi x_i$ and $o(b_i) = o(\phi x_i)$ for every $i$. Define $\psi$ to extend the correspondence $x_i \mapsto b_i$; owing to the conditions on orders and the structure of $X$, $\psi$ will in fact be a homomorphism $X \to B$. That it makes (1) commutative is evident.

Conversely, assume $X$ is a pure-projective group. By (30.1), there exists a diagram

$$X$$
$$\psi \diagup \quad \|$$
$$0 \to K \xrightarrow{\ \xi\ } G \xrightarrow{\ \eta\ } X \to 0$$

with pure-exact row where $G$ is a direct sum of cyclic groups. By assumption, there is a $\psi : X \to G$ such that $\eta \psi = 1_X$. Hence $\psi$ is an injection upon a direct summand of $G$. An application of (18.1) leads us to the conclusion that $X$ is likewise a direct sum of cyclic groups.□

Turning to the dual case, we define a group $Y$ *pure-injective* if any diagram

(2)

$$0 \to A \xrightarrow{\alpha} B \xrightarrow{\beta} C \to 0$$

with pure-exact row can be embedded in a commutative diagram by choosing $\psi : B \to Y$ properly. We prove a preliminary lemma, dual to (30.1).

**Lemma 30.3** (Łoś [1]). *Every group can be embedded as a pure subgroup in a direct product of cocyclic groups.*

Let $B_i$ ($i \in I$) be the family of all cocyclic quotient groups of the given group $A$, and let $B = \prod_{i \in I} B_i$. The epimorphisms $\eta_i : A \to B_i$ induce a homomorphism

$$\eta = [\prod \eta_i]\Delta : A \to B.$$

Since every $a \in A$, $a \neq 0$, is excluded from the kernel of some $\eta_i$ [see (25.2)], $\eta$ is a monomorphism. In order to verify the purity of Im $\eta$ in $B$, let us assume that $a \in A$, $a \notin p^n A$, and choose a subgroup $M$ of $A$ maximal with respect to the properties $p^n A \leq M$ and $a \notin M$. Then, by (25.2), $A/M$ is cocyclic, and so $p^n A \leq M$ implies $A/M = Z(p^k)$ with $k \leq n$. Since $A/M = B_i$ for some $i$, and $a + M$ is of height $k - 1$ in $A/M$, $\eta a \in p^n B$ is impossible. $\square$

The following dual of (30.2) describes the pure-injective groups.

**Theorem 30.4.** *A group is pure-injective exactly if it is a direct summand of a direct product of cocyclic groups.*

Let $Y$ be a direct summand of a direct product $G = \prod G_i$ where all $G_i$ are cocyclic. Let $\pi_i$, $\rho_i$ denote the coordinate projections and injections attached to this direct product, and let $\pi : G \to Y$, $\rho : Y \to G$ satisfy $\pi\rho = 1_Y$. If (2) has pure-exact row, then by (29.2) there exists a map $\psi_i : B \to G_i$ for every $i$, such that $\pi_i \rho\phi = \psi_i \alpha$. The $\psi_i$ yield the map $\psi = [\prod \psi_i]\Delta : B \to G$ where $\pi_i \psi = \psi_i$. Thus $\pi_i \rho\phi = \pi_i \psi\alpha$ for each $i$, that is, $\rho\phi = \psi\alpha$. Hence $\phi = \pi\rho\phi = \pi\psi\alpha$, and $\pi\psi : B \to Y$ is a desired homomorphism.

Next assume $Y$ is pure-injective. By (30.3), there is a pure-exact sequence $0 \to Y \to G \to H \to 0$ where $G$ is a direct product of cocyclic groups. By pure-injectivity, the identity map $Y \to Y$ factors through $G \to Y$, $Y \to G \to Y$, which implies that $Y$ is isomorphic to a direct summand of $G$. $\square$

Pure-injective groups are of considerable importance from several points of view. Their theory is expanded in Chapter VII.

In the light of (30.2) and (30.4), (30.1) and (30.3) get another interpretation. (30.1) states, namely, that every group $A$ can be embedded in a pure-exact sequence

(3)

$$0 \to G \to X \to A \to 0$$

where $X$ is pure-projective. [This is the complete analog of the familiar fact that every $A$ can be embedded in an exact sequence (3) with $X$ projective.] To this fact one may refer briefly by saying that " *there are enough pure-projectives.*" Similarly, given $A$, (30.3) ensures the existence of a pure-exact sequence

(4)                                                    $0 \to A \to Y \to H \to 0$

with pure-injective $Y$ [this being the analog of (24.1)], which states that "*there are enough pure-injectives.*"

EXERCISES

1.  Show that in the proof of (30.1), it suffices to select $A_i = \langle a_i \rangle$ only for a set $\{a_i\}_{i \in I}$ of generators of $A$, and in the proof of (30.3) one can restrict himself to $B_i$ where Ker $\eta_i$ excludes an element in a set of cogenerators of $A$.
2.  Give an example for a group which is a direct summand of a direct product of cocyclic groups, but is not a direct product of cocyclic groups. [*Hint*: $p$-adic integers.]
3.  Characterize the $p$-pure-projective groups. Are there enough $p$-pure-projectives?
4.  Characterize the $p$-pure-injective groups. Do there exist enough $p$-pure-injectives?
5.  A homomorphism $\alpha : A \to B$ is called *pure* if both Ker $\alpha$ is pure in $A$ and Im $\alpha$ is pure in $B$. Show that all abelian groups with pure homomorphisms as morphisms form a category which has enough projectives and injectives.
6.  Prove the analog of **14**, Ex. 11, for pure-projective groups $F^{(n)}$ and pure homomorphisms $\alpha_n$.
7.  Prove **24**, Ex. 9, for pure-injective groups $D_n$ and pure homomorphisms $\alpha_n$.

## 31.*  GENERALIZATIONS OF PURITY

In our study of pure subgroups, we have met numerous properties of purity whose significance will become more apparent in subsequent chapters. Due to the increasing importance of pure subgroups, recently much attention has been paid to their various generalizations.

These generalizations follow two main directions. The first is based on the simple observation that pure subgroups, under certain "finiteness" conditions, behave like direct summands. Consequently, notions analogous to purity will emerge if these finiteness conditions are replaced by suitable other conditions. The second direction came into existence when it was noticed that, in the group of extensions, the pure-extensions formed an important functorial subgroup, namely, the first Ulm subgroup [cf. (53.3)], and it turned out that another choice of functorial subgroups leads to a new type of purity. In this section we wish to discuss generalizations of purity, only of the first kind, in a rather general setting, based on the ideas developed by C. P. Walker [1].

As in **6**, we start with a nonempty class $\mathfrak{X}$ of groups $X$. With every $A \in \mathscr{A}$ we associate two families of subgroups of $A$ as follows:

$$_A\mathfrak{X} = \{\text{Ker } \phi \mid \phi : A \to X \in \mathfrak{X}\}$$

and

$$\mathfrak{X}_A = \{\text{Im } \psi \mid \psi : X \to A, \, X \in \mathfrak{X}\}.$$

Clearly, $A \in {}_A\mathfrak{X}$ and $0 \in \mathfrak{X}_A$, thus neither ${}_A\mathfrak{X}$ nor $\mathfrak{X}_A$ is empty. It is also clear that the $\mathfrak{X}_A$ are functorial in the sense that, for any $\alpha : A \to B$, $\alpha\mathfrak{X}_A$ is a subset of $\mathfrak{X}_B$, while the sets ${}_A\mathfrak{X}$ enjoy a property of continuity type: for every $B' \in {}_B\mathfrak{X}$ there is an $A' \in {}_A\mathfrak{X}$ such that $\alpha A' \leq B'$.

The correspondence $\mathfrak{X}_* : A \mapsto \mathfrak{X}_A \, (A \in \mathscr{A})$ defines $\mathfrak{X}_*$-purity in the following way. A subgroup $G$ of $A$ is called $\mathfrak{X}_*$-*pure* in $A$, if $G$ is a direct summand in every subgroup $B$ of $A$ that contains $G$ and $B/G \in \mathfrak{X}_{A/G}$. If $\mathfrak{X}$ is the set of all finitely generated [or all cyclic] groups, then $\mathfrak{X}_*$-purity means just ordinary purity.

It is evident that, for every class $\mathfrak{X}$, direct summands are $\mathfrak{X}_*$-pure, and if $G$ is $\mathfrak{X}_*$-pure in $A$, then $G$ is $\mathfrak{X}_*$-pure in every group between $G$ and $A$. The analog of (26.1) holds true in this general situation:

**Lemma 31.1** (C. P. Walker [1]). *Let $B$, $C$ be subgroups of $A$ satisfying $C \leq B \leq A$. Then*

(i) *if $C$ is $\mathfrak{X}_*$-pure in $B$ and $B$ is $\mathfrak{X}_*$-pure in $A$, then $C$ is $\mathfrak{X}_*$-pure in $A$;*
(ii) *if $B$ is $\mathfrak{X}_*$-pure in $A$, then $B/C$ is $\mathfrak{X}_*$-pure in $A/C$;*
(iii) *if $C$ is $\mathfrak{X}_*$-pure in $A$ and $B/C$ is $\mathfrak{X}_*$-pure in $A/C$, then $B$ is $\mathfrak{X}_*$-pure in $A$.*

Assume $C$ is $\mathfrak{X}_*$-pure in $B$ and $B$ is $\mathfrak{X}_*$-pure in $A$, and let $C \leq G \leq A$ with $G/C \in \mathfrak{X}_{A/C}$, i.e., there exist an $X \in \mathfrak{X}$ and epimorphism $\psi : X \to G/C$. If $\psi$ is composed with the natural epimorphism $G/C \to (G + B)/B$, then we find $(G + B)/B \in \mathfrak{X}_{A/B}$, and so $G + B = B \oplus H$ for some subgroup $H \leq A$. Let $B'$ be the image of $G$ under the projection onto $B$; this induces an epimorphism $G/C \to B'/C$ showing that $B'/C \in \mathfrak{X}_{B/C}$. Hence $B' = C \oplus K$ for some subgroup $K$. We find $G \leq B' \oplus H = C \oplus K \oplus H$, and so $G = C \oplus [G \cap (K \oplus H)]$, $C$ is $\mathfrak{X}_*$-pure in $A$.

To prove (ii), let $B$ be $\mathfrak{X}_*$-pure in $A$, and let $B/C \leq G/C \leq A/C$ with $(G/C)/(B/C) \cong G/B$ an epimorphic image of an $X \in \mathfrak{X}$. Thus $B$ is a direct summand of $G$ and $B/C$ is a direct summand of $G/C$. This proves (ii).

Finally, assume the hypotheses of (iii), and let $B \leq G \leq A$ with $G/B \in \mathfrak{X}_{A/B}$.

Because of $G/B \cong (G/C)/(B/C)$, $B/C$ is a direct summand of $G/C$, $G/C = B/C \oplus H/C$. The second summand is $\cong G/B$; thus by the $\mathfrak{X}_*$-purity of $C$ in $A$ we have $H = C \oplus K$ for some $K \leq H$. Consequently, $B + K = B + C + K = B + H = G$ and $B \cap K = B \cap H \cap K = C \cap K = 0$ imply $G = B \oplus K$, establishing (iii). $\square$

It is easy to see that not all of the results on purity carry over to the general case of $\mathfrak{X}_*$-purity, e.g., it is in general not an inductive property.

Call the exact sequence

(1) $$0 \to A \xrightarrow{\alpha} B \xrightarrow{\beta} C \to 0$$

$\mathfrak{X}_*$-*pure-exact* if $\alpha A$ is an $\mathfrak{X}_*$-pure subgroup of $B$. A group which has the projective [injective] property relative to $\mathfrak{X}_*$-pure exact sequences is said to be $\mathfrak{X}_*$-*pure-projective* [$\mathfrak{X}_*$-*pure-injective*].

**Theorem 31.2** (C. P. Walker [1]). *For every group $A$ and for every set $\mathfrak{X}$ there is an epimorphism $\phi : G \to A$ with $G$ $\mathfrak{X}_*$-pure-projective and* Ker $\phi$ *$\mathfrak{X}_*$-pure in $G$.*

*A group is $\mathfrak{X}_*$-pure-projective if and only if it is a direct summand of a direct sum of infinite cyclic groups and quotient groups of groups in $\mathfrak{X}$.*

If $Y$ is a quotient group of a group $X \in \mathfrak{X}$, if (1) is $\mathfrak{X}_*$-pure-exact, and $\eta : Y \to C$, then $\eta$ factors through $B \to C$, $\alpha A$ being a direct summand of $\beta^{-1}\eta Y$. Hence $Y$ is $\mathfrak{X}_*$-pure-projective, and so is every direct summand of a direct sum of $\mathfrak{X}_*$-pure-projectives. This proves the "if" part of our second assertion.

To prove the first statement, for every $A_i \in \mathfrak{X}_A$ let us select a monomorphism $\phi_i : Y_i \to A$ such that $Y_i$ is a quotient group of some $X_i \in \mathfrak{X}$ and Im $\phi_i = A_i$. These $\phi_i$, together with an epimorphism $\phi_0 : F \to A$ where $F$ is projective, give rise to an epimorphism

$$\phi = \nabla\left(\bigoplus_i \phi_i \oplus \phi_0\right) : G = \bigoplus_i Y_i \oplus F \to A.$$

By the preceding paragraph, $G$ is $\mathfrak{X}_*$-pure-projective. Let $K = $ Ker $\phi$ and $K \leq H \leq G$ with $H/K \in \mathfrak{X}_{G/K}$. In view of $G/K \cong A$, there is an epimorphism $\sigma : H \to Y_j$ for some index $j$, with Ker $\sigma = K$, such that

$$\begin{array}{ccc} H & \xrightarrow{\lambda} & G \\ {\scriptstyle\sigma}\downarrow & & \downarrow{\scriptstyle\phi} \\ Y_j & \xrightarrow{\phi_j} & A \end{array}$$

commutes; here $\lambda : H \to G$ [and later $\mu : K \to H$] denotes the obvious injection map. If $\rho_j : Y_j \to G$ is the injection map, then $\lambda - \rho_j\sigma : H \to G$ satisfies

$$\phi(\lambda - \rho_j\sigma) = \phi\lambda - \phi\rho_j\sigma = \phi\lambda - \phi_j\sigma = 0.$$

Thus there is a homomorphism $\tau : H \to K = \text{Ker } \phi$ such that $\lambda\mu\tau = \lambda - \rho_j\sigma$, and so $\lambda\mu\tau\mu = \lambda\mu - \rho_j\sigma\mu = \lambda\mu - \rho_j 0 = \lambda\mu$. Here $\lambda\mu$ is monic, whence $\tau\mu = 1_K$, $K$ is a direct summand of $H$, and $K$ is $\mathfrak{X}_*$-pure in $G$.

Turning to the " only if " part of the second assertion, assume that $H$ is $\mathfrak{X}_*$-pure-projective. By what has been proved, there is an $\mathfrak{X}_*$-pure-exact sequence $0 \to K \to G \to H \to 0$ with $G$ a direct sum of quotient groups of groups in $\mathfrak{X}$ and a free group. By our assumption on $H$, the identity map $1_H$ factors through $G \to H$, whence the isomorphism $G \cong K \oplus H$ follows.☐

The dual notion of $\mathfrak{X}$-purity, yielded by the correspondence $_*\mathfrak{X} : A \mapsto {}_A\mathfrak{X}$ is now easy to formulate: $G$ is $_*\mathfrak{X}$-*pure* in $A$ if $G/B$ is a direct summand of $A/B$ for every subgroup $B$ of $G$ such that $G/B \in {}_G\mathfrak{X}$. Notice that ordinary purity corresponds to the choice of the class of cocyclic groups for $\mathfrak{X}$. (31.1) can be verified for $_*\mathfrak{X}$-purity, and the dual of (31.2) also holds true. The proofs of these are left to the reader.

Two important special cases are worthwhile mentioning.

Let $\mathfrak{X}(\mathfrak{m})$ denote the class of groups whose cardinality is less than a fixed infinite cardinal $\mathfrak{m}$. Then the $\mathfrak{X}(\mathfrak{m})_*$-purity of $G$ in $A$ means that $G$ is a direct summand of any group $B$ such that $G \le B \le A$ and $|B/G| < \mathfrak{m}$; in this case $G$ is simply called $\mathfrak{m}$-*pure* in $A$ (Gacsályi [2]). It is clear that if $\mathfrak{m} > \mathfrak{n}$ then $\mathfrak{m}$-purity implies $\mathfrak{n}$-purity, and that $\aleph_0$-purity is equivalent to ordinary purity.

It is easy to give an example of an $\aleph_0$-pure subgroup which is not $\aleph_1$-pure. For instance, if $F$ is free such that $F/H \cong Q$, then $H$ is $\aleph_0$-pure, but not $\aleph_1$-pure. An example for an $\aleph_1$-pure subgroup which is not a direct summand is given by the subgroup $H$ of the free group $F$ if $F/H$ is the direct product of countably many infinite cyclic groups. This follows from (19.2), where it was shown that $F/H$ is $\aleph_1$-free, but not free.

The other important special case is when $\mathfrak{X}$ is the class of all cyclic groups of prime orders. Now $G$ is $_*\mathfrak{X}$-pure in $A$ if $B < G$, $G/B \cong Z(p)$ implies that $G/B$ is a direct summand of $A/B$. It is not difficult to see that this is equivalent to the direct summand character of $G/pG$ in $A/pG$, for every $p$, which in turn is equivalent to

$$pG = G \cap pA \qquad \text{for every prime} \quad p.$$

If this holds, $G$ is called a *neat* subgroup of $A$. This notion has been studied by Honda [1].

EXERCISES

1.  Prove (31.1) and the dual of (31.2) for $_*\mathfrak{X}$-purity.
2.  Relate $\mathfrak{X}_*$- and $\mathfrak{Y}_*$-purity, $_*\mathfrak{X}$- and $_*\mathfrak{Y}$-purity for classes $\mathfrak{X}$ contained in $\mathfrak{Y}$.

3. Show that $\mathfrak{X}_*$-purity [$_*\mathfrak{X}$-purity] does not change if $\mathfrak{X}$ is replaced by the class of all quotient groups [subgroups] of groups in $\mathfrak{X}$.

4. (C. P. Walker [1]) A group is $\mathfrak{m}$-pure-projective exactly if it is a direct sum of groups of cardinality $< \mathfrak{m}$ provided $\mathfrak{m}$ is of the form $\mathfrak{m} = \aleph_{\sigma+1}$ [i.e., $\mathfrak{m}$ is not a limit cardinal]. [*Hint*: (31.2) and (9.10).]

5. $G$ is $\mathfrak{m}$-pure in $A$ exactly if every system of equations over $G$ with a set of unknowns of power $< \mathfrak{m}$ is solvable in $G$ whenever it admits a solution in $A$. [*Hint*: (28.5).]

6. If $p^\sigma A$ is defined for ordinals $\sigma$ as in **37**, then

$$p^\sigma G = G \cap p^\sigma A$$

holds for an $\mathfrak{m}$-pure subgroup $G$ of $A$ and for any ordinal $\sigma$ of cardinality less than $\mathfrak{m}$. [*Hint*: by induction on $\sigma$, prove that $a \in p^\sigma A$ is equivalent to the solvability of a certain system of equations with not more than $|\sigma|$ unknowns.]

7. A subgroup $B$ of a group $A$ can be embedded in an $\aleph_\sigma$-pure subgroup of cardinality $\leq |B|^{\aleph_\rho}$ where $\rho = \sigma - 1$ or $\sigma$, according as $\sigma$ is an isolated or a limit ordinal. [*Hint*: argue as in (26.2) and refer to Ex. 5.]

8. A directly indecomposable torsion-free group of cardinality greater than $\aleph_{\sigma+1}$ contains an $\aleph_\sigma$-pure subgroup which is not a direct summand. [*Hint*: Ex. 7; assume the generalized continuum hypothesis.]

9. Prove (9.2), part (c), for $B$ $\mathfrak{m}$-pure in $A$ under the assumption $|V:U| < \mathfrak{m}$, and show that this characterizes $\mathfrak{m}$-purity.

10. (Honda [1]) (a) Give an example for a neat subgroup which is not pure.
    (b) In torsion-free groups, neatness and purity are equivalent.
    (c) Neatness is an inductive property.

11. If $B$ is neat in $A$, and if either $B$ or $A/B$ is an elementary $p$-group, then $B$ is a direct summand of $A$.

12. (Honda [1]) (a) If $B \leq A$ and $C (\leq A)$ is $B$-high, then $C$ is neat in $A$.
    (b) For a subgroup $B$ of $A$ to be an absolute direct summand of $A$, it is necessary and sufficient that, if $C$ is neat in $A$ and disjoint from $B$, then $B + C$ must again be neat in $A$.

13. A group does not contain neat subgroups other than the trivial ones if and only if it is of rank $\leq 1$.

14. (Dlab [1]) Let $A$ be of finite rank. A subgroup $B$ is neat exactly if $r(B) + r(A/B) = r(A)$.

15. Describe the neat-projective and neat-injective groups.

16. (a) Let $E$ be the divisible hull of $A$. $B$ is a neat subgroup of $A$ if and only if $B = A \cap D$ for some divisible subgroup $D$ of $E$.
    (b) Assertion (a) is not true if $E$ is not a minimal divisible group containing $A$.

17. Every subgroup $B$ of $A$ can be embedded in a neat subgroup of $A$ which is minimal neat containing $B$. [*Hint*: Ex. 16.]

18. A group of infinite rank $\mathfrak{r}$ contains $2^{\mathfrak{r}}$ neat subgroups. [*Hint*: maximal independent system and Ex. 17.]

19. (Rangaswamy [2]) A subgroup $G$ of $A$ is the intersection of neat subgroups of $A$ if and only if $A[p]$ is not contained in $G$ whenever $(A/G)[p] \neq 0$.

20. (a) (Kulikov [3]) Call a subgroup $C$ of a $p$-group $A$ *isotype* if

$$p^\sigma C = C \cap p^\sigma A \qquad \text{for every ordinal } \sigma.$$

Try to prove (31.1) for isotype subgroups.

(b) (Irwin and Walker [2]) Show that $C$ is isotype in $A$ if and only if

$$(p^\sigma C)[p] = C \cap (p^\sigma A)[p] \qquad \text{for every } \sigma.$$

## NOTES

The observation that purity can be generalized in two basically different ways [corresponding to the properties listed in **27** and **28**, respectively] was the starting point in C. P. Walker's paper [1]; she calls them copurity and purity, respectively. Other authors use more specified versions of purity for modules, like $\alpha N = N \cap \alpha M$ for every $\alpha \in R$ or $LN = N \cap LM$ for every left [right or twosided] ideal $L$ of $R$ to define the purity of $N$ in $M$.

A natural way of introducing purity for modules is *via* the impressive (29.5). First notice that for an exact sequence $0 \to N \to M \to P \to 0$ of R-modules the following conditions are equivalent:

(a) it is the direct limit of splitting exact sequences;
(b) every finitely presented R-module has the projective property relative to it;
(c) every finite system of equations over $N$ which is solvable in $M$ has a solution in $N$;
(d) for every right R-module $J$, the induced sequence

$$0 \to J \otimes_R N \to J \otimes_R M \to J \otimes_R P \to 0$$

is exact [cf. (60.4)];

(e) the induced sequence

$$0 \to \operatorname{Hom}_Z(P, Q/Z) \to \operatorname{Hom}_Z(M, Q/Z) \to \operatorname{Hom}_Z(N, Q/Z) \to 0$$

is splitting exact.

Purity in terms of (d) was defined by P. M. Cohn [*Math. Z.* **71** (1959), 380–398]; so far this is the most useful definition. Under this definition, all exact sequences of R-modules are pure-exact if and only if $R$ is regular in von Neumann's sense.

Notice that pure submodules of injective modules need not be injective; they are if and only if the ring is left Noetherian.

Purity can also be dealt with from the point of view of relative homological algebra. In this connection, Nunke [3] is most interesting, and so is the general discussion of purity by B. Stenström [*J. Algebra* **8** (1968), 352–361].

Pure-projectivity and pure-injectivity make sense as soon as a definition of purity has been accepted. (30.1) and (30.2) generalize to suitable versions of purity; cf., e.g., D. J. Fieldhouse [*Queen's Preprint Series*, **No. 14** (1967)]. Comments on pure-injectivity will appear in the Notes to Chapter VII. (27.5) can be extended to "bounded" quasi-injective modules, see L. Fuchs [to appear in *Ann. Scuola Norm. Sup. Pisa*].

*Problem 13.* Find conditions on a subgroup of $A$ to be the intersection of a finite number of pure ($p$-pure) subgroups of $A$.

For infinite intersections in $p$-groups see Megibben [1].

*Problem 14.* Investigate pure-high subgroups (i.e., maximal disjoint among pure subgroups).

*Problem 15.* (a) For a subfunctor $F$ of the identity, call the exact sequence $0 \to A \to B \to C \to 0$ *F-pure-exact*, if the induced sequence $0 \to F(A) \to F(B) \to F(C) \to 0$ is exact. Develop a theory for $F$-purity.

(b) Dually, for $F^*(A) = A/F(A)$.

*Problem 16.* Give conditions on the classes $\mathfrak{X}$ and $\mathfrak{Y}$ of groups for the equivalence of (a) $\mathfrak{X}_*$- and $\mathfrak{Y}_*$-purities; (b) $_*\mathfrak{X}$- and $_*\mathfrak{Y}$-purities; (c) $\mathfrak{X}_*$- and $_*\mathfrak{Y}$-purities.

*Problem 17.* (a) Develop properties of *quasi-pure-injective* groups $A$ (i.e., every homomorphism $G \to A$ with $G$ pure in $A$ is induced by an endomorphism of $A$).

(b) Dually for *quasi-pure-projectives*.

*Problem 18.* For which ordinals $\sigma$ are there $\aleph_\sigma$-pure subgroups which fail to be $\aleph_{\sigma+1}$-pure?

The answer is affirmative if $\sigma$ is an isolated ordinal not exceeding the first inaccessible ordinal.

*Problem 19.* State conditions on an isotype subgroup of a $p$-group to be a direct summand.

*Problem 20.* Find the cardinality of the set of (a) all [nonisomorphic] isotype subgroups of a $p$-group; (b) all $\aleph_\sigma$-pure subgroups.

# VI

## BASIC SUBGROUPS

As we know, a reduced group does not always decompose into a direct sum of cyclic groups. But every group contains direct sums of cyclic groups, and if we focus our attention on largest subgroups which are direct sums of cyclic groups of order infinity and powers of prime, then there is some weak uniqueness as is shown by the ranks. For $p$-groups, Kulikov [2] pointed out the importance of largest pure subgroups which are direct sums of cyclic groups. They proved to be invariants of $p$-groups, and the theory of $p$-groups is based, to a considerable extent, on our knowledge of how these so-called basic subgroups are located in the groups themselves.

It turned out that some results of the theory of basic subgroups can be extended to arbitrary groups: for every prime $p$, one can define $p$-basic subgroups where, in the definition, the prime $p$ plays a distinguished role. We shall see that these $p$-basic subgroups are useful in various contexts, especially in the next chapters.

### 32. $p$-BASIC SUBGROUPS

Let $A$ be an arbitrary group and $p$ any fixed prime. A system $\{a_i\}_{i \in I}$ of elements of $A$, not containing 0, is called $p$-independent, if for every finite sub-system $a_1, \cdots, a_k$ and for any positive integer $r$,

(1) $$n_1 a_1 + \cdots + n_k a_k \in p^r A \qquad (n_i a_i \neq 0, n_i \in \mathbf{Z})$$

implies

(2) $$p^r \,|\, n_i \qquad (i = 1, \cdots, k).$$

Thus, by definition, $p$-independence is of finite character. Hence, every $p$-independent system of $A$ can be expanded to a maximal one.

135

Notice that *every p-independent system is necessarily independent.* For, if $a_1, \cdots, a_k$ are $p$-independent, and if $n_1 a_1 + \cdots + n_k a_k = 0$ with $n_i a_i \neq 0$, then (1), and so (2) holds for every $r$, hence $n_i = 0$, and the arising contradiction establishes independence.

Also, observe that *a p-independent system contains only elements of infinite order and of orders which are powers of the given prime p.* For, if $a$ is of finite order $m$ in a $p$-independent system, and if $p^s$ is the highest power of $p$ dividing $m$, then $p^s a$ is contained in every $p^r A$, thus $p^r \mid p^s$ for all $r$, unless $p^s a = 0$.

Next we verify the following technical lemma.

**Lemma 32.1** (Fuchs [9]). *A subgroup generated by a p-independent system in A is p-pure in A.*

*If an independent system containing but elements of p-power and infinite order generates a p-pure subgroup, then it is p-independent.*

Let $C$ be the subgroup generated by a $p$-independent system $\{a_i\}_{i \in I}$. Assume $c \in C \cap p^r A$, that is, $c = n_1 a_1 + \cdots + n_k a_k \in p^r A$, where $n_i a_i \neq 0$. By $p$-independence, there are integers $m_i$ with $n_i = p^r m_i$, and thus $c = p^r(m_1 a_1 + \cdots + m_k a_k) \in p^r C$, $C$ is $p$-pure in $A$.

In order to verify the second assertion, let $\{a_i\}_{i \in I}$ denote an independent system in $A$, such that $o(a_i)$ is $\infty$ or a power of the given $p$. Set

$$C = \langle \cdots, a_i, \cdots \rangle = \bigoplus_{i \in I} \langle a_i \rangle$$

which is assumed to be $p$-pure. If $n_1 a_1 + \cdots + n_k a_k \in p^r A$ with $n_i a_i \neq 0$, then the $p$-purity of $C$ implies $n_1 a_1 + \cdots + n_k a_k = p^r(m_1 a_1 + \cdots + m_k a_k)$ for some integers $m_i$. On account of independence, $n_i a_i = p^r m_i a_i$ for $i = 1, \cdots, k$. From the hypothesis on the orders of the $a_i$ we infer $p^r \mid n_i$. $\square$

We have come to the definition of *p-basic subgroups.* By a *p-basic subgroup B of A* we mean a subgroup of $A$ satisfying the following three conditions:

(i) $B$ is a direct sum of cyclic $p$-groups and infinite cyclic groups;
(ii) $B$ is $p$-pure in $A$;
(iii) $A/B$ is $p$-divisible.

[We emphasize that $p$ denotes the same fixed prime throughout.] According to this definition, $B$ possesses a basis which is said to be a *p-basis of A.*

If $A$ is equipped with the $p$-adic topology, then conditions (i)–(iii) imply that $B$ is Hausdorff in its $p$-adic topology, which is the same as the one induced by the $p$-adic topology of $A$, and $B$ is dense in $A$. [Cf. Kaloujnine [1].] Evidently, $A$ is a $p$-basic subgroup of itself if, and only if (i) holds for $A$, and 0 is a $p$-basic subgroup if and only if $A$ is $p$-divisible.

**Lemma 32.2.** $\{a_i\}_{i \in I}$ *is a p-basis of A exactly if it is a maximal p-independent system of A.*

In view of the definition of the $p$-basis and an italicized remark above, we need to consider only elements of infinite and $p$-power orders.

First suppose $\{a_i\}_{i \in I}$ is a $p$-basis. From (32.1) and from conditions (i), (ii) we obtain the $p$-independence of the set $\{a_i\}_{i \in I}$. From (iii) it follows that to every nonzero $g \in A$ there is a relation of the form $g + n_1 a_1 + \cdots + n_k a_k \in pA$. Therefore, if we enlarge the system $\{a_i\}_{i \in I}$ by adjoining $g$ to it, the arising system is no longer $p$-independent.

Conversely, let $\{a_i\}_{i \in I}$ be a maximal $p$-independent system in $A$. Then it is independent, and (i) holds evidently. (32.1) implies (ii). To prove (iii), let $g \neq 0$ be in $A$. By maximality, some relation $n_0 g + n_1 a_1 + \cdots + n_k a_k \in p^r A$ holds where $n_0 g \neq 0$ $n_i a_i \neq 0$ for $i = 1, \cdots, k$ and $p^r \nmid n_j$ for some $j$ $(0 \leq j \leq k)$. By the $p$-independence of the $a_i$, we have certainly $p^r \nmid n_0$. Write $n_0 = p^s m_0$ with $0 \leq s < r$, $(m_0, p) = 1$. Now $n_1 a_1 + \cdots + n_k a_k \in p^s A$ implies $n_i = p^s m_i$ with suitable integers $m_i$ $(i = 1, \cdots, k)$. Thus

$$p^s(m_0 g + m_1 a_1 + \cdots + m_k a_k) = p^r b$$

for some $b \in A$, whence $p^s(m_0 g - p^{r-s} b + m_1 a_1 + \cdots + m_k a_k) = 0$. Suppose that we have already proved the divisibility of the elements of $A$ by $p$ mod $B = \langle \cdots, a_i, \cdots \rangle$ provided that their orders are powers of $p$ and less than $p^r$. Then it follows that $m_0 g - p^{r-s} b + m_1 a_1 + \cdots + m_k a_k + B$, and hence $m_0 g + B$ is divisible by $p$. Since $(m_0, p) = 1$, also $g + B$ is divisible by $p$, and so, by induction, every element of finite order in $A$ is divisible by $p$, mod $B$. If $g$ is of infinite order, then we argue in the same way to show that $m_0 g + B$, and hence $g + B$ is divisible by $p$.□

**Theorem 32.3** ([Fuchs [9]]). *Every group contains p-basic subgroups, for every prime p.*

There exists a maximal $p$-independent system in the group. By the preceding lemma, it generates a $p$-basic subgroup.□

That a group may contain a number of $p$-basic subgroups, will be seen in **35**, but (35.2) will show that all these are isomorphic.

Let $B$ be a $p$-basic subgroup of $A$. We collect the cyclic direct summands of the same order in a decomposition of $B$, and form their direct sums to obtain

(3) $$B = B_0 \oplus B_1 \oplus \cdots \oplus B_n \oplus \cdots \quad \text{where}$$

(4) $$B_0 = \oplus Z \quad \text{and} \quad B_n = \oplus Z(p^n) \quad (n = 1, 2, \cdots).$$

For each $n \geq 1$, $B_1 \oplus \cdots \oplus B_n$ is $p$-pure and hence pure in $A$, and since it is a bounded $p$-group, in view of (27.5), it is a direct summand of $A$:

(5) $$A = B_1 \oplus \cdots \oplus B_n \oplus A_n.$$

However, $B_0$ is, in general, not a direct summand of $A$. We show that the subgroups $A_n$ of $A$ in (5) can be chosen such that

$$A_n = B_{n+1} \oplus A_{n+1} \qquad (n = 1, 2, \cdots).$$

Moreover, we have the following result:

**Theorem 32.4** (R. Baer; Boyer [1]). *Assume that $B$ is a subgroup of a group $A$ such that (3) and (4) hold. For $B$ to be a p-basic subgroup of $A$ it is necessary and sufficient that*

(a) $B_0$ *is p-pure in $A$; and*
(b) *with $B_n^* = B_0 \oplus B_{n+1} \oplus B_{n+2} \oplus \cdots$, we have*

$$A = B_1 \oplus \cdots \oplus B_n \oplus (B_n^* + p^n A) \qquad (n = 1, 2, \cdots).$$

For a $p$-basic subgroup $B$, (a) is obviously satisfied. By virtue of (iii), every $a \in A$ is of the form $a = b + p^n g$ ($b \in B$, $g \in A$) whatever $n$ is, thus $B_1, \cdots, B_n, B_n^*$ and $p^n A$ together generate $A$. If $a \in B_n^* + p^n A$ belongs to $B_1 \oplus \cdots \oplus B_n$ too, then write $a = c + p^n h$ ($c \in B_n^*$, $h \in A$). It follows that $p^n h \in B$, moreover, $p^n h \in B_n^*$ because of the $p$-purity of $B$ in $A$ and the structure of $B$. Consequently, $a \in B_n^* \cap (B_1 \oplus \cdots \oplus B_n) = 0$, and (b) is valid.

Conversely, if $B$ satisfies (a) and (b), then first of all (i) is trivial. (a) implies $B_0$ is $p$-pure in $B_n^* + p^n A$, and (b) implies $B_1 \oplus \cdots \oplus B_n \oplus B_0$ is $p$-pure in $A$, and so is $B$ as the union of an ascending chain of $p$-pure subgroups of $A$. Finally, to prove (iii), write $a \in A$ in the form $a = b_1 + \cdots + b_n + c + p^n g$ with $b_i \in B_i$, $c \in B_n^*$, $g \in A$. Then $a + B = p^n g + B = p^n(g + B)$, and (iii) follows.☐

EXERCISES

1. (Kulikov [2]) A $p$-group $A$ is a direct sum of cyclic groups if and only if it contains a $p$-independent set $S$ such that

$$\langle S \rangle [p] = A[p].$$

2. (a) A $p$-basic subgroup of $J_p$ is $Z$.
   (b) Show that $B_0$ in (3) need not be a direct summand of $A$.
3. (a) If $B_i$ is a $p$-basic subgroup of $A_i$ for $i \in I$, then $\oplus B_i$ is a $p$-basic subgroup of $\oplus A_i$.
   (b) In which cases is $\oplus B_i$ $p$-basic in $\prod A_i$?
4. If $B$ is a $p$-basic subgroup of $A$, then, for every $n > 0$, $nB$ is $p$-basic in $nA$.
5. A $p$-basis of $A$ is a basis of $A/pA$. If $A$ is torsion-free, every basis of $A/pA$ can be lifted to a $p$-basis of $A$.
6. A subgroup $C$ of $A$ is $p$-pure in $A$ exactly if a $p$-basic subgroup of $C$ is $p$-pure in $A$.

7.  Every $p$-basic subgroup of a $p$-pure subgroup of $A$ is a direct summand of some $p$-basic subgroup of $A$.
8.  The subgroups $B_n^* + p^n A$ $(n = 1, 2, \cdots)$ in (32.4) are absolute direct summands of $A$.
9.  (B. Charles) Show that $p^n A$ is an essential subgroup of $B_n^* + p^n A$.


## 33.   BASIC SUBGROUPS OF $p$-GROUPS

We now focus our attention on $p$-groups where $p$-basic subgroups are particularly important. If $A$ is a $p$-group and $q$ is a prime $\neq p$, then evidently $A$ has only one $q$-basic subgroup, namely 0. Therefore, in $p$-groups we may refer to the $p$-basic subgroups simply as *basic subgroups* (Kulikov [2]), without danger to confusion.

Occasionally it is useful to speak of a basic subgroup of a torsion group $A$, too. Thereby, we mean the direct sum $\oplus_p B_p$ of the basic subgroups $B_p$ of the $p$-components of $A$. Thus a basic subgroup $B$ of a torsion group $A$ is defined by the conditions:

(i)   $B$ is a direct sum of cyclic groups of prime power orders;
(ii)  $B$ is pure in $A$;
(iii) $A/B$ is divisible.

The following example, due to Kulikov [2], is typical.

*Example.* Let $B_n$ be a direct sum of cyclic groups of the same order $p^n$, and $A$ the torsion part of the direct product of the $B_n$ $(n = 1, 2, \cdots)$. Then *the direct sum $B = \oplus_{n=1}^{\infty} B_n$ is a basic subgroup of $A$.* Recall that $A$ consists of all vectors $a = (b_1, \cdots, b_n, \cdots)$ with $b_n \in B_n$ such that there is an integer $k$ with $p^k b_n = 0$ for every $n$, while $B$ consists of all $a \in A$ with almost all $b_n = 0$. Conditions (i) and (ii) are clearly satisfied, so it suffices to prove that every $a \in A$ is divisible by $p$ mod $B$. Since $p^k b_n = 0$, $a - (b_1 + \cdots + b_k) = (0, \cdots, 0, b_{k+1}, b_{k+2}, \cdots)$ is certainly divisible by $p$, because of $p \mid b_{k+i}$ in $B_{k+i}$ $(i = 1, 2, \cdots)$.

The following method of finding a basic subgroup in a $p$-group $A$ points out an interesting relation with the socle $A[p]$. For every $n \geq 0$, let $S_n$ denote the socle of $p^n A$. Since $S_0 \geq S_1 \geq \cdots \geq S_n \geq \cdots$ and every $S_n$ is an elementary $p$-group, there exist subgroups $P_n$ $(n = 1, 2, \cdots)$ such that

$$S_n = P_{n+1} \oplus S_{n+1} \qquad (n = 0, 1, \cdots).$$

The nonzero elements of $P_{n+1}$ are clearly of height $n$ in $A$; we write

$$P_{n+1} = \oplus_i \langle c_{n+1, i} \rangle$$

and select $a_{n+1,i} \in A$ satisfying $p^n a_{n+1,i} = c_{n+1,i}$. We let $B_{n+1} = \oplus_i \langle a_{n+1,i} \rangle$, and claim that $B = \sum B_n$ is a basic subgroup of $A$. Clearly, $B = \oplus B$, so (i) is satisfied. Also, $B$ is pure in $A$, because the elements of $B[p] = \oplus B_n[p] = \oplus P_n$ have the same height in $B$ as in $A$. Finally, $A/B$ is divisible, for if $a \in S_0 = P_1 \oplus \cdots \oplus P_{n+1} \oplus S_{n+1}$, $a = a_1 + a_2$ ($a_1 \in P_1 \oplus \cdots \oplus P_{n+1}$, $a_2 \in S_{n+1}$), then $a + B = a_2 + B$ shows that $a$ is divisible by $p^n$ mod $B$, i.e., $a + B$ is of infinite height in $A/B$. The purity of $B$ in $A$ guarantees that every coset of order $p$ in $A/B$ can be represented by an element of order $p$ in $A$ [cf. (28.1)], therefore the divisibility of $A/B$ results from **20**(C). This leads to the sufficiency part of the following:

**Theorem 33.1** (Charles [1]). *Let* $B = \oplus_{n=1}^\infty B_n$, $B_n$ *a direct sum of groups* $Z(p^n)$, *be a subgroup of the p-group* $A$. *A necessary and sufficient condition for* $B$ *to be a basic subgroup of* $A$ *is that, for every* $n \geqq 0$,

(1)                                        $S_n = B_{n+1}[p] \oplus S_{n+1}$.

To establish necessity, let $B$ be a basic subgroup of $A$. Then for $A_n = B_n^* + p^n A$ we clearly have $p^n A = p^n A_n$ [see (32.4)], whence $S_n = (p^n A)[p] \leqq A_n[p]$. If we had proper inclusion here, then choosing some $a \in A_n[p]$ of height $r < n$ we could write $a = b^* + p^n g$ ($b^* \in B_n^*$, $g \in A$) where $h(b^*) = r$. From $pb^* + p^{n+1} g = 0$ we would obtain $h(pb^*) \geqq n + 1$, which would be incompatible with $h(b^*) = r < n$ in a direct sum $B_n^*$ of cyclic groups of order $\geqq p^{n+1}$. Hence $S_n = A_n[p]$, and $A_n = B_{n+1} \oplus A_{n+1}$ implies (1) at once. $\square$

Another useful characterization is contained in :

**Theorem 33.2** (Szele [7]). *A subgroup* $B = \oplus_{n=1}^\infty B_n$, $B_n$ *a direct sum of groups* $Z(p^n)$, *is a basic subgroup of a p-group* $A$ *if and only if for every positive integer* $n$, $B_1 \oplus \cdots \oplus B_n$ *is a maximal* $p^n$-*bounded direct summand* [*or equivalently, a* $p^{n+1}A$-*high subgroup*] *of* $A$.

If $B$ is basic subgroup of $A$, then $A_n[p]$ has no element of order $p$ and of height $< n$, hence $A_n$ has no direct summand of order $\leqq p^n$ [cf. the proofs of (33.1) and (27.2)]. Conversely, if $B$ satisfies the stated condition, then (i) and (ii) are obvious. If $A/B$ were not divisible, then by (27.2) it would have a direct summand $C/B = \langle c^* \rangle \cong Z(p^m)$. By (28.2), $C = B \oplus \langle c \rangle$ ($c \in c^*$), and $C$ also satisfies (i) and (ii). But $B_1 \oplus \cdots \oplus B_m \oplus \langle c \rangle$ would be a $p^m$-bounded pure subgroup, hence a direct summand of $A$ larger than $B_1 \oplus \cdots \oplus B_m$, a contradiction.

The equivalence of the condition in brackets with what has been shown is evident by (27.7). $\square$

The last result enables us to find a basic subgroup in a direct product of $p$-groups.

**Corollary 33.3.** *Let $A$ be the torsion part of the direct product $\prod_i A_i$ of $p$-groups $A_i$, and let*

$$B_i = \bigoplus_{n=1}^{\infty} B_{in} \qquad with \quad B_{in} = \bigoplus Z(p^n)$$

*be a basic subgroup of $A_i$ $(i \in I)$. Then*

$$B = \bigoplus_{n=1}^{\infty} B_n \qquad with \quad B_n = \prod_i B_{in}$$

*is a basic subgroup of $A$.*

For any $n$, $A_i = B_{i1} \oplus \cdots \oplus B_{in} \oplus A_{in}^*$ where $A_{in}^*$ has no cyclic direct summand of order $\leq p^n$. Clearly, $A = B_1 \oplus \cdots \oplus B_n \oplus A_n^*$ where $A_n^*$ is the torsion part of $\prod_i A_{in}^*$, and it is clear that $A_{in}^*$ cannot have any cyclic direct summand of order $\leq p^n$. An application of (33.2) concludes the proof.☐

The next theorem gives a characterization of subgroups of basic subgroups. Its similarity with Kulikov's theorem (17.1) is apparent.

**Theorem 33.4** (Kovács [1]).   *A subgroup $C$ of a $p$-group $A$ is contained in a basic subgroup of $A$ if and only if $C$ is the union of an ascending chain of subgroups*

$$C_1 \leq C_2 \leq \cdots \leq C_n \leq \cdots$$

*such that the heights of the elements of $C_n$ [taken in $A$] are bounded for every integer $n$.*

If $C$ can be embedded in a basic subgroup $B = \bigoplus_{n=1}^{\infty} B_n$, $B_n = \bigoplus Z(p^n)$, then the subgroups $C_n = C \cap (B_1 \oplus \cdots \oplus B_n)$ satisfy the conditions.

To prove sufficiency, we may without loss of generality assume that $C_n \cap p^n A = 0$ for every $n$. We select ascending chains $G_1 \leq G_2 \leq \cdots \leq G_n \leq \cdots$ of subgroups of $A$ subject to the following conditions:

$$\text{(a)} \quad C_n \leq G_n, \qquad \text{(b)} \quad G_n \cap p^n A = 0$$

for every $n$. If we introduce a partial order in the set of all such chains in the obvious manner [cf. the proof of (17.1)], then by Zorn's lemma we conclude that there is a chain which is maximal in our present sense. Without danger of confusion, we may denote this again by $G_n$ $(n = 1, 2, \cdots)$. We claim that $B = \bigcup_{n=1}^{\infty} G_n$ will be a basic subgroup of $A$.

The main step in the proof of this is to show that $G_n$ is $p^n A$-high in $A$. By way of contradiction, assume that some $a \in A$, with $a \notin G_n$, $pa \in G_n$, satisfies $\langle G_n, a \rangle \cap p^n A = 0$. By the maximal choice of the chain of the $G_n$, there is a first $m \geq n+1$ such that $\langle G_m, a \rangle$ intersects $p^m A$, say, $a + g_m = p^m b \neq 0$ $(g_m \in G_m, b \in A)$. Now $pa + pg_m = p^{m+1} b \in G_m \cap p^m A = 0$ implies $pg_m = -pa \in G_n$, but $g_m \notin G_n$, for otherwise $a + g_m = p^m b \in \langle G_n, a \rangle \cap p^n A = 0$. Define the index $k$ so as to satisfy $g_m \in G_{n+k} \backslash G_{n+k-1}$; then $1 \leq k \leq m - n$.

Since $G_{n+k-1}$ is maximal with respect to the property (b) within $G_{n+k}$, $\langle G_{n+k-1}, g_m \rangle$ intersects $p^{n+k-1}A$, say $g_{n+k-1} + g_m = p^{n+k-1}c \neq 0$ for some $g_{n+k-1} \in G_{n+k-1}$ and $c \in A$. Thus

$$a - g_{n+k-1} = p^m b - p^{n+k-1}c \in \langle G_{n+k-1}, a \rangle \cap p^{n+k-1}A = 0$$

in view of $n + k - 1 < m$ and the choice of $m$. Hence $a \in G_{n+k-1} \leqq G_m$, and $\langle G_m, a \rangle = G_m$ does not intersect $p^m A$.

Now that we know $G_n$ is $p^n A$-high in $A$, we see that $G_n$ is a direct summand of $A$ [cf. (27.7)], whence $G_{n+1} = G_n \oplus B_{n+1}$ for some subgroup $B_{n+1}$ of $A$. The rest follows at once from (33.2).☐

If we start with $C_n = 0$ for all $n$, then the last proof yields another method of constructing basic subgroups.

The existence of basic subgroups in $p$-groups can be interpreted as follows: every $p$-group $A$ can be obtained from a pure subgroup $B$ which is a direct sum of cyclic groups, and from a divisible quotient group $A/B$. Since $B$ has a basis and $A/B$ is a direct sum of copies of $Z(p^\infty)$ [thus, also $A/B$ has an easily describable generating system], it is natural to combine these generating systems to obtain one for $A$.

We write

$$B = \bigoplus_{i \in I} \langle a_i \rangle \qquad \text{and} \qquad A/B = \bigoplus_{j \in J} C_j^* \qquad \text{where} \qquad C_j^* = Z(p^\infty).$$

If $C_j^*$ is generated by the cosets $c_{j1}^*, \cdots, c_{jn}^*, \cdots$ mod $B$ such that $pc_{j1}^* = 0^*$, $pc_{j,n+1}^* = c_{jn}^* (n = 1, 2, \cdots)$, then, by the purity of $B$ in $A$, we can pick out $c_{jn} \in c_{jn}^*$ of the same order as $c_{jn}^*$. Then we get the following set of relations:

$$(2) \qquad pc_{j1} = 0, \qquad pc_{j,n+1} = c_{jn} - b_{jn} \qquad (n \geq 1; b_{jn} \in B)$$

where $b_{jn}$ must be of order $\leqq p^n$, since $o(c_{jn}) = p^n$.

We shall call the set $\{a_i, c_{jn}\}_{i \in I, j \in J, n=1, 2, \cdots}$ a *quasibasis* of $A$. This terminology is justified by:

**Proposition 33.5** (Fuchs [2]). *If $\{a_i, c_{jn}\}$ is a quasibasis of the $p$-group $A$, then every $a \in A$ can be written in the following form:*

$$(3) \qquad a = s_1 a_{i_1} + \cdots + s_m a_{i_m} + t_1 c_{j_1 n_1} + \cdots + t_r c_{j_r n_r}$$

*where $s_i$ and $t_j$ are integers, no $t_j$ is divisible by $p$, and the indices $i_1, \cdots, i_m$ as well as $j_1, \cdots, j_r$ are distinct. (3) is unique in the sense that it uniquely defines the terms $sa_i$ and $tc_{jn}$.*

Given $a \in A$, we first express the coset $a + B$ in terms of $c_{jn}^*$, $a^* = t_1 c_{j_1 n_1}^* + \cdots + t_r c_{j_r n_r}^*$ where none of the $t \in Z$ is divisible by $p$. Hence the terms $tc_{jn}^*$ and so $tc_{jn}$ are uniquely determined by $a$. If we express $a - t_1 c_{j_1 n_1} - \cdots - t_r c_{j_r n_r}$ as a linear combination of basis elements $a_i$ of $B$, we get (3), where the $sa_i$ are again uniquely determined, $B$ being a direct sum of cyclic groups.☐

A $p$-group can be defined in a rather simple way in terms of a quasibasis, because the defining relations belonging to the generators $\{a_i, c_j\}$ will have the form $p^m a_i = 0$ and (2), where the $b_{jn}$ are replaced by linear combinations of the generators $a_i$:

EXERCISES

1.  (Bourbaki [1]) Call the system $\{a_i\}_{i \in I}$ in $A$ *pure-independent*, if it is independent and generates a pure subgroup of $A$. Prove that:
    (a) $\{a_i\}$ is pure-independent exactly if $mb = n_1 a_1 + \cdots + n_k a_k$ implies $n_j a_j = m n'_j a_j$ $(j = 1, \cdots, k)$.
    (b) Every pure-independent system is embeddable in a maximal one.
    (c) A pure-independent system $S$ in a $p$-group $A$ is maximal if and only if $A/\langle S \rangle$ is divisible.
    (d) A maximal pure-independent system is a basis of some basic subgroup of the $p$-group $A$.
2.  If a basic subgroup $B$ of a $p$-group $A$ is a characteristic subgroup of $A$, then either $B = A$ or $B = 0$.
3.  (a) Every countable subgroup of a $p$-group $A$ without elements of infinite height is contained in a basic subgroup of $A$.
    (b) This fails to hold if countability is removed.
4.  (a) Let $D$ be a divisible group containing the $p$-group $A$ and $B = \oplus_i \langle a_i \rangle$ a basic subgroup of $A$. Embed each $\langle a_i \rangle$ in a quasicyclic subgroup $D_i$ of $D$, and show that $E = A + \sum_i D_i$ is a minimal divisible group containing $A$.
    (b) Use (a) to prove that every $p$-group can be embedded in a minimal divisible group.
5.  (Irwin [1]) If $A$ is a $p$-group, every $A^1$-high subgroup of $A$ contains a basic subgroup of $A$. [*Hint*: **26**, Ex. 12.]
6.  Define neat-independence [cf. Ex. 1] and prove:
    (a) Every neat-independent system can be embedded in a maximal one.
    (b) Every element in the socle of the quotient group mod a subgroup generated by a maximal neat-independent system is divisible by any prime.
    (c) A maximal neat-independent system of a $p$-group need not be pure-independent.
7.  If $A$ has 0 $p$-component, then $A/B$ has the same property for every $p$-basic subgroup $B$ of $A$.

## 34.   FURTHER RESULTS ON $p$-BASIC SUBGROUPS

In this section we prove a number of useful results on $p$-basic subgroups. Most of them have been proved for $p$-groups by Kulikov; they immediately extend to the general case.

Throughout this section, $A$ denotes an arbitrary group and $B$ a $p$-basic subgroup of $A$.

(A) *For every integer $n \geq 0$, $A = B + p^n A$.*
This is an immediate consequence of the $p$-divisibility of $A/B$.

(B) *For every integer $n \geq 0$, $B/p^n B \cong A/p^n A$.*
In view of (A) and the second isomorphism theorem, we have $A/p^n A = (B + p^n A)/p^n A \cong B/(B \cap p^n A)$. Here $B \cap p^n A = p^n B$, since $B$ is $p$-pure in $A$.

(C) *For every integer $n \geq 0$, $p^n A/p^n B \cong A/B$.*
The proof is similar to the previous one.

(D) *A $p$-basic subgroup $B'$ of a $p$-basic subgroup $B$ of $A$ is necessarily a $p$-basic subgroup of $A$.*
The nontrivial part of the proof is to show that $A/B'$ is $p$-divisible. This follows from the $p$-divisibility of $A/B$ and $B/B'$.

(E) *If $C$ is a $p$-pure subgroup of $A$, then $A/C$ is $p$-divisible exactly if a [moreover, every] $p$-basic subgroup of $C$ is $p$-basic in $A$.*
If $A/C$ is $p$-divisible and $B$ is a $p$-basic subgroup of $C$, then it is straightforward to check that $B$ is $p$-basic in $A$. Conversely, if $C$ contains a $p$-basic subgroup $B$ of $A$, then $A/C$ is an epimorphic image of the $p$-divisible group $A/B$.

(F) *Let $A_0 = A/A^1$ be the 0th Ulm factor of $A$. Then. the image $B_0$ of a $p$-basic subgroup $B$ of $A$ under the canonical epimorphism $\phi : A \to A_0$ is a $p$-basic subgroup of $A_0$, and $\phi$ induces an isomorphism of $B$ with $B_0$.*
Since the elements of $A^1$ are divisible by every power of $p$, $B \cap A^1 = 0$. Thus $\phi | B$ is an isomorphism of $B$ with a subgroup $B_0$ of $A_0$; in particular, $B_0$ is a direct sum of cyclic groups of infinite order and orders of powers of $p$. If $\{a_i\}_{i \in I}$ is a basis of $B$, then $\{\phi a_i\}_{i \in I}$ is a basis of $B_0$ which must be $p$-independent in $A_0$, since if $n_1 \phi a_1 + \cdots + n_k \phi a_k \in p^r A_0$ with $n_i \phi a_i \neq 0$, then $n_1 a_1 + \cdots + n_k a_k \in p^r A$, and so $p^r | n_i$ $(i = 1, \cdots, k)$. Therefore $B_0$ is $p$-pure in $A_0$ [see (32.1)]. Finally, $A_0/B_0$ is $p$-divisible, being isomorphic to $[A/A^1]/[(B + A^1)/A^1] \cong A/(B + A^1)$, which is an epimorphic image of $A/B$.

(G) *Let $C$ be a $p$-pure subgroup of $A$, $\{a_i\}$ a $p$-basis of $C$ and $\{b_j^*\}$ a $p$-basis of $A/C = A^*$. If $b_j \in A$ is a representative of the coset $b_j^*$ of the same order as $b_j^*$, then $\{a_i, b_j\}$ is a $p$-basis of $A$.*
The orders of the elements $b_j^*$ are powers of $p$ or infinite, therefore the $p$-purity of $C$ in $A$ guarantees the existence of elements $b_j \in A$ of the desired kind. To prove the $p$-independence of the system $\{a_i, b_j\}$, let

$$n_1 a_1 + \cdots + n_s a_s + m_1 b_1 + \cdots + m_t b_t = p^r g \qquad (g \in A),$$

and assume that here no summand vanishes. Passing mod $C$, this yields $m_1 b_1^* + \cdots + m_t b_t^* = p^r g^*$ with no vanishing summand, whence $p^r \mid m_j$ for every $j$. Hence, we obtain $n_1 a_1 + \cdots + n_s a_s = p^r h$ for some $h \in A$, and so $p^r \mid n_i$ for every $i$, establishing the $p$-independence of $\{a_i, b_j\}$. For every $a \in A$, there is a relation $a + m_1 b_1 + \cdots + m_l b_l = c + pg$ with $c \in C$, $g \in A$, where for $c$ we have a relation $c + n_1 a_1 + \cdots + n_k a_k = pd$ with $d \in C$. Hence $a$ plus some linear combination of the $a_i$ and $b_j$ belongs to $pA$, proving that $\{a_i, b_j\}$ is maximal $p$-independent.

It follows that under the same hypothesis, *a $p$-basic subgroup of $A$ is isomorphic to the direct sum of $p$-basic subgroups of $C$ and $A/C$.*

(H)   Under certain rather general circumstances, endomorphisms are completely determined by their effects on a $p$-basic subgroup.

**Proposition 34.1.** *Let $A$ have no $p$-divisible subgroup $\neq 0$, and let $B$ be a $p$-basic subgroup of $A$. If $\chi$, $\eta$ are endomorphisms of $A$ inducing the same map on $B$, then $\chi = \eta$.*

The kernel of $\chi - \eta$ includes $B$, hence $\mathrm{Im}(\chi - \eta)$ is an epimorphic image of $A/B$. Thus $\mathrm{Im}(\chi - \eta)$ is a $p$-divisible subgroup of $A$, and so it is $0$.☐

(I)   For $p$-groups we have:

**Theorem 34.2** (Fuchs [4]). *If $G$ is an epimorphic image of a $p$-group $A$, then every basic subgroup of $G$ is an epimorphic image of every basic subgroup of $A$.*

Let $B$, $C$ denote basic subgroups of $A$, $G$. The restriction of the epimorphism $\phi : A \to G$ to $p^n A$ is an epimorphism $\phi_n : p^n A \to p^n G$ which induces an epimorphism $\phi_n^* : p^n A/p^{n+1} A \to p^n G/p^{n+1} G$. As in (B), it follows that $p^n A/p^{n+1} A \cong p^n B/p^{n+1} B$ and $p^n G/p^{n+1} G \cong p^n C/p^{n+1} C$. Take into account that $p^n B/p^{n+1} B$ is a direct sum of groups of order $p$, where the number of summands equals the number of summands in $B$ that are of order $\geq p^{n+1}$, and the same holds for $C$. The existence of an epimorphism $p^n B/p^{n+1} B \to p^n C/p^{n+1} C$ implies an inequality between the cardinalities of sets of components of order $\geq p^{n+1}$ in $B$ and $C$, for $n = 0, 1, \cdots$. Considering that $B$ and $C$ are direct sums of cyclic groups, the existence of an epimorphism $B \to C$ follows.☐

In particular, *if a direct sum of cyclic groups is an epimorphic image of a $p$-group $A$, then the same direct sum is an epimorphic image of a basic subgroup of $A$* (Szele [7]).

(J)   The following theorem gives an upper estimation for the cardinality of a group in terms of its $p$-basic subgroups.

**Theorem 34.3.** *If $A$ is a reduced group and if, for every prime $p$, $B_p$ denotes a $p$-basic subgroup of $A$, then*

$$|A| \leq \left( \sum_p |B_p| \right)^{\aleph_0}.$$

Let $B$ be the subgroup of $A$ generated by all $B_p$ [one for each prime $p$]. Then $A/B$ is $p$-divisible for every $p$, hence it is divisible, and so we may write $A/B = \bigoplus_i (C_i/B)$ with $C_i/B \cong Q$ or $Z(p^\infty)$. If $C_i/B \cong Q$, then there are elements $c_{i1}, \cdots, c_{in}, \cdots \in C_i$ such that $c_{in} = (n+1)c_{i,n+1} + b_{in}$ for some $b_{in} \in B$. If for $i \neq j$, the sequences $b_{i1}, \cdots, b_{in}, \cdots$ and $b_{j1}, \cdots, b_{jn}, \cdots$ were equal, then the elements

$$c_{in} - c_{jn} = (n+1)(c_{i,n+1} - c_{j,n+1}) \qquad (n = 1, 2, \cdots)$$

would generate a subgroup $\cong Q$ in $A$, contrary to the reducedness of $A$. Hence the number of groups $\cong Q$ among $C_i/B$ does not exceed the cardinality of the set of sequences $b_{i1}, \cdots, b_{in}, \cdots$ in $B$, which is obviously $|B|^{\aleph_0}$. The same argument applies to the number of $C_i/B$ isomorphic to $Z(p^\infty)$. Now the estimate follows from $|B| \leq \sum_p |B_p|$ [where we take into account that $B$ can be finite only if $A$ is finite.$\square$

For $p$-groups, in particular, we have

**Corollary 34.4** (Kulikov [3]). *If $B$ is a basic subgroup of a reduced $p$-group $A$, then*

$$|A| \leq |B|^{\aleph_0}. \square$$

Notice that it may happen that $|B| < |A|$ and $|A| = |B|^{\aleph_0}$ hold simultaneously. In fact, if $A$ is the torsion part of the direct product of cyclic groups $Z(p^n)$ $(n = 1, 2, \cdots)$, then $B$ is countable, while $A$ is of the power of the continuum [cf. example in 33].

**Corollary 34.5.** *For a reduced group $A$ and for its 0th Ulm factor $A_0$, the following inequality holds:*

$$|A| \leq |A_0|^{\aleph_0}.$$

Keeping the notation of (34.3), we obtain $\sum_p |B_p| \leq |A_0|$ from (F). Combining this with (34.3), the stated inequality becomes obvious.$\square$

It is a remarkable fact that an upper estimate for the cardinality of $A$ can be given in terms of the 0th Ulm factor of $A$.

EXERCISES

1. If $B$ is a $p$-basic subgroup of $A$, then:
   (a) $p^n A = p^n B + p^{n+k} A$ for all nonnegative integers $n, k$;
   (b) $p^n B / p^{n+k} B \cong p^n A / p^{n+k} A$ for $n$ and $k \geq 0$.

2. If $B$ denotes a $p$-basic subgroup of $A$, then:
   (a) $(A/B)[p] \cong A[p]/B[p]$;
   (b) $(p^n A/p^n B)[p] \cong (p^n A)[p]/(p^n B)[p]$ for every $n \geq 0$;
   (c) $(p^n A)[p] = (p^n B)[p] + (p^{n+1} A)[p]$ for every $n \geq 0$;
   (d) $(p^{n+1} B)[p] = (p^n B)[p] \cap (p^{n+1} A)[p]$ for every $n \geq 0$;
   (e) $(p^n B)[p]/(p^{n+1} B)[p] \cong (p^n A)[p]/(p^{n+1} A)[p]$ for every $n \geq 0$.

3. Which of (a)–(e) in Ex. 2 hold if $[p]$ is throughout replaced by $[p^k]$ with an integer $k \geq 2$?

4. Give an example of a reduced group $A$ and a nontrivial subgroup $B$, such that $B$ is $p$-basic in $A$ for every prime $p$.

5. Not every homomorphism of a basic subgroup of a $p$-group $A$ into $A$ can be extended to an endomorphism of $A$.

6. Let $0 \to A \to G \to C \to 0$ be a $p$-pure-exact sequence and $A_p$, $C_p$ $p$-basic subgroups of $A$, $C$. There exists a $p$-basic subgroup $B_p$ of $G$ such that, with the induced mappings, $0 \to A_p \to B_p \to C_p \to 0$ is splitting exact. [*Hint*: see (G).]

7. Generalize (34.2) to $p$-basic subgroups of any groups. [*Hint*: cf. second part of proof in (35.2).]

8. If $F$ is the Frattini subgroup of the reduced group $A$, then

$$|A| \leq |A/F|^{\aleph_0}.$$

9. (E. A. Walker [3]) Prove the following generalization of (34.3): If $A$ is a reduced group and $B$ is a subgroup of $A$ such that $A/B$ is divisible, then $|A| \leq |B|^{\aleph_0}$.

## 35. DIFFERENT $p$-BASIC SUBGROUPS

First of all we show that a group contains, in general, many different $p$-basic subgroups.

**Lemma 35.1.** *Let* $A = \bigoplus_{k=1}^{\infty} \langle a_k \rangle$, *where either* $o(a_k) = p^{n_k}$ *with* $n_1 < \cdots < n_k < \cdots$ *or* $o(a_k) = \infty$ *for every* $k$. *Then* $A$ *contains a $p$-basic subgroup $B$ different from* $A$.

Define

$$b_k = a_k - p^{n_{k+1} - n_k} a_{k+1} \qquad (k = 1, 2, \cdots)$$

where $n_1 < \cdots < n_k < \cdots$ is any increasing sequence in case all $a_k$ are of infinite order. It is straightforward to show that $\{b_k\}_{k=1,2,\ldots}$ is a $p$-independent system and the subgroup $B$ generated by these $b_k$ does not contain $a_1$. Since $A/B$ is evidently $p$-divisible [namely, isomorphic to $Z(p^\infty)$ or to $Q^{(p)}$], $B$ is $p$-basic in $A$. □

If we start with $A$ as described in the lemma, then the $p$-basic subgroup $B$ obtained also satisfies the hypotheses, and so it contains a $p$-basic subgroup $B_2 < B$. Thus proceeding, we obtain an infinite properly descending chain

$$A > B > B_2 > \cdots > B_n > \cdots$$

where, in view of **34**(D), the groups $B_n$ are all $p$-basic subgroups of $A$. We have, however:

**Theorem 35.2** (Kulikov [2], Fuchs [9]). *For a given prime $p$, all $p$-basic subgroups of a group are isomorphic.*

Let $B = \bigoplus_{n=0}^{\infty} B_n$ with $B_0 = \oplus Z$, $B_n = \oplus Z(p^n)$ for $n \geq 1$, be a $p$-basic subgroup of $A$. The number of the summands $Z(p^n)$ in $B_n$ is equal to the number of the summands $Z(p^n)$ in $B/p^k B$ for any $k > n$. From **34**(B) we know that $B/p^k B \cong A/p^k A$, and so the cardinal number in question does not depend on the particular choice of $B$ in $A$.

In order to prove the unicity of $B_0$ up to isomorphism, we begin with establishing the equality

(1)                                    $$pB_0 = B_0 \cap (T_p + pA)$$

where $T_p$ is the $p$-component of $A$. The inclusion $\leq$ being obvious, let $b_0 \in B_0 \cap (T_p + pA)$, and write $b_0 = c + pa$ with $c \in T_p$, $a \in A$. For some $n$, $p^n b_0 = p^{n+1} a \in B_0 \cap p^{n+1} A = p^{n+1} B_0$, that is, $p^n b_0 = p^{n+1} b$ for some $b \in B_0$. Owing to the torsion-freeness of $B_0$, $b_0 = pb \in pB_0$, proving (1). We infer

$$B_0/pB_0 = B_0/[B_0 \cap (T_p + pA)] \cong (B_0 + T_p + pA)/(T_p + pA) = A/(T_p + pA)$$

where we have made use of the fact that $B_0 + T_p + pA \geq B + pA = A$. We conclude that the rank of $B_0$, which is obviously equal to the rank of $B_0/pB_0$, is independent of the special choice of $B_0$. $\square$

The preceding theorem shows that a $p$-basic subgroup $B$, considered as an abstract group, is an invariant of the group $A$. This yields a system $\{ \mathfrak{m}_0, \mathfrak{m}_1, \cdots, \mathfrak{m}_n, \cdots \}$ of cardinal numbers as a system of invariants for $A$, namely,

$$B = \bigoplus_{n=0}^{\infty} B_n \quad \text{with} \quad B_0 = \bigoplus_{\mathfrak{m}_0} Z, \quad B_n = \bigoplus_{\mathfrak{m}_n} Z(p^n) \quad \text{for} \quad n \geq 1.$$

Thus with every group $A$ and with every prime $p$, we can associate a countable system of cardinal numbers which are invariants of $A$. [This, however, fails to be a complete system of invariants.]

The question naturally arises: which are the groups containing only one $p$-basic subgroup? There are only a few of them:

**Theorem 35.3.**  *A group $A$ contains exactly one $p$-basic subgroup if and only if $A$ is either of the following types:*

(a)  *$A$ is $p$-divisible;*

(b)  *$A = B \oplus C$, where $B$ is a bounded $p$-group and $C$ is $p$-divisible without elements of order $p$;*

(c)  *$A = B \oplus C$, where $B$ is a bounded $p$-group and $C$ is a free group of finite rank.*

Assume $A$ has only one $p$-basic subgroup $B'$. Then $B'$ cannot contain a free group of infinite rank or an unbounded $p$-group as direct summands, because otherwise we get a contradiction from (35.1) and **34**(D) to the uniqueness of $B'$. Thus $B' = B_0 \oplus B_1$ with $B_0$ free of finite rank and $B_1$ a bounded $p$-group. If $B_0 = B_1 = 0$, we get (a). If $B_0 = 0$, but $B_1 \neq 0$, then $B' = B_1$ [as a bounded pure subgroup] is a direct summand of $A$, $A = B' \oplus C$. Here $C$ is $p$-divisible, but contains no element of order $p$, since if $0 \neq c \in C[p]$, and if $b$ is one of the basis elements of $B'$, then $b$ can be replaced by $b + c$ to obtain a $p$-basic subgroup different from $B'$. Finally, if $B_0 \neq 0$, then $B' < A$ is impossible, because we can replace a basis element $b$ of $B_0$ by $b + pc$ where $c \in A$ does not depend on $B'$, and we get another $p$-basic subgroup. This shows the necessity of (a), (b), (c).

If (a) holds, then $0$ is the only $p$-basic subgroup. In case (b), $B$ is a $p$-basic subgroup of $A$, and by (35.2) every $p$-basic subgroup $B'$ of $A$ must be contained in the $p$-component $B$ of $A$. Now $B/B' = 0$ as a bounded $p$-divisible group; thus $B$ is the only $p$-basic subgroup of $A$. Finally, $A$ is its own $p$-basic subgroup if it is of type (c). If $B'$ is any $p$-basic subgroup of $A$, then for some $n$, $p^n(A/B') = A/B'$ is an epimorphic image of the finitely generated group $C$; thus $A/B' = 0$, $B' = A$. $\square$

**Corollary 35.4** (Kulikov [2]).  *A $p$-group has only one basic subgroup exactly if it is divisible or bounded.* $\square$

The notion of basic subgroup can be extended in a rather obvious fashion to $p$-adic modules. Since all $p$-adic modules are $q$-divisible for primes $q \neq p$, only the $p$-basic submodules may yield something nontrivial. Now a basic submodule $B$ of a $p$-adic module $A$ is defined just as basic subgroups in **33** with condition (i) replaced by:

(i′)  $B$ is the direct sum of cyclic $p$-adic modules;

i.e., the components are $J_p$ or $Z(p^k)$ for $k = 1, 2, \cdots$. The reader can check it for himself that (32.1)–(32.3) as well as (35.2) prevail for $p$-adic modules. For the sake of convenient reference let us display this as:

**Theorem 35.5.**  *Every $p$-adic module contains basic submodules which are all isomorphic.* $\square$

In the rest of this section we confine ourselves to $p$-groups.

Let $A$ be an unbounded reduced $p$-group. If $B$ is a basic subgroup of $A$, then it is unbounded, and hence it contains basic subgroups different from $B$ itself [see (35.1)]. Therefore, $A/B$ is not in general an invariant of $A$, not even for direct sums of cyclic $p$-groups of unbounded order. Since $A/B$ is a direct sum of groups $Z(p^\infty)$, $r(A/B)$ already characterizes $A/B$, and since every non-void set of cardinal numbers contains a minimal cardinal, we may select a basic subgroup $B_u$ of $A$ for which $r(A/B_u)$ is minimal. Such a $B_u$ is called an *upper basic subgroup* of $A$. Though for upper basic subgroups $B_u$ of $A$, both $B_u$ and $A/B_u$ are unique up to isomorphism, $B_u$ is not unique as a subset; moreover, if $r(A/B_u)$ is infinite, then an upper basic subgroup contains another one.

The next example shows that $r(A/B_u) = 1$ is possible for reduced $p$-groups $A$.

*Example* (Prüfer [2]). Let $A = \langle a_0, a_1, \cdots, a_n, \cdots \rangle$ be defined by the defining relations

$$pa_0 = 0, \qquad pa_1 = a_0, \qquad p^2 a_2 = a_0, \qquad \cdots, \qquad p^n a_n = a_0, \qquad \cdots.$$

Then $o(a_n) \leq p^{n+1}$, and if $C = Z(p^\infty)$ is defined as in 3, (1), then $a_n \mapsto c_{n+1}$ $(n = 0, 1, 2, \cdots)$ gives rise to an epimorphism $A \to C$, showing that $o(a_n) = p^{n+1}$. Since $a_0 \neq 0$ is of infinite height, $A$ is not a direct sum of cyclic groups. Notice that every $a \in A$ may be written in the form $a = t_0 a_0 + t_1 a_1 + \cdots + t_n a_n$ with $0 \leq t_i < p^i$ $(i = 1, \cdots, n)$ and $0 \leq t_0 < p$. If here $t_n a_n \neq 0$ $(n \geq 1)$, then the correspondence $a_0, \cdots, a_{n-1} \mapsto 0$, $a_m \mapsto c_m$ [for $m \geq n$] extends to an epimorphism $\phi: A \to C$ where $\phi a \neq 0$, showing that $a = 0$ only if $t_0 a_0 = \cdots = t_n a_n = 0$. It is now routine to show that $A$ is reduced and $a_1 - pa_2, \cdots$, $a_n - pa_{n+1}, \cdots$ is a $p$-basis of $A$. The subgroup $B$ generated by them must be an upper basic subgroup, since clearly $A/B \cong Z(p^\infty)$. Thus $r(A/B) = 1$, as we wished to have.

There is an upper bound for the ranks $r(A/B)$ of quotient groups $A/B$ mod basic subgroups $B$, namely, $r(A)$. Hence, there is a least upper bound for the $r(A/B)$; this is easy to describe. Define the *final rank* (Szele [7]) of a $p$-group $A$ as the infimum of the cardinals $r(p^n A)$ for $n = 1, 2, \cdots$:

$$\operatorname{fin} r(A) = \inf_n r(p^n A).$$

By virtue of 34(C), $r(A/B) = r(p^n A/p^n B) \leq r(p^n A)$, whence the inequality $r(A/B) \leq \operatorname{fin} r(A)$ follows. Call a basic subgroup $B_l$ of $A$ a *lower basic subgroup*, if $r(A/B_l) = \operatorname{fin} r(A)$. The existence of lower basic subgroups is established in:

**Theorem 35.6** (Fuchs [2]). *Every basic subgroup of a $p$-group contains a lower basic subgroup.*

If $A$ is a $p$-group, and if fin $r(A)$ is finite, then fin $r(B) \leq$ fin $r(A)$ holds for every basic subgroup $B$ of $A$, and hence fin $r(B) = 0$, because $B$ is a direct sum of cyclic groups. This means $B$ is bounded; thus every basic subgroup of $A$ is bounded and hence lower.

Next assume fin $r(A)$ is infinite. If the basic subgroup $B$ of $A$ is not lower, then $r(A/B) <$ fin $r(A)$, and from **34**(C), we obtain

$$r(p^n A/p^n B) < \text{fin } r(A) \leq r(p^n A).$$

Hence $r(p^n B) = r(p^n A)$. This equality implies that the cardinality of the set of cyclic direct summands $\langle a_i \rangle$ of order $> p^n$ in $B$ is at least fin $r(A)$ for every integer $n$. We may therefore decompose $B$ into a direct sum of groups $C_j$ where each $C_j$ is a direct sum of cyclic groups of unbounded order, and $j$ runs over an index set of cardinality fin $r(A)$. An application of (35.1) shows that each $C_j$ contains a basic subgroup $B_j < C_j$. The direct sum $B'$ of all these $B_j$ is evidently a basic subgroup of $B$, and hence of $A$. This $B'$ is lower, since by construction $r(A/B') \geq r(B/B') = \sum_j r(C_j/B_j) \geq$ fin $r(A)$.☐

EXERCISES

1. Show that in a $p$-group $A = B \oplus C$, where $B$ is finite and $C$ is divisible of finite rank $m$, the number of different basic subgroups is $|B|^m$.
2. Characterize the groups with a finite number of $p$-basic subgroups.
3. (Khabbaz and Walker [1], Hill [1]) If $A$ is a $p$-group with infinitely many different basic subgroups $B$, then the cardinality of the set of different basic subgroups of $A$ is $|A|^{|B|}$.
4. For every basic subgroup $B$ of a $p$-group $A$, $r(A/B) \geq r(A^1)$, where $A^1$ is the first Ulm subgroup of $A$.
5. Given a cardinal $m$, construct a reduced $p$-group $A$ such that $r(A/B_u) = m$ for an upper basic subgroup $B_u$ of $A$.
6. If $A$ is the torsion part of $\prod_{n=1}^{\infty} Z(p^n)$, then every basic subgroup of $A$ is both upper and lower.
7. If $n$ is a cardinal such that $r(A/B_u) \leq n \leq r(A/B_l)$ for an upper and for a lower basic subgroup $B_u$, $B_l$ of the $p$-group $A$, then $A$ contains a basic subgroup $B$ such that $r(A/B) = n$.
8. (Scott [1]) (a) Let $A$ be a $p$-group of infinite rank $m$. Then $A$ has an epimorphic image isomorphic to $\oplus_m C$, where $C$ is either quasicyclic or of order $p$. [*Hint*: consider $A/B_l$ or $A/pA$.]
   (b) There are $2^m$ different subgroups $G$ of $A$ such that

$$A/G \cong \bigoplus_m C.$$

9. (a) If $A$ is a $p$-group, then the intersection of all basic subgroups of $A$ is either $A$ or 0.

(b) If $A$ is a $p$-group with a basic subgroup $B$ satisfying $r(B) \leq \text{fin } r(A)$, then $A$ contains two disjoint basic subgroups.

(c) (Mitchell and Mitchell [1]) If $|A| = \text{fin } r(A)$, then $A$ contains two disjoint basic subgroups.

10.  For a $p$-group $A$ and an infinite cardinal $\mathfrak{m}$, fin $r(A) \geq \mathfrak{m}$ if and only if $A$ contains a subgroup isomorphic to

$$\bigoplus_{\mathfrak{m}} G \qquad \text{where} \quad G = \bigoplus_{n=1}^{\infty} Z(p^n).$$

11.  (Fuchs [3]) Every $p$-group $A$ can be written in the form $A = A' \oplus A''$ where $A'$ is bounded and $r(A'') = \text{fin } r(A'')$. [*Hint*: if fin $r(A) = r(p^m A)$, and $B = \oplus B_n$, $B_n = \oplus Z(p^n)$ is basic in $A$, write $A' = B_1 \oplus \cdots \oplus B_m$.]

12.  Describe all $p$-groups in which all maximal $p$-independent systems are maximal independent.

13.  Prove the analog of (35.3) for $p$-adic modules.

## 36.  BASIC SUBGROUPS ARE ENDOMORPHIC IMAGES

In this section we restrict our attention to $p$-groups $A$ and their basic subgroups $B$.

If $A$ is reduced, then—as is shown by (34.1)—every endomorphism of $A$ is determined by its restriction to $B$. Another outstanding property of basic subgroups has been discovered by T. Szele:

**Theorem 36.1** (Szele [7]). *A basic subgroup of a $p$-group $A$ is an endomorphic image of $A$.*

Let $B = \bigoplus_{n=1}^{\infty} B_n$, $B_n = \oplus Z(p^n)$, be a basic subgroup of $A$. If fin $r(B) = r(p^m B)$, then with this $m$ we decompose $B = B_1 \oplus \cdots \oplus B_m \oplus B_m^*$. The cyclic summands $\langle a_i \rangle$ of $B_m^*$ are manifestly of order $> p^m$; we use them in order to establish a correspondence $f \colon \langle a_i \rangle \mapsto \langle a_j \rangle$ in the set $\mathfrak{D}$ of direct summands in a fixed direct decomposition of $B_m^*$. $f$ is subject to the following conditions:

(i)  $f$ is one-to-one between $\mathfrak{D}$ and a subset $\mathfrak{D}'$ of $\mathfrak{D}$;

(ii)  if $f \colon \langle a_i \rangle \mapsto \langle a_j \rangle$, where $e(a_i) = k_i$, $e(a_j) = k_j$, then $k_j \geq 2k_i$.

The choice of $m$ guarantees that for every $k \geq m$, $\mathfrak{D}$ contains fin $r(B)$ many summands $\langle a_i \rangle$ of order $\geq p^k$, so such an $f$ does exist. Now define a mapping $\varepsilon$ on a subset of $B$ as follows:

(a)  $\varepsilon a = a$        if $a \in B_1 \oplus \cdots \oplus B_m$ ;

(b)  $\varepsilon a_j = a_i$      if $\langle a_j \rangle \in \mathfrak{D}'$ and $f \colon \langle a_i \rangle \mapsto \langle a_j \rangle$;

(c)  $\varepsilon a_j = 0$        if $\langle a_j \rangle \notin \mathfrak{D}'$.

Evidently, $\varepsilon$ induces a well-defined endomorphism of $B$ onto itself which may again be denoted by $\varepsilon$. This extends to an endomorphism $\eta \colon A \to B$ as follows.

If $a \in A$ is of order $p^r$, then write $A = B_1 \oplus \cdots \oplus B_n \oplus A_n$ $(n \geq 2r)$, and $a = b + c$ with $b \in B_1 \oplus \cdots \oplus B_n$, $c \in A_n$, and define $\eta a = \varepsilon b$. This definition does not depend on the choice of $n$ ($\geq 2r$), for the $B_s$-component of $a$ in the decomposition $B_1 \oplus \cdots \oplus B_s \oplus A_s$ is divisible by $p^{s-r}$, and therefore it is carried into 0 by $\varepsilon$ whenever $s \geq 2r$. That $\eta$ preserves sums is evident, hence $\eta$ is an endomorphism of $A$ onto $B$.☐

**Corollary 36.2.** *If $C$ is a pure subgroup of a p-group $A$, and if it is a direct sum of cyclic groups, then $C$ is an endomorphic image of $A$.*

A basis of $C$ is $p$-independent in $A$, hence it extends to a $p$-basis of $A$. By (36.1), the subgroup $B$ generated by this $p$-basis is an endomorphic image of $A$, and so is $C$ as a direct summand of $B$.☐

EXERCISES

1.  (Szele [7]) Every $p$-group of cardinality $\leq \mathfrak{m}$ (an infinite cardinal) is an epimorphic image of a $p$-group $A$ if and only if $\mathfrak{m} \leq \mathrm{fin}\ r(B)$ where $B$ is a basic subgroup of $A$.
2.  Let $A$ be torsion-free. Its $p$-basic subgroup $B$ is an endomorphic image of $A$ if and only if $A$ has a direct summand which is a free group of rank $r(B)$.
3.  Under which conditions is a basic submodule of a $p$-adic module $A$ an endomorphic image of $A$? [*Hint*: cf. Ex. 2.]
4.  State a necessary and sufficient condition for the existence of an epimorphism $B \rightarrow A$, where $B$ is basic in the $p$-group $A$.

## 37.  THE ULM SEQUENCE

We have already introduced the Ulm subgroups and the Ulm factors of a group [see 6], but we have failed to state some important results on them which have numerous consequences in our further developments and which at the same time shed more light on the structure of groups in general. Since we are now in a position to establish the most important properties of the Ulm sequence, we turn our attention to this important invariant of a group.

Let us recall that, for an ordinal $\sigma$, the $\sigma$th Ulm subgroup $A^\sigma$ of a group $A$ was defined inductively as follows: $A^0 = A$, $A^{\sigma+1} = \bigcap_n nA^\sigma$ and, for a limit ordinal $\sigma$, $A^\sigma = \bigcap_{\rho < \sigma} A^\rho$. The *Ulm length* $u(A)$ of $A$ is the least ordinal $\tau$ such that $A^{\tau+1} = A^\tau$. Clearly, this $\tau$ exists and does not exceed $|A|$. It is also evident that $A^\tau$ is the maximal divisible subgroup of $A$. Thus, if we restrict ourselves to reduced groups $A$—which we may do without loss of generality— then $A^\tau = 0$, and we arrive at a well-ordered descending chain of subgroups of $A$:

$$A = A^0 > A^1 > A^2 > \cdots > A^\sigma > A^{\sigma+1} > \cdots > A^\tau = 0.$$

For every $\sigma < \tau$, we form the quotient group $A_\sigma = A^\sigma/A^{\sigma+1}$ which has been called the $\sigma$th Ulm factor of $A$. The well-ordered sequence

$$A_0, A_1, \cdots, A_\sigma, \cdots \qquad (\sigma < \tau)$$

is said to be the *Ulm sequence* of $A$; this is manifestly an invariant of $A$.

Clearly, for a reduced group $A$, $u(A) = 1$ is equivalent to being Hausdorff in the $Z$-adic topology. The example of **35** is of Ulm length 2. We shall see that every ordinal is the Ulm length of some group.

Occasionally, it is useful to have the following notation at our disposal. For an ordinal $\sigma$, $p^\sigma A$ is defined as follows: $p^0 A = A$,

$$p^{\sigma+1} A = p(p^\sigma A),$$

and $p^\sigma A = \bigcap_{\rho < \sigma} p^\rho A$ whenever $\sigma$ is a limit ordinal. The minimal ordinal $\tau$ such that $p^{\tau+1} A = p^\tau A$ will be called the *p-length* of $A$; clearly, $p^\tau A$ is the maximal $p$-divisible subgroup of $A$.

Notice that for every ordinal $\sigma$ we have

$$A^\sigma = \bigcap_p p^{\omega\sigma} A.$$

In connection with the subgroups $p^\sigma A$, we introduce the *generalized p-height* $h_p^*(a)$ of an element $a \in A$. Given $a \notin p^\tau A$, there is a unique ordinal $\sigma$ such that $a \in p^\sigma A \backslash p^{\sigma+1} A$, and we put

$$h_p^*(a) = \sigma.$$

For $a \in p^\tau A$, we let $h_p^*(a)$ be equal to $\tau$. Obviously, this is a finer notion than the height $h_p(a)$ introduced in **1**; namely, $h_p^*(a) = h_p(a)$ if $h_p(a)$ is finite, but $h_p^*(a)$ renders it possible to distinguish between elements of infinite height. It is readily checked that, for all $a, b \in A$,

  (i) $h_p^*(pa) \geqq h_p^*(a) + 1$;
  (ii) $h_p^*(a + b) \geqq \min(h_p^*(a), h_p^*(b))$;
  (iii) if $h_p^*(a) \neq h_p^*(b)$, then in (ii) equality holds;
  (iv) if $A = A' \oplus A''$ and $a \in A', b \in A''$, then again equality holds in (ii);
  (v) $h_p^*$ does not diminish under homomorphisms.

Of great importance are the Ulm-Kaplansky invariants of a group $A$. For a fixed prime $p$, the $\sigma$th *Ulm-Kaplansky invariant of $A$* is the cardinal number

$$f_\sigma(A) = r((p^\sigma A)[p]/(p^{\sigma+1} A)[p]).$$

[To simplify notation, we omit the reference to $p$.] These invariants for $\sigma < \omega$ characterize the basic subgroup of $A$ in case $A$ is a $p$-group [see Ex. 9].

Let us turn our attention to the Ulm sequence. A preliminary lemma is proved first of all.

**Lemma 37.1.** *Let* $\alpha: A \to B$ *be an epimorphism such that* $\operatorname{Ker} \alpha$ *contains only elements of generalized p-heights* $\geqq \sigma$. *Then*

$$h_p^*(\alpha a) = h_p^*(a),$$

*whenever the latter is* $< \sigma$.

In view of (v), only $h_p^*(\alpha a) \leqq h_p^*(a)$ needs a verification in case $h_p^*(a) < \sigma$. First, assume $h_p^*(\alpha a) < \sigma$, and—as a basis of induction—the stated inequality holds for every $a'$ with a smaller $h_p^*(\alpha a')$. [Note that if $h_p^*(\alpha a) = 0$, there is nothing to prove in view of (v).] If $h_p^*(\alpha a) = \rho + 1$, then there is an $a' \in A$ such that $h_p^*(\alpha a') = \rho$ and $\alpha a = p\alpha a'$. Thus $a = pa' + x$ with $x \in \operatorname{Ker} \alpha$. From (i) and (ii), we infer $h_p^*(a) \geqq \min(h_p^*(a') + 1, \sigma)$, and so $h_p^*(a) \geqq \rho + 1 = h_p^*(\alpha a)$, by induction hypothesis. If $h_p^*(\alpha a) = \rho$ is a limit ordinal, then for every $\rho' < \rho$ and integer $n > 0$, there is an $a' \in A$ such that $h_p^*(\alpha a') \geqq \rho'$ and $\alpha a = p^n \alpha a'$. As above, we obtain $h_p^*(a) \geqq \rho' + n$, whence $h_p^*(a) \geqq \rho = h_p^*(\alpha a)$. In a similar fashion one can show that $h_p^*(\alpha a) \geqq \sigma$ leads to the contradiction $h_p^*(a) \geqq \sigma$.□

This lemma has the important consequence that *in an Ulm factor* $A_\sigma$ *every element* $\neq 0$ *is of finite p-height for at least one prime p*. In fact, by (37.1), every element of $A^\sigma$ not in $A^{\sigma+1}$ is mapped under the natural epimorphism $A^\sigma \to A^\sigma/A^{\sigma+1} = A_\sigma$ upon an element of the same p-height for every prime p.

**Lemma 37.2.** *If for some* $\sigma$, *the p-component of* $A_\sigma$ *is bounded, then* $A^{\sigma+1}$ *is p-divisible.*

Let $p^n$ be an upper bound for the orders of elements in the p-component of $A_\sigma$. Then every equation $px = a \in A^{\sigma+1}$ is solvable in $A^{\sigma+1}$, since $p^{n+1}y = a$ is, by the definition of $A^{\sigma+1}$, solvable in $A^\sigma$ and $p^n y = x \in A^{\sigma+1}$.□

**Lemma 37.3.** *All the Ulm factors* $A_\sigma$ *of* $A$—*with the possible exception of the last one if this exists*—*have at least one unbounded p-component.*

If all the p-components of $A_\sigma$ are bounded, then, by (37.2), $A^{\sigma+1}$ is divisible, and hence $A_\sigma$ is the last Ulm factor.□

**Lemma 37.4.** *For any ordinal* $\sigma$, $A_\sigma, A_{\sigma+1}, \cdots$ *is the Ulm sequence of* $A^\sigma$ *and* $A_0, A_1, \cdots, A_\rho, \cdots$ ($\rho < \sigma$) *is the Ulm sequence of* $A/A^\sigma$.

The first assertion is evident. To prove the second one, notice that the natural map $\eta: A \to A/A^\sigma = C$ preserves, by (37.1), the generalized p-heights insofar as they are less than $\omega\sigma$. Hence, it follows that, for every $\rho \leqq \sigma$, $A^\rho$ is the complete inverse image of $C^\rho$, and so we find

$$A_\rho = A^\rho/A^{\rho+1} \cong (A^\rho/A^\sigma)/(A^{\rho+1}/A^\sigma) \cong C^\rho/C^{\rho+1} = C_\rho \qquad \text{for} \quad \rho < \sigma.□$$

**Lemma 37.5.** *Let* $A = \bigoplus_{i \in I} A_i$ *be a direct sum. Then for the Ulm factors* $A_\sigma$ *of* $A$ *one has* $A_\sigma = \bigoplus_{i \in I} A_{i\sigma}$ *where* $A_{i\sigma}$ *is the* $\sigma$*th Ulm factor of* $A_i$ *if* $\sigma < u(A_i)$ *and* $A_{i\sigma} = 0$ *otherwise. The same holds for the direct product.*

From (iv), it follows easily that $A^\sigma = \oplus A_i^\sigma$ for the $\sigma$th Ulm subgroups, whence the assertion follows at once. $\square$

The basic result on the Ulm sequence is the following theorem [proved for $p$-groups by Kulikov [3], Fuchs [2]].

**Theorem 37.6.** *Let $A$ be a reduced group and $A_0, A_1, \cdots, A_\sigma, \cdots (\sigma < \tau)$ the Ulm sequence of $A$. We denote by $A_{\sigma,p}, B_{\sigma,p}$ the $p$-component and a $p$-basic subgroup of $A_\sigma$, respectively. Then*

(a)   *the first Ulm subgroup of $A_\sigma$ vanishes, for every $\sigma < \tau$;*
(b)   $\sum_{0 \leq \sigma < \tau} |A_\sigma| \leq |A| \leq \prod_{0 \leq n < \min(\omega, \tau)} |A_n|$;
(c)   $\sum_{\rho \leq \sigma < \tau} |A_\sigma| \leq |A_\rho|^{\aleph_0}$      *for every $0 \leq \rho < \tau$;*
(d)   $r(B_{\sigma+1, p}) \leq \text{fin } r(A_{\sigma, p})$      *for every $\sigma + 1 < \tau$ and prime $p$;*
(e)   $\sum_{\rho < \sigma < \tau} |B_{\sigma, p}| \leq |B_{\rho, p} \cap A_{\rho, p}|^{\aleph_0}$   *for every $0 \leq \rho < \tau$ and prime $p$.*

Statement (a) is evident by the definition of $A_\sigma$. In (b), the first inequality follows from $A = \bigcup_{\sigma < \tau} (A^\sigma \backslash A^{\sigma+1})$ and from $|A_\sigma| = |A^\sigma / A^{\sigma+1}| \leq |A^\sigma \backslash A^{\sigma+1}|$. To establish the second inequality, first let $\tau = n$ be a natural integer; then

$$|A| = |A/A^1| \, |A^1/A^2| \cdots |A^{n-1}/A^n| = |A_0| \, |A_1| \cdots |A_{n-1}|.$$

If $\tau \geq \omega$, then setting $\min_{n < \omega} |A_n| = |A_m|$, we obtain

$$|A| = |A/A^m| \, |A^m| = |A_0| \, |A_1| \cdots |A_{m-1}| \, |A^m|.$$

Applying (34.5) to $A^m$ whose 0th Ulm factor is $A_m$, one gets

$$|A^m| \leq |A_m|^{\aleph_0} \leq |A_m| \, |A_{m+1}| \cdots$$

whence the second part of (b) follows.

The inequality (c) can be verified at once, since for $A^\rho$ we get by (b) and (34.5)

$$\sum_{\rho \leq \sigma < \tau} |A_\sigma| \leq |A^\rho| \leq |A_\rho|^{\aleph_0}.$$

To prove (d), write $B_{\sigma+1, p} = \oplus_{i \in I} \langle a_i \rangle$. By (37.4), $A_{\sigma+1}$ is the first Ulm subgroup of $A^\sigma / A^{\sigma+2}$, and therefore, for each $n > 0$, the equation $p^n x = a_i$ has a solution $x_i$ in $A^\sigma / A^{\sigma+2}$. We prove that, for different indices $i_1, \cdots, i_k$, the elements $x_{i_1}, \cdots, x_{i_k}$ are independent mod $A^{\sigma+1}$. For, otherwise, $m_1 x_{i_1} + \cdots + m_k x_{i_k} = c \in A_{\sigma+1} = A^{\sigma+1}/A^{\sigma+2}$ where no generality is lost in supposing $m_j = p^{n-1} m_j', (m_j', p) = 1$ for $j = 1, \cdots, k$; then $pc = m_1' a_{i_1} + \cdots + m_k' a_{i_k} \in B_{\sigma+1, p}$ is a contradiction to the $p$-purity of $B_{\sigma+1, p}$ in $A_{\sigma+1}$. We conclude that $A_{\sigma, p}$ contains an independent set of elements of order $p^n$ whose cardinality is not less than the rank of $B_{\sigma+1, p}$. This yields (d) at once.

Turning to (e), from **34(F)** it follows that $B_{\sigma, p}$ is isomorphic to a $p$-basic subgroup of $A^\sigma$. This being $p$-pure in $A^\sigma$, its elements are contained in

$p^{\omega\sigma}A \; (\geqq A^{\sigma})$ but not in $p^{\omega(\sigma+1)}A$. Therefore it intersects trivially the last group, and so the inequality

$$|B_{\sigma,p}| \leqq |p^{\omega\sigma}A/p^{\omega(\sigma+1)}A|$$

holds for every $\sigma$. Hence

$$\sum_{\rho<\sigma<\tau} |B_{\sigma,p}| \leqq \sum_{\rho<\sigma<\tau} |p^{\omega\sigma}A/p^{\omega(\sigma+1)}A| \leqq |p^{\omega(\rho+1)}A/p^{\omega\tau}A|,$$

the latter inequality being a variant of the first part of (b). If (34.3) is formulated for groups without $p$-divisible subgroups $\neq 0$, then we obtain

$$|p^{\omega(\rho+1)}A/p^{\omega\tau}A| \leqq |B_{\rho+1,p}|^{\aleph_0}.$$

Now (d) implies $|B_{\rho+1,p}| \leqq |A_{\rho,p}|$ whence, in view of (34.4),

$$|B_{\rho+1,p}| \leqq |B_{\rho,p} \cap A_{\rho,p}|^{\aleph_0}. \; \square$$

EXERCISES

1.   If $C$ is a subgroup of the reduced group $A$, then $C^{\sigma} \leqq A^{\sigma}$ for every ordinal $\sigma$ and $u(C) \leqq u(A)$. Also, $p^{\sigma}C \leqq p^{\sigma}A$ for every $\sigma$.
2.   A quotient group of $A$ may have an Ulm length greater than $A$, even if it is taken mod a pure subgroup.
3.   If $A$ is as in (37.5), then $u(A)$ is the least ordinal $\tau$ such that $u(A_i) \leqq \tau$ for every $i$.
4.   Let $C$ be a subgroup of $A^{\sigma}$ such that $A$ and $A/C$ are reduced, and $A/C$ is of Ulm length $\leqq \sigma$. Then $C = A^{\sigma}$.
5.   If $C$ is a subgroup of $A$ such that $C/(A^{\sigma} \cap C)$ is divisible, then $C \leqq A^{\sigma}$.
6.   Determine the Ulm-Kaplansky invariants of the example in 35.
7.   Let $A$ be a $p$-group and $A^{\sigma}$ its $\sigma$th Ulm subgroup. Show that there is an epimorphism $A \rightarrow C$ such that $C_{\rho} \cong A_{\rho}$ for $\rho < \sigma$, and $C_{\sigma}$ is isomorphic to a basic subgroup of $A^{\sigma}$. [Hint: combine (36.1) and (37.1).]
8.   If $A = \bigoplus_i A_i$, then for the Ulm-Kaplansky invariants one has

$$f_{\sigma}(A) = \sum_i f_{\sigma}(A_i).$$

9.   If $A$ is a $p$-group and $f_n(A) = \mathfrak{m}_n \; (n < \omega)$, then any basic subgroup $B$ of $A$ satisfies

$$B \cong \bigoplus_{n=0}^{\infty} \bigoplus_{\mathfrak{m}_n} Z(p^{n+1}).$$

[Hint: 34, Ex. 2 (e).]

10.   If $A$ is a $p$-group, then its Ulm-Kaplansky invariants completely determine the basic subgroups of the Ulm subgroups and Ulm factors of $A$.

11.* Show that to every group $G$ there exists a group $A$ whose first Ulm subgroup is isomorphic to $G$. [*Hint*: if $G$ is cyclic $= \langle a_0 \rangle$, apply the construction as in the example of **35** to make $a_0$ of infinite $p$-height; pass to direct sums of cyclic groups; for arbitrary $G$ use a $p$-basis of $G$, for every $p$.]

### NOTES

The material of Chapter VI is based essentially on two papers: the pioneering article by Kulikov [2] where basic subgroups of $p$-groups were introduced and a wealth of details about them was published, and a paper by the author (Fuchs [9]) who recognized that a certain method of introducing basic subgroups works in the general situation, provided a prime $p$ has been fixed, and the arising $p$-basic subgroups still enjoy most of the advantageous properties of basic subgroups of $p$-groups. A major difference is that Szele's beautiful theorem (36.1) fails to hold in general. Kulikov [3] also develops basic submodules over two rings: the $p$-adic integers and the rational numbers with denominators prime to $p$. So far, $p$-basic submodules have not been investigated in a general setting.

Ulm subgroups and sequences were introduced by Ulm [1] for $p$-groups. Their importance for the theory of $p$-groups cannot be overestimated [cf. Chapter XII]. These concepts for arbitrary abelian groups did not receive full recognition until it turned out that the Ulm factors of cotorsion groups are algebraically compact [see (54.3)]. The construction of groups with given Ulm sequences will be discussed in Vol. II.

*Problem 21.* When is a subgroup of $A$ (a) contained in a $p$-basic subgroup of $A$? (b) disjoint from some $p$-basic subgroup?

For the primary case, (33.4) is the answer to (a).

*Problem 22.* Give conditions on a pure subgroup to be an endomorphic image of the group.

For $p$-groups, a sufficient condition is to be a direct sum of countable groups [this follows from (36.2)].

# VII

## ALGEBRAICALLY COMPACT GROUPS

In Chapter IV we saw that a group is a direct summand of every group containing it if, and only if, it is divisible. This chapter is devoted to the study of groups that are direct summands of every group in which they are contained as pure subgroups.

These groups, known under the name "algebraically compact" groups, possess a number of remarkable properties which are characteristic for them. These properties are not all group-theoretic in character; some of them are of topological or of homological nature. The algebraically compact groups will turn out to be identical with the pure-injective groups introduced in **30**. Our discussion culminates in the structure theorem establishing a complete and independent system of cardinal invariants for algebraically compact groups. The reader will notice that in many respects the theory of algebraically compact groups runs parallel to the theory of divisible groups—a fact which was first pointed out by Maranda [1].

We shall often meet algebraically compact groups in subsequent discussions.

## 38. ALGEBRAIC COMPACTNESS

A group $A$ is called *algebraically compact*, if $A$ is a direct summand in every group $G$ that contains $A$ as a pure subgroup. By (21.2), every divisible group, and by (27.5), every bounded group is algebraically compact. In particular, cocyclic groups are algebraically compact.

It follows at once that a direct summand of an algebraically compact group is again algebraically compact, and a group is algebraically compact exactly if its reduced part is algebraically compact.

It is convenient to begin with the result which gives various characterizations of algebraically compact groups; the equivalence of (a) with (c) in (38.1) has already been observed in (30.4).

**Theorem 38.1.**  *The following conditions on a group A are equivalent:*

(a)  *A is pure-injective;*
(b)  *A is algebraically compact;*
(c)  *A is a direct summand of a direct product of cocyclic groups;*
(d)  *A is algebraically a direct summand of a group that admits a compact topology;*
(e)  *if every finite subsystem of a system of equations over A has a solution in A, then the whole system is solvable in A.*

The proof is cyclic. First assume (a) and let $A$ be pure in $G$. From the pure-exact sequence $0 \to A \to G \to G/A \to 0$ and from (a), we conclude that the identity map $1_A : A \to A$ can be factored as $A \to G \to A$. Thus $A$ is a direct summand of $G$, proving (b).

Now assume (b). By (30.3), $A$ can be embedded as a pure subgroup in a direct product of cocyclic groups. Hence (b) implies (c).

Next suppose (c). Since the direct product of compact groups is again compact in the product topology, and since the property of being a direct summand is transitive, (d) will be established as soon as we can show that cocyclic groups share property (d). But this is evident, since $Z(p^k)$ $(k < \infty)$ is compact in the discrete topology, while $Z(p^\infty)$ is a direct summand of the group of reals mod 1 [which is a wellknown compact group].

In order to derive (e) from (d), let us start with a system

(1)                    $$\sum_{j \in J} n_{ij} x_j = a_i \qquad (a_i \in A, \quad i \in I)$$

where $n_{ij}$ are integers such that, for a fixed $i$, almost all of them vanish; and assume every finite subsystem of (1) admits a solution in $A$. By (d), there is a group $B$ such that $A \oplus B = C$ admits a compact topology; we may equally well consider (1) as a system over $C$. A solution of the $i$th equation of (1) can be regarded as an element $(\cdots, c_j, \cdots)$ of $C^J = \bar{C}$ such that $x_j = c_j$ $(j \in J)$ satisfies the $i$th equation. The set of all solutions of the $i$th equation is thus a subset $X_i$ of the compact space $\bar{C}$; moreover, $X_i$ is closed, since it is defined in terms of an equation. The hypothesis that every finite subsystem of (1) is solvable amounts to the condition that every finite subset of the set of the $X_i$ has a nonempty intersection. By the compactness of $\bar{C}$, the intersection $\bigcap X_i$ of all the $X_i$ is not vacuous, thus the whole system (1) admits a solution in $C$. The $A$-components of a solution yield a solution in $A$.

Finally, assume (e). Let $0 \to B \to C \to C/B \to 0$ be a pure-exact sequence and $\eta : B \to A$. Let $c_j$ $(j \in J)$ be a generating system of $C$ mod $B$ and

$$\sum_{j \in J} n_{ij} c_j = b_i \in B \qquad (i \in I)$$

all the relations between these $c_j$ and elements of $B$. We consider the system

(2)
$$\sum_j n_{ij} x_j = \eta b_i \quad (\in A) \quad (i \in I).$$

A finite subsystem of (2) contains but a finite number of unknowns $x_{j_1}, \cdots, x_{j_k}$ explicitly. Owing to purity, $B$ is a direct summand of $B' = \langle B, c_{j_1}, \cdots, c_{j_k} \rangle$, $B' = B \oplus C'$ [cf. (28.4)], and the images of the $B$-components of $c_{j_1}, \cdots, c_{j_k}$ under $\eta$ yield a solution in $A$. Consequently, (2) satisfies the hypotheses of (e). We infer there exists a solution $x_j = a_j$ in $A$ of the whole system (2). The correspondence $c_j \mapsto a_j$ gives rise to an extension of $\eta$ to a homomorphism $C \to A$. Hence (a) follows.☐

For a reduced group $A$, condition (c) can be improved:

**Corollary 38.2.** *A reduced algebraically compact group is a direct summand of a direct product of cyclic p-groups.*

If $A$ is reduced algebraically compact, then, for some group $B$, $A \oplus B = C = C_1 \oplus C_2$, where $C_1$ is a direct product of cyclic groups $Z(p^k)$ and $C_2$ is a direct product of quasicyclic groups. Obviously, $C_2$ is the maximal divisible subgroup of $C$. Its full invariance implies $C_2 = (A \cap C_2) \oplus (B \cap C_2)$ [cf. (9.3)] where the first summand must vanish. Thus $C_2 \leqq B$ and $A \oplus (B/C_2) \cong C_1$.☐

In particular, $A^1 = 0$ holds for the first Ulm subgroup of a reduced algebraically compact group $A$.

For the sake of easy reference let us formulate the following immediate consequence of (c) or (d).

**Corollary 38.3.** *A direct product of groups is algebraically compact if and only if every component is algebraically compact.*☐

Combining (30.3) with the theorem above we conclude:

**Corollary 38.4.** *Every group can be embedded as a pure subgroup in an algebraically compact group.*☐

For an improvement of this result we refer to (41.7).

We shall refer to the following weaker form of the definition of algebraically compact groups.

**Proposition 38.5** (E. Sąsiada). *A group $A$ is algebraically compact if and only if $A$ is a direct summand of every group $G$ such that $A$ is pure in $G$ and $G/A$ is isomorphic to $Q$ or to some $Z(p^\infty)$.*

Only the "if" part needs a verification. Assume $A$ has the stated property. Then $A$ must be a direct summand in every group $G$ in which $A$ is pure and $G/A$ is divisible; in fact, this follows from our hypothesis and (9.4), (23.1). Next, we consider the case in which $A$ is pure in $G$ and $G/A$ is torsion. If $B/A$ is a basic subgroup of $G/A$, then from (28.2) we infer $B = A \oplus B'$ for some $B' \leqq B$. Now $G/B \cong (G/A)/(B/A)$ is divisible; therefore, $G/B'$ contains

$B/B' \cong A$ as a pure subgroup [$B$ is pure in $G$] with divisible quotient group, thus $G/B' = B/B' \oplus G'/B'$ for some $G' \leq G$. Since $G = B + G' = A + B' + G' = A + G'$, and $A \cap G' = (A \cap B) \cap G' = A \cap (B \cap G') = A \cap B' = 0$, we have $G = A \oplus G'$.

If $G/A$ is torsion-free, then by (24.6) there is a group $H$, such that $G \leq H$ and $H/A$ is torsion-free divisible. By the first part of this proof, $A$ is a direct summand in $H$, and *a fortiori* in $G$. Finally, if $G/A$ is arbitrary, and if $T/A$ is the torsion part of $G/A$, then $A$ is a direct summand in $T$, $T = A \oplus T'$, and since $T/T' \cong A$ is a direct summand in $G/T'$, it follows that $G = A \oplus G'$ for some $G' \leq G$.☐

EXERCISES

1. Prove that $Z$ is not algebraically compact, and more generally, free groups are never algebraically compact unless 0.

2. A direct product of copies of $p$-adic integers is algebraically compact.

3. If $A$ is algebraically compact and $B$ is a pure subgroup in $A$, then $A/B$ is algebraically compact. [*Hint*: (24.6) and (26.1).]

4. Prove that $A$ is algebraically compact if it has the injective property relative to the class of pure-exact sequences $0 \to B \to C \to D \to 0$ with countable $D$. [*Hint*: (38.1) and (38.5).]

5. $A$ is algebraically compact if it satisfies condition (e) of (38.1) for systems of equations only with countably many unknowns.

6. A group can be embedded as a pure subgroup in a reduced algebraically compact group if and only if its first Ulm subgroup vanishes.

7. If $A$ is algebraically compact, then its first Ulm subgroup coincides with its maximal divisible subgroup.

8. Every group is isomorphic to a pure subgroup of a suitable group which admits a compact topology.

9. A group $A$ need not be algebraically compact if both its subgroup $C$ and the quotient group $A/C$ are algebraically compact. [*Hint*: embed the 0th Ulm factor of the example in **35** in a reduced algebraically compact group (Ex. 6), apply (24.6), Ex. 7.—Cf. cotorsion groups.]

10. Every linearly compact group is algebraically compact. [*Hint*: verify (e) as in the proof of (38.1), the solutions of the $i$th equation form a coset mod a closed subgroup; cf. **13**, Ex. 6.]

## 39.  COMPLETE GROUPS

In this section we shall examine more closely the groups which are *complete* in their $Z$-adic or $p$-adic topologies. This is not a new class of groups; namely, we shall see that just the reduced algebraically compact groups are complete in their $Z$-adic topologies.

All the groups which are discrete in their Z-adic topologies [that is, the bounded groups] are evidently complete. Further examples may be obtained by taking direct summands of direct products of complete groups—as is suggested by (13.4).

The principal result on complete groups is the next theorem, which is essentially due to Kaplansky [2].

**Theorem 39.1.** *A group is complete in the Z-adic topology if, and only if, it is a reduced algebraically compact group.*

Assume first $A$ is reduced and algebraically compact. From (38.2) we know that it is a direct summand of a direct product of cyclic groups $Z(p^k)$. Each $Z(p^k)$ being complete in its own Z-adic topology, a repeated application of (13.4) shows $A$ complete in its Z-adic topology.

Next suppose $A$ is complete in its Z-adic topology. Then $A$ is Hausdorff and, hence, reduced. Let $G$ contain $A$ as a pure subgroup such that $G/A$ is divisible; to complete the proof, we need only show that $A$ is a direct summand of $G$ [cf. (38.5)]. If $G$ is Hausdorff in its Z-adic topology, then by the density of $A$ in $G$, every $g \in G$ is the limit of some sequence in $A$. The relative topology of $A$ being the Z-adic topology of $A$, this sequence is Cauchy in $A$, whence $g \in A$ and $A = G$. If $G$ is not Hausdorff, then $G/G^1$ is Hausdorff and contains $(A + G^1)/G^1 \cong A$ as a pure subgroup [notice $A \cap G^1 = 0$]. From what has already been shown, we infer $A + G^1 = G$, whence $G = A \oplus G^1$.☐

*Remark.* The " only if " part of (39.1) can slightly be improved by observing that *A is complete in the Z-adic topology whenever it is complete in some (Hausdorff) topology coarser than the Z-adic topology.* For, let $a_1, \cdots, a_n, \cdots$ be a Cauchy sequence in the Z-adic topology, where we may assume that $a_n - a_{n+1} = n!g_n$ (for some $g_n \in A$) holds for every $n$. Define

$$b_{nk} = g_n + (n + 1)g_{n+1} + \cdots + (n + 1)\cdots(n + k)g_{n+k};$$

then the sequence $b_{n0}, \cdots, b_{nk}, \cdots$ is again Cauchy in the Z-adic topology in view of $k! | (n + 1)\cdots(n + k)$, and we have, for every $n$ and $k$, $a_n = n!b_{nk} + a_{n+k+1}$. Every Cauchy sequence in the Z-adic topology is Cauchy in a coarser topology, hence letting $k \to \infty$ we obtain $a_n = n!b_n + a$ for the limits $a$ and $b_n$ of $\{a_n\}$ and $\{b_{nk}\}$, respectively. Thus $a - a_n \in n!A$, and $a$ is the limit of $\{a_n\}$ in the Z-adic topology, too.

Since the Z-adic topology on $A$ induces a topology on a subgroup $B$ coarser than the Z-adic topology of $B$, from (39.1) and our remark, together with (13.1) and (13.2), we are led to:

**Corollary 39.2.** *If $A$ is reduced algebraically compact and $B$ is a subgroup such that $(A/B)^1 = 0$, then both $B$ and $A/B$ are reduced algebraically compact.*☐

The following fact deserves special mention.

**Corollary 39.3.** *If $\theta$ is an endomorphism of a complete group $A$, then both* Ker $\theta$ *and* Im $\theta$ *are complete groups.*

Apply (39.2) with $B = \operatorname{Ker} \theta$ and notice that $A/\operatorname{Ker} \theta$ is complete because of (13.2).□

A result of special interest is:

**Proposition 39.4.** *The inverse limit of reduced algebraically compact groups is again reduced and algebraically compact. In particular, $J_p$ is algebraically compact.*

The inverse limit is, by **12** (e), a closed subgroup of the direct product in the product and hence in the box topology. The direct product is, because of the hypothesis and (38.3), reduced algebraically compact; thus (39.2) concludes the proof.□

Next we investigate the embedding of a group $A$ in its $Z$-adic completion $\hat{A}$. As was seen in **13**, this can be performed either by Cauchy sequences or by inverse limits.

**Theorem 39.5.** *For any group $A$,*

$$\hat{A} = \varprojlim_n A/nA$$

*is a complete group. The canonical map $\mu : a \mapsto (\cdots, a + nA, \cdots) \in \hat{A}$ has $A^1$ for its kernel and $\mu A$ is pure in $\hat{A}$ with $\hat{A}/\mu A$ divisible.*

Since the $A/nA$ are bounded and hence algebraically compact, the first part follows from (39.4). Clearly, $\mu a = 0$ amounts to $a \in nA$ for every $n$, hence $\operatorname{Ker} \mu = A^1$. If $\hat{a} = (\cdots, a_n + nA, \cdots)$ $(a_n \in A)$ satisfies $m\hat{a} = \mu a$ for some $a \in A$, then $ma_n - a \in nA$ for every $n$, in particular, for $n = m$, whence $a \in mA$ and purity follows. To prove $\hat{A}/\mu A$ divisible, it suffices to show that the induced topology of $\hat{A}$ is the $Z$-adic, since then the density of $\mu A$ in $\hat{A}$ will at once imply $\hat{A}/\mu A$ divisible.

Because the groups $A/nA$ are bounded and hence discrete, the neighborhoods of $\hat{A}$ are $U_n = \pi_n^{-1}0$ where $\pi_n$ is the $n$th coordinate projection $\hat{A} \to A/nA$; we show $U_n = n\hat{A}$. The inclusion $n\hat{A} \leq U_n$ being trivial, assume $\hat{a} \in U_n$, i.e., $\hat{a} = (\cdots, a_m + mA, \cdots)$ with $a_n = 0$. For every $m$, there is an element $c_m \in A$ satisfying $a_{(m+1)!n} - a_{m!n} = m!nc_m$. Define $b_1 = 0$ and $b_{m+1} = b_m + m!c_m$ for $m > 1$. Then by induction we have $a_{m!n} = nb_m$ for every $m \geq 1$, and $\hat{b} = (\cdots, b_m + m!A, \cdots) \in \hat{A}$. This $\hat{b}$ satisfies $n\hat{b} = \hat{a}$, proving $U_n \leq n\hat{A}$.□

**Corollary 39.6.** *Under the canonical map $\mu$, a $p$-basic subgroup of $A$ is mapped isomorphically upon a $p$-basic subgroup of $\hat{A}$, for any prime $p$.*

Since $\operatorname{Ker} \mu = A^1$, **34** (F) implies that $\mu$ maps a $p$-basic subgroup of $A$ isomorphically upon a $p$-basic subgroup of $\mu A$. Owing to the divisibility of $\hat{A}/\mu A$, and the purity of $\mu A$ in $\hat{A}$, **34** (E) completes the proof.□

The importance of the canonical map $\mu : A \to \hat{A}$ lies in the fact that it is functorial [which justifies its name]:

**Proposition 39.7.** *If* $\alpha : A \to B$ *is a homomorphism, then it induces a unique homomorphism* $\hat{\alpha} : \hat{A} \to \hat{B}$ *making the diagram*

(1)
$$
\begin{array}{ccc}
A & \xrightarrow{\ \alpha\ } & B \\
{\scriptstyle \mu}\downarrow & & \downarrow{\scriptstyle \nu} \\
\hat{A} & \xrightarrow{\ \hat{\alpha}\ } & \hat{B}
\end{array}
$$

*commute* [*the vertical maps are canonical*].
The proof is given in the comments made after (13.8).□

The group $\hat{A}$ [which is unique up to continuous isomorphisms, cf. **13**] is called a *Z-adic completion of A*. Because of (39.7), the correspondence $A \mapsto \hat{A}$, $\alpha \mapsto \hat{\alpha}$ is a functor on $\mathscr{A}$ to the category of complete groups.

As we know from (39.5), this completion can be achieved by using inverse limits. By making use of this fact, we shall show that it carries pure-exact sequences into splitting exact sequences:

**Theorem 39.8.** *Let* $0 \to A \xrightarrow{\ \alpha\ } B \xrightarrow{\ \beta\ } C \to 0$ *be a pure-exact sequence. Then the sequence*

(2)
$$
0 \to \hat{A} \xrightarrow{\ \hat{\alpha}\ } \hat{B} \xrightarrow{\ \hat{\beta}\ } \hat{C} \to 0
$$

*is splitting exact.*

If the given sequence is pure-exact, then the induced sequences $0 \to A/nA \to B/nB \to C/nC \to 0$ are exact for all $n$, because of (29.1). From (39.5) and (12.3), we conclude that for the exactness of (2) it suffices to verify that $\hat{\beta}$ is epic. Im $\hat{\beta} \le \hat{C}$ and (39.2) guarantee the completeness of Im $\hat{\beta}$. Since Im $\hat{\beta}$ contains the canonical image of $C$ which is dense in $\hat{C}$, necessarily Im $\hat{\beta} = \hat{C}$. What remains to prove is the purity of Im $\hat{\alpha} \cong \hat{A}$ in $\hat{B}$. Now the map $a + A^1 \mapsto \alpha a + B^1$ carries $A/A^1$ onto a pure subgroup of $B/B^1$ which, together with the purity of $\nu B$ in $\hat{B}$, shows $\nu\alpha A$ pure in $\hat{B}$. In view of $\nu\alpha = \hat{\alpha}\mu$ [cf. (1)] and the divisibility of $\hat{\alpha}\hat{A}/\hat{\alpha}\mu A$, we infer that $\hat{\alpha}\hat{A}$ must be pure in $\hat{B}$.□

In the remainder of this section we focus our attention on direct decompositions of complete groups. We wish to prove first of all:

**Theorem 39.9.** *Assume that the complete group A is contained in the direct sum* $\bigoplus_{i \in I} C_i$ *of groups* $C_i$ *such that* $C_i^1 = 0$ *for every i. Then there exists an integer* $n > 0$ *such that nA is contained in the direct sum of a finite number of the* $C_i$.

If the conclusion fails, then there exist a properly increasing sequence of natural numbers $n_1, \cdots, n_j, \cdots$ with $n_j | n_{j+1}$ and groups $B_j$, each being a direct sum of a finite number of $C_i$, such that the $B_j$ generate their direct sum in $\bigoplus C_i$ and

$$
n_j A \cap \bigoplus_{k=1}^{j-1} B_k < n_j A \cap \bigoplus_{k=1}^{j} B_k \qquad (j = 1, 2, \cdots).
$$

Let $a_j$ be in the right, but not in the left member; then $a_{j-1}$ has $0$ and $a_j$ has nonzero coordinate in $B_j$. Therefore, the Cauchy sequence $a_1, \cdots, a_j, \cdots \in A$ cannot have a limit in $\bigoplus_i C_i$. $\square$

**Corollary 39.10.** *If* $A = \bigoplus_{i \in I} C_i$ *is a direct decomposition of a complete group* $A$, *then all the* $C_i$ *are complete groups, and there is an integer* $n > 0$ *such that* $nC_i = 0$ *for almost all* $i \in I$.

The first statement results from (13.4), while the second is an immediate consequence of (39.9). $\square$

EXERCISES

1.  Let $m > 0$ be an integer. Show that $A$ is complete if and only if $mA$ is complete.
2.  (Kaplansky [2]) If $C$ is a pure subgroup of a complete group $A$, then the closure of $C$ in $A$ is a direct summand of $A$.
3.  Describe the $Z$-adic completions of the following groups:

$$Z, \quad Q_p, \quad Q^{(p)}, \quad \bigoplus_{n=1}^{\infty} Z(p^n), \quad \bigoplus_p Z(p).$$

4.  The $Z$-adic completion of a direct product is the direct product of the $Z$-adic completions of the components.
5.  (a) Show that the $p$-adic completion of a group $A$ is isomorphic to $\varprojlim_k A/p^k A$.
    (b) Prove that the $p$-adic completion is always a $p$-adic module.
    (c) If $A$ is complete in its $p$-adic topology, then $qA = A$ for every prime $q \neq p$.
6.  Prove (39.7) and (39.8) for $p$-adic completions [replace pure-exactness by $p$-pure-exactness].
7.  Let $B = \bigoplus_n B_n$ where $B_n = \bigoplus Z(p^n)$. Show that if $B^* = \prod_n B_n$ and $C/B$ is the maximal divisible subgroup of $B^*/B$, then $C$ is a $p$-adic completion of $B$.
8.  If $A$ is complete in its $p$-adic topology and $B$ is a $p$-basic subgroup of $A$, then every homomorphism $B \to A$ extends uniquely to an endomorphism of $A$.
9.  If $B$ is a basic submodule of the $p$-adic module $A$, then every finitely generated direct summand of $B$ is a direct summand of $A$, too.
10. Prove that neither $J = \prod_p J_p$ nor $\prod_p Z(p)$ is a direct sum of infinitely many subgroups $\neq 0$.
11. Let $A_i$ $(i \in I)$ be algebraically compact groups. Find a necessary and sufficient condition for $\bigoplus_i A_i$ to be algebraically compact.

12.   The inverse limit of nonreduced algebraically compact groups need not
      be algebraically compact. [*Hint*: use the idea of **21**, Ex. 6 to get an un-
      bounded *p*-group for inverse limit.]
13.   (Fuchs [13]) The inverse limit of groups satisfying the minimum con-
      dition is algebraically compact. [*Hint*: **13**, Exs. 6, 7 and **38**, Ex. 10.]

## 40.   THE STRUCTURE OF ALGEBRAICALLY COMPACT GROUPS

We have seen a number of useful properties of algebraically compact
groups, and now we have come to the structure theorem on algebraically
compact groups. Since the structure of divisible groups can be described in
terms of cardinal invariants, there is no loss of generality in restricting our
discussion to the reduced case.

A further reduction is possible in view of:

**Proposition 40.1** (Kaplansky [2]). *For a reduced group $A$ to be algebraically
compact, it is necessary and sufficient that it is of the form*

$$(1) \qquad\qquad\qquad A = \prod_p A_p,$$

[*the product being extended over all distinct primes p*] *where each $A_p$ is complete
in its p-adic topology. The $A_p$ are uniquely determined by A.*

First, let $A$ be algebraically compact and $A \oplus B = C$ a direct product of
cyclic *p*-groups [see (38.2)]. Collect the summands $Z(p^k)$ belonging to the
same *p* and form their direct product $C_p$. This is a subgroup of $C$, and clearly,
$C = \prod_p C_p$. The $C_p$ are fully invariant in $C$, hence (9.3) implies $C_p = A_p \oplus B_p$
where $A_p = A \cap C_p$ and $B_p = B \cap C_p$. Thus, $C = \prod_p A_p \oplus \prod_p B_p$. Obviously,
the $A_p$ with varying *p* generate their direct sum $A_0$ in $A$. If $C$ is regarded as a
topological group in the *Z*-adic topology, then it is clear that the closure of
$A_0$ in $C$ must contain $\prod_p A_p$ because of the divisibility of $\prod_p A_p / A_0$. Since
$B^1 = 0$, $A$ is closed in $C$, and therefore we obtain the inclusion $\prod_p A_p \leq A$.
Analogously, $\prod_p B_p \leq B$, and consequently, $A = \prod_p A_p$ and $B = \prod_p B_p$. As
a direct summand of a complete group, $A_p$ must be complete in its *Z*-adic
topology, which is now identical with the *p*-adic topology, for the *Z*-adic and
*p*-adic topologies coincide on $Z(p^k)$.

Conversely, let $A_p$ be a group complete in its *p*-adic topology. The following definition makes $A_p$ in a natural way into a module over the ring $Q_p^*$ of the
*p*-adic integers. Let $a \in A_p$ and $\pi = s_0 + s_1 p + \cdots + s_n p^n + \cdots \in Q_p^*$. The
sequence

$$s_0 a, \qquad (s_0 + s_1 p)a, \qquad \cdots, \qquad (s_0 + s_1 p + \cdots + s_n p^n)a, \qquad \cdots$$

is Cauchy in $A_p$, hence it tends to a limit in $A_p$ which we define as $\pi a$. It is
straightforward to show that, in this way, $A_p$ has actually become a unital

$Q_p^*$-module. Hence we conclude $qA_p = A_p$ for all primes $q \neq p$, that is, the Z-adic and $p$-adic topologies of $A_p$ are the same. By (39.1), $A_p$ is algebraically compact, and (38.3) implies the algebraic compactness of their direct product $A$.

Finally, the uniqueness of the components $A_p$ in (1) follow at once from the relation

$$A_p = \bigcap_{q \neq p} q^k A \qquad (k = 1, 2, \cdots),$$

where $q$ again denotes primes. This is a consequence of the relations $qA_p = A_p$ ($q \neq p$) and $\bigcap_k p^k A_p = 0$. $\square$

The group $A_p$ of the preceding theorem will be called the *p-adic component of A*. It is a *p-adic algebraically compact* group in the sense that it is complete in its $p$-adic topology. As is shown by the preceding proof, it must be a $p$-adic module. It can be characterized completely *via* basic subgroups, as is indicated by the following theorem.

**Theorem 40.2.** *There is a one-to-one correspondence between the p-adic algebraically compact groups $A_p$ and direct sums $B_p$ of cyclic p-adic modules: given $A_p$, we let its basic submodule $B_p$ correspond to $A_p$; to a given $B_p$, there corresponds its p-adic completion.*

It follows immediately from (35.5) that the correspondence $A_p \mapsto B_p$ with $A_p$ a $p$-adic module and $B_p$ its basic submodule is single-valued [$B_p$ is now regarded as an abstract group]. On the other hand, the completion $\hat{B}$ of a direct sum $B$ of cyclic $p$-adic modules, with respect to the $p$-adic topology, does exist and is unique up to isomorphism; $\hat{B}$ is algebraically compact [cf. (39.1)]. If we identify $B$ with a subgroup of $\hat{B}$ under the natural embedding, then it is sufficient to refer simply to (39.6) in order to conclude that $B$ is basic in $\hat{B}$. Therefore, we infer that if the correspondence $B_p \mapsto A_p$ is followed by the correspondence $A_p \mapsto B_p$, then we get back to the original direct sum of cyclic $p$-adic modules.

We still need to verify the isomorphy of two $p$-adic algebraically compact groups $A_p$ and $A_p'$ under the hypothesis of having isomorphic basic submodules $B_p$. We start with the diagram

$$0 \to B_p \longrightarrow A_p \to A_p/B_p \to 0$$

where the row is pure-exact and $\rho$ is the injection $B_p \to A_p'$. By the pure-injectivity of $A_p'$, there is a homomorphism $\phi : A_p \to A_p'$ making the diagram commutative. Analogously, we get a homomorphism $\psi : A_p' \to A_p$ that extends the injection map $B_p \to A_p$. Hence $\psi\phi : A_p \to A_p$, whose restriction to $B_p$ is the identity, thus $\psi\phi = 1_{A_p}$ by (34.1). Similarly, $\phi\psi$ is the identity map of $A'_p$ whence $A_p \cong A'_p$. $\square$

In view of the foregoing theorem, the $p$-adic algebraically compact groups $A_p$ can be characterized by the same systems of invariants as their basic submodules $B_p$, namely, by the cardinal numbers $\mathfrak{m}_0$ and $\mathfrak{m}_k$ $(k = 1, 2, \cdots)$ of the sets of components [in a direct decomposition of $B_p$] isomorphic to $J_p$ and to $Z(p^k)$, respectively. This countable system of cardinals is a complete, independent system of invariants for $A_p$, from which we can reconstruct $A_p$ by forming first the obvious direct sum and then $p$-adic completion:

$$A_p \cong p\text{-adic completion of } \left[\bigoplus_{\mathfrak{m}_0} J_p \oplus \bigoplus_{k=1}^{\infty} \bigoplus_{\mathfrak{m}_k} Z(p^k)\right].$$

If $A$ is an arbitrary algebraically compact group, then the invariants of its maximal divisible subgroup along with the invariants of its $p$-adic components $A_p$ constitute a complete, independent system of invariants for $A$. We are therefore justified to claim that the structure of algebraically compact groups is known.

*Example 1.* Let $\mathfrak{m}$ be an infinite cardinal. Then

$$J_p^{\mathfrak{m}} \cong p\text{-adic completion of } \bigoplus_{2^{\mathfrak{m}}} J_p.$$

For, it is a torsion-free $p$-adic algebraically compact group whose basic submodule is of the same rank as $J_p^{\mathfrak{m}}/pJ_p^{\mathfrak{m}} \cong Z(p)^{\mathfrak{m}}$.

*Example 2.* Let $\mathfrak{m}_k$ $(k = 1, 2, \cdots)$ be any cardinals. Then

$$\prod_{k=1}^{\infty}\left[\bigoplus_{\mathfrak{m}_k} Z(p^k)\right] \cong p\text{-adic completion of } \left[\bigoplus_{\mathfrak{m}} J_p \oplus \bigoplus_{k=1}^{\infty}\bigoplus_{\mathfrak{m}_k} Z(p^k)\right]$$

where $\mathfrak{m} = \prod \mathfrak{m}_k$. In fact, the group $A$ in question is $p$-adic algebraically compact whose basic submodule has the indicated torsion part, by (33.3). Clearly, $A/pA \cong \prod_k \bigoplus_{\mathfrak{m}_k} Z(p)$ and the torsion subgroup of $A$ is represented by the elements of the direct sum in this direct product. From the proof of (35.2), we get $\mathfrak{m} = \prod \mathfrak{m}_k$.

In view of the above theorem, we are now able to derive the following elementary, useful corollaries.

**Corollary 40.3.** *If a reduced algebraically compact group is torsion, then it is bounded.*

In a decomposition (1) of such a group $A$, only a finite number of $A_p \neq 0$ may occur. By (40.2), each $A_p$ is the $p$-adic completion of a direct sum $B_p$ of cyclic $p$-groups. It is straightforward to check that this completion contains elements of infinite order, unless $B_p$ is bounded when $B_p = A_p$. $\square$

**Corollary 40.4** (Kaplansky [2]). *Every reduced algebraically compact group $\neq 0$ contains a direct summand isomorphic to $J_p$ or $Z(p^k)$ $(k = 1, 2, \cdots)$ for some $p$.*

If the group $A \neq 0$ is algebraically compact and reduced, and if $p$ is a prime such that $A_p \neq 0$ [see (40.1)], then the basic submodule $B_p$ of $A_p$ is not 0, and hence has a direct summand of the indicated type. This is pure in $A$ and algebraically compact, thus it is a summand of $A$, too.☐

It follows that the directly indecomposable algebraically compact groups are: $J_p$, $Q$, and subgroups of $Z(p^\infty)$. [That $J_p$ is indecomposable will also be shown in (Chapter XIII).]

EXERCISES

1.  Every reduced algebraically compact group is a unital module over the ring $\prod_p Q_p^*$.
2.  If, for each prime $p$, $A_p$ is a group complete in its $p$-adic topology, then the product and box topologies coincide in $A = \prod_p A_p$.
3.  (a) A countable algebraically compact group is the direct sum of a divisible and a bounded group.
    (b) There exist no countable reduced algebraically compact torsion-free groups.
4.  Determine the invariants of the following algebraically compact group:

$$\left[ \prod_{k=1}^{\infty} Z(p^k) \right]^{\mathfrak{m}} \oplus J_p^{\mathfrak{n}} \qquad (\mathfrak{m}, \mathfrak{n} \text{ infinite cardinals}).$$

5.  If $A$ and $A'$ are algebraically compact groups, each isomorphic to a pure subgroup of the other, then $A \cong A'$.
6.  If $A$ is an algebraically compact group such that $A \oplus J_p \cong A' \oplus J_p$, then $A \cong A'$.
7.  If $A$ is algebraically compact and $A \oplus A \cong A' \oplus A'$, then $A \cong A'$.
8.  (Mader [1]) A reduced algebraically compact group $A$ is the $Z$-adic completion of a direct sum of cyclic groups exactly if the rank of $A/[pA + T(A)]$ is independent of the prime $p$.

## 41.  PURE-ESSENTIAL EXTENSIONS

The striking analogy between divisible and algebraically compact groups can be pushed further by pointing out the analog of the divisible hull.

We first introduce the following concept. Let $G$ be a pure subgroup of $A$, and let $K(G, A)$ denote the set of all subgroups $H \leq A$ such that

(i)   $G \cap H = 0$      and      (ii)   $(G + H)/H$ is pure in $A/H$.

Since $G + H$ is the direct sum of $G$ and $H$, condition (ii) amounts to the fact that if $nx = g + h$ ($g \in G$, $h \in H$) is solvable in $A$, then there exists a solution of $ny = g$ in $G$. Hence $K(G, A)$ contains, along with $H$, all subgroups of $H$. The assumed purity of $G$ in $A$ implies $0 \in K(G, A)$, i.e., $K(G, A)$ is not empty.

Following Maranda [1], we call the group $A$ a *pure-essential extension* of its subgroup $G$ if $G$ is pure in $A$, and if $K(G, A)$ consists of 0 only. That is to say, if $G$ is pure in $A$ and if $\phi$ is a homomorphism of $A$ inducing a monomorphism on $G$ such that $\phi G$ is pure in $\phi A$, then $\phi$ is a monomorphism. The analogy with essential extensions is apparent.

Let us begin by introducing some lemmas.

**Lemma 41.1** (Maranda [1]). *If $G$ is a pure subgroup of $A$, then there exists a homomorphism $\phi$ of $A$ which induces a monomorphism on $G$ and for which $\phi A$ is a pure-essential extension of $\phi G$.*

The set $K(G, A)$ of subgroups $H$ of $A$ is inductive. In fact, if $\{H_i\}_{i \in I}$ is a chain in $K(G, A)$, and if $H$ is the union of this chain, then (i) is obvious, and if $nx = g + h \, (g \in G, h \in H)$ is solvable in $A$, then for some $i, h \in H_i \in K(G, A)$, whence $ny = g$ is solvable in $G$, and $H \in K(G, A)$. By Zorn's lemma, $K(G, A)$ contains a maximal member $B$. Define $\phi : A \to A/B$ as the canonical epimorphism; then $\phi \mid G$ is a monomorphism, and $\phi G = (G + B)/B$ is pure in $\phi A = A/B$. The maximal choice of $B$ implies $K(\phi G, \phi A) = \{0\}$.□

**Lemma 41.2.** *Let $A_i \, (i \in I)$ be a chain of groups and $A$ their union. If each $A_i$ is a pure-essential extension of $G$, then $A$ has the same property.*

It is immediately clear that if $G$ is pure in every $A_i$, then it is pure in $A$, too. If some $H \neq 0$ belonged to $K(G, A)$, then for some $i$, $H \cap A_i \neq 0$, and one would have $H \cap A_i \in K(G, A_i)$, a contradiction.□

**Lemma 41.3** (Maranda [1]). *If $C$ is a pure-essential extension of $G$, and if $A$ is an algebraically compact group containing $G$ as a pure subgroup, then the identity map of $G$ can be extended to a monomorphism of $C$ into $A$.*

Since $0 \to G \to C$ is pure-exact and $A$ is pure-injective, there is a homomorphism $\phi : C \to A$ which induces the identity map on $G$. Since $G$ is pure in $A$ and hence in $\phi C$, $\phi$ must be a monomorphism.□

The group $A$ will be called a *maximal pure-essential extension* of $G$ if it is a pure-essential extension of $G$, and if $A'$ with $A < A'$ is never a pure-essential extension of $G$. The existence of such maximal extensions may easily be established:

**Proposition 41.4** (Maranda [1]). *Every pure-essential extension of a group $G$ is contained in a maximal pure-essential extension of $G$.*

Let $C_0$ be a pure-essential extension of $G$. By way of contradiction, suppose that $G$ has no maximal pure-essential extension containing $C_0$. On account of (41.2), this means that, for every ordinal $\sigma$, there is a well-ordered, properly ascending chain of groups

$$C_0 < C_1 < \cdots < C_\sigma$$

such that for every $\rho \leqq \sigma$, $C_\rho$ is a pure-essential extension of $G$. Let $A$ be an algebraically compact group containing $G$ as a pure subgroup. (41.3) guarantees the existence of a monomorphism $C_\sigma \to A$, which is clearly impossible if $\sigma$ is of larger cardinality than $A$. $\square$

The maximal pure-essential extensions can easily be described:

**Proposition 41.5.** *A maximal pure-essential extension of a group $G$ is a minimal algebraically compact group containing $G$ as a pure subgroup.*

Let $A$ be a maximal pure-essential extension of $G$ and $B$ a group containing $A$ as a pure subgroup. By (41.1), there is a homomorphism $\phi$ of $B$ which induces a monomorphism on $G$ such that $\phi B$ is a pure-essential extension of $\phi G$. Hence $\phi G$ is pure in $\phi A$, and the definition of pure-essential extensions implies that $\phi$ is monic on $A$, too, so $\phi A \cong A$ is a maximal pure-essential extension of $\phi G \cong G$. Thus $\phi B = \phi A$, and $\phi$ followed by the inverse of $\phi \mid A$ is a projection of $B$ onto $A$. Hence $A$ is a direct summand of $B$, and $A$ is algebraically compact.

If $A'$ is an algebraically compact group such that $G \leqq A' \leqq A$, then by (41.3) there exists a monomorphism $\psi : A \to A'$ which induces the identity on $G$. Obviously, $\psi A$ is again a maximal pure-essential extension of $G$, and therefore $\psi A \leqq A' \leqq A$ implies $\psi A = A$. Hence $A' = A$, and $A$ is minimal. $\square$

**Proposition 41.6.** *A minimal algebraically compact group containing $G$ as a pure subgroup is a maximal pure-essential extension of $G$.*

If $A$ is a minimal algebraically compact group containing $G$ as a pure subgroup, and $B$ is a maximal pure-essential extension of $G$, then, in view of (41.3), there is a monomorphism $\phi : B \to A$ which is the identity on $G$. Evidently, $\phi B$ is, by (41.6), an algebraically compact group containing $G$, hence $\phi B = A$ because of the minimality of $A$. $\square$

Now we have arrived at the main result of this section.

**Theorem 41.7** (Maranda [1]). *For every group $G$, there exists a minimal algebraically compact group $A$ that contains $G$ as a pure subgroup. $A$ is a maximal pure-essential extension of $G$ and is unique up to isomorphism over $G$.*

Given $G$, by (41.4) there is a maximal pure-essential extension $A$ of $G$. (41.5) and (41.6) imply the first two assertions.

Let both $A$ and $A'$ be minimal algebraically compact groups containing $G$ as a pure subgroup. The identity map of $G$ can be extended, owing to (41.3), to a monomorphism $\phi : A \to A'$. Here $\phi A$ is minimal algebraically compact containing $\phi G = G$ as a pure subgroup, whence $\phi A = A'$, and $\phi$ is an isomorphism. $\square$

We may call a minimal algebraically compact group containing a given group $G$ as a pure subgroup the *pure-injective hull* of $G$. The last part of the preceding proof shows that *every algebraically compact group containing*

*G as a pure subgroup contains a pure-injective hull of G.*
The following result provides a useful criterion.

**Lemma 41.8.** *An algebraically compact group A containing G as a pure subgroup is the pure-injective hull of G exactly if the following conditions hold:*

(i) *the maximal divisible subgroup D of A is the injective hull of $G^1$;*
(ii) *the quotient group A/G is divisible.*

First, assume $A$ is the pure-injective hull of $G$. Write $A = D \oplus C$ with $D$ divisible and $C$ reduced. In view of the purity of $G$ in $A$,

$$G^1 = \bigcap nG = \bigcap (G \cap nA) = G \cap A^1 = G \cap D.$$

If $D$ were not the divisible hull of $G^1$, then we could write $D = D_1 \oplus D_2$ with $G^1 \leqq D_1$ and $D_2 \neq 0$. Here $D_2 \in K(G, A)$, which is absurd. If $A/G$ were not divisible, then let $E/G = (A/G)^1 < A/G$. Clearly, $D \leqq E$, thus $E = D \oplus (C \cap E)$. From $C/(C \cap E) \cong (C + E)/E = A/E \cong (A/G)/(E/G)$, it follows that $C \cap E$ is closed in the complete group $C$, thus is itself complete. Hence $E$ would be algebraically compact, properly contained in $A$. This proves (ii).

To prove sufficiency, assume (i) and (ii). By what has been noticed before the lemma, there is a pure-injective hull $E$ of $G$ in $A$, $G \leqq E \leqq A$. From the first part of the proof we know that $E/G$ is divisible, whence $E$ is pure in $A$, so $E$ is a direct summand of $A$, $A = E \oplus E_0$. By (ii), $E_0$ is divisible, $E_0 \leqq D$, and therefore $E_0 \cap G = 0$ and (i) imply $E_0 = 0$.□

Now it is easy to characterize the pure-injective hull of a given group $G$.

**Theorem 41.9.** *The pure-injective hull of a group G is isomorphic to the direct sum of the injective hull of $G^1$ and the completion $\hat{G}$ of G.*
Let $D$ be the injective hull of $G^1$, and let $\phi : G \to D$ extend the inclusion map $G^1 \to D$. If $\mu : G \to \hat{G}$ is the canonical map, then the homomorphism $(\phi \oplus \mu)\Delta : G \to D \oplus \hat{G}$ is obviously monic, and the purity of $\mu G$ in $\hat{G}$ implies $\mathrm{Im}\, (\phi \oplus \mu)\Delta$ pure in $D \oplus \hat{G}$. Combining (39.5) with (41.8), the assertion is clear.□

Recall that the divisible hull of $G^1$ is determined by the ranks of $G^1$, while the completion $\hat{G}$ is determined by the $p$-basic subgroups of $G$. Hence, these invariants serve to characterize the pure-injective hull of $G$.

EXERCISES

1. Let $G$ be pure in $A$ and let $H \in K(G, A)$. Then $K \in K(G, A)$, if $K/H \in K((G + H)/H, A/H)$.
2. Prove that $H \in K(G, A)$, if $H$ is contained in the first Ulm subgroup $A^1$ of $A$ and $H \cap G = 0$.

3. Let $A$ be a $p$-group and $B$ a basic subgroup of $A$. Then $K(B, A)$ is the set of all subgroups of $A^1$.

4. Assume $C$ is an algebraically compact group containing $G$ as a pure subgroup. Prove that the pure-injective hull $A$ of $G$ in $C$ is a direct summand of $C$.

5. Let $B$ be a pure-essential extension of $C$ and $A$ a pure-injective hull of $B$. Prove that $A$ is a pure-injective hull of $C$, too. [*Hint*: $A$ contains a pure-injective hull $A'$ of $C$ such that $B \leq A'$; $A = A' \oplus A''$ with $A''$ divisible; from (41.8), conclude $A'' = 0$.]

6. If $B$ is a pure-essential extension of $C$, then $B/C$ is divisible. [*Hint*: (41.8) and Ex. 5.]

7. A pure-essential extension of a pure-essential extension is a pure-essential extension. [*Hint*: Ex. 5.]

8. Let $C$ be pure in $B$ and $B$ pure in $A$, such that $A$ is a pure-essential extension of $C$. Then $B$ is a pure-essential extension of $C$ and $A$ is a pure-essential extension of $B$.

9. Characterize the pure-injective hull of the example in **35**.

10. If $A_i$ is the pure-injective hull of $G_i$ ($i \in I$), then $\prod A_i$ is not always the pure-injective hull of $\prod G_i$.

## 42.*   MORE ABOUT ALGEBRAICALLY COMPACT GROUPS

In our subsequent developments, we shall frequently meet algebraically compact groups; our knowledge on their structure will enable us to get a better insight into the structure of certain groups. There are some results on algebraically compact groups which do not fit properly into our future discussions; since they are worthwhile noticing, we intend to discuss them in this section.

The first result will show that algebraically compact groups are abundant among quotient groups of direct products.

Let $G_i$ ($i \in I$) be a family of groups with an arbitrary index set $I$, and let **K** be an ideal in the Boolean algebra **I** of all subset of $I$. Furthermore, let **K**\* be the $\sigma$-ideal generated by **K**, i.e., it consists of all subsets of unions of countable many members of **K** [it is easy to see that these members of **K** may be assumed to be mutually disjoint]. We are concerned with the **K**- and **K**\*-direct sums of the $G_i$:

**Theorem 42.1** (Fuchs [11]). *If* **K** *and* **K**\* *are defined as above, then the quotient group*

(1) $$A = \bigoplus_{\mathbf{K}^*} G_i \Big/ \bigoplus_{\mathbf{K}} G_i$$

*is algebraically compact.*

Ignoring a trivial case, we suppose $\mathbf{K} \neq \mathbf{K}^*$. From (38.5) and (38.1), it follows that it suffices to establish a solution in $A$ for the system

(2)                    $$\sum_k n_{jk} x_k = \bar{a}_j \qquad (\bar{a}_j \in A; \quad j = 1, 2, \cdots)$$

with a countable number of unknowns and equations, under the hypothesis that every finite subsystem of (2) admits a solution in $A$. We first pick out representatives $a_j \in \bar{a}_j$, where $a_j \in \bigoplus_{\mathbf{K}^*} G_i$, and focus our attention on the index set

$$\xi = \bigcup_{j=1}^{\infty} s(a_j);$$

since all the supports $s(a_j)$ belong to $\mathbf{K}^*$, manifestly $\xi \in \mathbf{K}^*$. If $\xi \in \mathbf{K}$, then $x_k = \bar{0}$ is a solution. Therefore, suppose $\xi \notin \mathbf{K}$. By hypothesis, the subsystem $(2_m)$ of (2) consisting of the first $m$ equations of (2) is solvable in $A$; assigning 0 to the unknowns not occurring in $(2_m)$ explicitly, we obtain values $x_k = \bar{g}(k, m) \in A$ for every $m$, which satisfy the first $m$ equations in (2).

We proceed to the system

(3)                    $$\sum_k n_{jk} x_k = a_j \qquad (a_j \in \bigoplus_{\mathbf{K}^*} G_i; \quad j = 1, 2, \cdots),$$

and select some representatives $g(k, m) \in \bar{g}(k, m)$ in $\bigoplus_{\mathbf{K}^*} G_i$, with 0 as a representative of $\bar{0}$. The system $(3_m)$ consisting of the first $m$ equations of (3) need not be satisfied by $x_k = g(k, m)$, but the set $\xi_m$ of indices $i \in I$ for which the $i$th components of $\sum_k n_{jk} g(k, m)$ and $a_j$ [as elements of $\bigoplus_{\mathbf{K}^*} G_i$] are different for $j = 1, 2, \cdots, m$ is certainly a member of $\mathbf{K}$. Starting with the sets $\xi_1, \cdots, \xi_m, \cdots \in \mathbf{K}$, we define successively the sets $\zeta_0, \zeta_1, \cdots, \zeta_m, \cdots \in \mathbf{K}$ so as to be pairwise disjoint and to satisfy

$$\zeta_0 = \xi_1, \qquad \zeta_1 \supseteq \xi_2 \backslash \zeta_0, \qquad \cdots, \qquad \zeta_m \supseteq \xi_{m+1} \backslash (\zeta_0 \cup \cdots \cup \zeta_{m-1}), \qquad \cdots$$

and, in addition,

$$\bigcup_{m=0}^{\infty} \zeta_m = \xi \cup \xi_1 \cup \cdots \cup \xi_m \cup \cdots.$$

This can be done easily.

In order to conclude the proof, set $g_k = (\cdots, g_{ki}, \cdots)$ where $g_{ki}$ is the $i$th component of $g(k, m)$ if $i \in \zeta_m$ and $g_{ki} = 0$ otherwise. Then, obviously, $g_k \in \bigoplus_{\mathbf{K}^*} G_i$. As the $i$th components of $\sum_k n_{jk} g_k$ and $a_j$ coincide for $i \in \zeta_m$ whenever $j \leq m$, it is clear that $x_k = \bar{g}_k$ yield a solution for the whole system (2). $\square$

If the index set $I$ is countable and $\mathbf{K}$ is the ideal of all finite subsets of $I$, then $\mathbf{K}^* = \mathbf{I}$, and we get the interesting special case:

**Corollary 42.2** (Hulanicki [3]). *For every countable family of groups* $G_n$ $(n = 1, 2, \cdots)$, *the quotient group*

$$\prod_{n=1}^{\infty} G_n \Big/ \bigoplus_{n=1}^{\infty} G_n$$

*is algebraically compact.*☐

For the invariants of the quotient group see Golema and Hulanicki [1].

A rather interesting feature of algebraically compact groups is the existence of generalized limits—a phenomenon discovered by Łoś [3].

Let $\aleph_\sigma$ denote an infinite cardinal number and $\omega_\sigma$ the first ordinal of power $\aleph_\sigma$. Sequences $a_0, a_1, \cdots, a_\rho, \cdots$ $(\rho < \omega_\sigma)$ of elements in a group $A$ will be denoted briefly as $\{a_\rho\}_{\rho < \omega_\sigma}$. We restrict ourselves to well-ordered sequences of the same type $\omega_\sigma$.

We shall say that a group $A$ admits $\omega_\sigma$-*limits* if to *each* sequence $\{a_\rho\}_{\rho < \omega_\sigma}$ in $A$ there is assigned a welldefined element $a \in A$, denoted as

$$a = \omega_\sigma\text{-}\lim a_\rho,$$

subject to the following postulates:

(i) $\omega_\sigma\text{-}\lim(a_\rho + b_\rho) = \omega_\sigma\text{-}\lim a_\rho + \omega_\sigma\text{-}\lim b_\rho$;
(ii) $\omega_\sigma\text{-}\lim a_\rho = a$ if $a_\rho = a$ for all $\rho < \omega_\sigma$;
(iii) $\omega_\sigma\text{-}\lim a_\rho = \omega_\sigma\text{-}\lim b_\rho$ if $a_\rho = b_\rho$ for $\rho > \rho_0$ with some fixed $\rho_0 < \omega_\sigma$.

In other words, the $\omega_\sigma$-limit is additive, assigns the constant to a constant sequence, and makes no distinction between sequences which are ultimately the same. In spite of these natural assumptions, the definition may at first glance appear rather strange, because all sequences of type $\omega_\sigma$ are required to have $\omega_\sigma$-limits. As might be expected, this yields a Hausdorff topology only in the trivial case $A = 0$, but otherwise it seems to be of interest in the light of the next theorem.

**Theorem 42.3** (Łoś [3]). *If a group $A$ admits $\omega_0$-limits, then it is algebraically compact. If $A$ is algebraically compact, then it admits $\omega_\sigma$-limits for every $\sigma$.*

As is apparent from the definition, the existence of $\omega_0$-limits amounts to the existence of a homomorphism

$$\eta : A^{\aleph_0} \to A$$

of the direct product of countably many copies of $A$ into $A$, such that

$$\eta(a, \cdots, a, \cdots) = a \qquad \text{and} \qquad \eta(\bigoplus A) = 0.$$

Hence, we may consider $\eta$ as a homomorphism of $A^* = A^{\aleph_0}/\bigoplus A$ into $A$. Let $\delta : a \mapsto (a, \cdots, a, \cdots) + (\bigoplus A)$ be the diagonal map of $A$ into the group

$A^*$. Evidently, $\delta\eta$ is an endomorphism of $A^*$ that leaves $\delta A$ elementwise fixed, i.e., $\delta A \cong A$ is a direct summand of $A^*$. We know from (42.2) that $A^*$ is algebraically compact, thus $A$ is also.

Conversely, let $A$ be algebraically compact, $\sigma$ an ordinal number, and define $G = \prod A / \bigoplus A$, where we use $\aleph_\sigma$ copies of $A$ to form the direct product and the direct sum. The diagonal map $\delta : A \to G$ [where $\delta a = (a, \cdots, a, \cdots) + (\bigoplus A)$] is a monomorphism, and, evidently, $\delta A$ is pure in $G$. By algebraic compactness, $G = \delta A \oplus H$ for some $H \leq G$. If given any $\omega_\sigma$-sequence $\{a_\rho\}_{\rho < \omega_\sigma}$, it may be regarded as an element of $A^{\aleph_\sigma}$, and the $\delta A$-component of its coset in $G$ gives rise to an $\omega_\sigma$-limit in $A$.☐

It must be noted that $\omega_\sigma$-limits are not uniquely chosen in $A$; in fact, the last part of the proof shows that every choice of $H$ yields a possible definition of $\omega_\sigma$-limits.

EXERCISES

1.  A group is algebraically compact if and only if it is a direct summand of a group of the form $\prod G / \bigoplus G$ with countably many factors.
2.  Prove that every group $G$ can be embedded as a pure subgroup in an algebraically compact group by taking the diagonal in $\prod G / \bigoplus G$ [countably many components].
3.  Given $G$, there is an algebraically compact group of cardinality $\leq |G|^{\aleph_0}$ that contains $G$ as a pure subgroup.
4.  Let $G_n$ ($n = 1, 2, \cdots$) be reduced groups such that there is no $m \in \mathbf{Z}$, $m > 0$, such that $mG_n = 0$ for almost all $n$. Prove that $\prod_n G_n / \bigoplus_n G_n$ is not reduced.
5.  (Baumslag and Blackburn [1]) If $G_n$ ($n = 1, 2, \cdots$) are reduced groups, then $\bigoplus_n G_n$ is a direct summand of $\prod_n G_n$ if, and only if, $mG_n = 0$ for some $m > 0$ and for almost all $n$. [*Hint*: for necessity, Ex. 4.]
6.  Let $H$ be the subgroup of $\prod_{i \in I} G_i$ consisting of elements with countable support. Prove that $H / \bigoplus G_i$ is a direct summand of $\prod G_i / \bigoplus G_i$.
7.  (Balcerzyk [3]) Let $A = \prod \mathbf{Z} / \bigoplus \mathbf{Z}$, countably many components. Prove that

$$A = \bigoplus_{\mathfrak{c}} Q \oplus \prod_p A_p$$

where $A_p$ is the $p$-adic completion of $\bigoplus_{\mathfrak{c}} J_p$, $\mathfrak{c} = 2^{\aleph_0}$. [*Hint*: the element $(s_1, \cdots, m! s_m, \cdots) \in \prod \mathbf{Z}$ is divisible by every $m \bmod(\bigoplus \mathbf{Z})$; select a set of power $\mathfrak{c}$ of sequences $s_1, \cdots, s_m, \cdots$ such that $s_m = 0$ or 1, and any two sequences differ in countably many places; in a similar fashion prove $|A/pA| = \mathfrak{c}$.]
8.  Let $A$ be a Hausdorff topological group in which $\omega_0$-limits exist and are limits in the topological sense. Then $A = 0$.

## NOTES

The problem of describing the algebraic structure of compact abelian groups led Kaplansky [2] to the discovery of algebraically compact groups. He phrased the definition in terms of the decomposition theorem (40.1) and proved that all compact groups are algebraically compact [the precise result appears in (47.3) and (47.4)]. A different line of investigation was followed by Łoś [1], [2], who considered groups with the property which served as our definition of algebraic compactness. He proved that these groups are exactly the direct summands of compact groups. Balcerzyk [2] noticed that the classes discussed by Kaplansky and Łoś are identical. The significant pure-injectivity was discovered by Maranda [1]. **38** follows the presentation in Fuchs [10].

The theory of complete groups is due to Kaplansky [2]; he gives credit to I. Fleischer for the torsion-free case.

Starting with a definition of purity for modules, that of algebraic compactness offers no difficulty. Surprisingly, (38.1) holds for Cohn's purity with the exception of (c), as was shown by R. B. Warfield, Jr. [*Pac. J. Math.* **28** (1969), 699–720]. Algebraic compactness over left Noetherian rings R was studied in L. Fuchs [*Ind. J. Math.* **9** (1967), 357–374]; most results hold for any R. It turns out that algebraic compactness splits into various $\mathfrak{m}$-compactness properties, depending on the cardinal number $\mathfrak{m}$ of unknowns in (38.1) (e), but this distinction disappears as soon as $|R|$ is reached (thus, for a countable ring, $\mathfrak{m}$-compactness is the same for every infinite $\mathfrak{m}$). For algebraic compactness of general algebraic systems see J. Mycielski [*Coll. Math.* **13** (1964), 1–9].

A remarkable fact underlining the importance of algebraically compact groups does not seem to be generally known: an injective module $M$ over any ring R must be algebraically compact as an abelian group [if $D$ is the divisible hull of the additive group of $M$, then $M$ is a submodule and hence a direct summand of $\text{Hom}_Z (R, D)$, which is by (47.7) an algebraically compact group].

*Problem 23.* Let $A_1, \cdots, A_n, \cdots$ be infinite cyclic groups. Does there exist a group topology on $\prod A_n$ such that $\oplus A_n$ is a closed subgroup and $\prod A_n / \oplus A_n$ is compact in the induced topology?

*Problem 24.* Characterize groups which are quotients of direct products modulo direct sums.

*Problem 25.* Describe the structure of ultraproducts.

If $A_i$ ($i \in I$) is a set of groups and **K** a prime ideal in the Boolean algebra of all subsets of $I$, then the ultraproduct is the quotient of the direct product $\prod A_i$ modulo the **K**-direct sum $\oplus_\mathbf{k} A_i$. Assume that the finite subsets of $I$ belong to **K**.

*Problem 26.* Which groups are inverse limits of groups, each of which is a finite direct sum of bounded and quasicyclic groups?

Cf. **39**, Ex. 13.

*Problem 27.* Given $\sigma \geq 1$, find the groups that are direct summands in every group in which they are $\aleph_\sigma$-pure.

*Problem 28.* (a) Give conditions on a group to be a (countable) direct sum of complete groups.

(b) Characterize these groups by invariants.

*Problem 29.* Which algebraically compact groups can be additive groups of injective modules?

# VIII

## HOMOMORPHISM GROUPS

This chapter is devoted to the study of homomorphism groups; that is, the elements of the groups are homomorphisms of a fixed group $A$ into a fixed group $C$. The fact that these homomorphisms form a group $\operatorname{Hom}(A, C)$ has proved to be extraordinarily profound; as a matter of fact, the homomorphism groups have many remarkable properties, and they are very useful in various contexts.

Our first aim is to find the elementary properties of homomorphism groups and to point out their functorial behavior. It is a surprising fact that $\operatorname{Hom}(A, C)$ is in some important cases algebraically compact; e.g., if $A$ is a torsion group or if $C$ is algebraically compact. In these cases, we can determine the invariants of $\operatorname{Hom}(A, C)$ in terms of those of $A$ and $C$. In particular, if $C$ is the additive group of reals mod 1—in which case, $\operatorname{Hom}(A, C)$ [with a certain topology] is the character group of $A$—our description leads to a complete characterization of the structure of compact (abelian) groups.

Finally, we discuss the Pontryagin duality theory. Since we wish to refrain from entering into an extensive study of this theory, we shall confine ourselves to the case when no deeper results are needed. In accordance with this, we develop the duality between discrete torsion groups and 0-dimensional compact groups.

## 43. GROUPS OF HOMOMORPHISMS

We have already noticed in **2** that, if $\alpha$ and $\beta$ are homomorphisms of a group $A$ into a group $C$, then their sum $\alpha + \beta$, defined as

$$(\alpha + \beta)a = \alpha a + \beta a \qquad (a \in A),$$

is again a homomorphism of $A$ into $C$. It is now routine to check that the homomorphisms of $A$ into $C$ form an abelian group [as a matter of fact, a subgroup of $C^A$] which we call the *homomorphism group* of $A$ into $C$ and

*If A is a p-group does ∃ $G_A$ ∋ Hom($A, G$) ≅ A?*

denote by $\mathrm{Hom}(A, C)$. The zero in this group is the trivial homomorphism of $A$ into $C$, and the inverse $-\alpha$ of $\alpha : A \to C$ maps $a \in A$ on $-\alpha a \in C$.

If $A = C$, the homomorphisms of $A$ into $C$ are endomorphisms of $A$, and the group $\mathrm{Hom}(A, A) = \mathrm{End}\, A$ is called the *endomorphism group* of $A$. [We can define a ring structure on $\mathrm{End}\, A$; see Chapter XV.]

*Example 1.* If $A = Z$, then each $\alpha : Z \to C$ is completely determined by $\alpha 1 = c\,(\in C)$, and evidently, to every $c \in C$ there exists a homomorphism $\alpha : Z \to C$ such that $\alpha 1 = c$. Since $\alpha 1 = c_1$ and $\beta 1 = c_2$ imply $(\alpha + \beta)1 = c_1 + c_2$, the correspondence $\alpha \mapsto c$ if $\alpha 1 = c$ turns out to be a natural isomorphism between $\mathrm{Hom}(Z, C)$ and $C$,

$$\mathrm{Hom}(Z, C) \cong C.$$

*Example 2.* If $A = Z(m)$, then, again, every $\alpha : Z(m) \to C$ is characterized by the image $\alpha 1 = c$ of 1, but here $mc = 0$ must hold, i.e., $c \in C[m]$. Conversely, each such $c$ gives rise to a homomorphism $\alpha : 1 \mapsto c$, and, as in the preceding example, $\alpha \mapsto c$ if $\alpha 1 = c$ is a natural isomorphism

$$\mathrm{Hom}(Z(m), C) \cong C[m].$$

*Example 3.* Next let $A$ be quasicyclic, say, $A = \langle a_1, \cdots, a_n, \cdots \rangle$ with the defining relations $pa_1 = 0$, $pa_{n+1} = a_n$ $(n \geq 1)$, and let $A = C$. If $\eta$ is an endomorphism of $A$, then write $\eta a_n = k_n a_n$ with an integer $k_n$ $(0 \leq k_n < p^n)$ for every $n$. Now $k_n a_n = \eta a_n = \eta(pa_{n+1}) = p\eta a_{n+1} = k_{n+1} a_n$ implies $k_n \equiv k_{n+1} \bmod p^n$, thus the sequence $k_1, \cdots, k_n, \cdots$ tends to a $p$-adic integer $\pi$. The correspondence $\eta \mapsto \pi$ between endomorphisms $\eta$ and $p$-adic integers $\pi$ is readily seen to be additive. If the endomorphisms $\eta_1, \eta_2$ of $A$ define the same $\pi$, then $\eta_1 - \eta_2$ maps each $a_n$ on 0, i.e., $\eta_1 - \eta_2 = 0$. If $\pi = s_0 + s_1 p + \cdots + s_n p^n + \cdots$ is any $p$-adic integer, then $a_n \mapsto (s_0 + s_1 p + \cdots + s_{n-1} p^{n-1})a_n$ [this element may be written as $\pi a_n$] extends uniquely to an endomorphism $\eta$ of $A$, such that $\eta \mapsto \pi$. We conclude:

$$\mathrm{End}\, Z(p^\infty) \cong J_p.$$

*Example 4.* Now let $A = Q^{(p)}$, the group of rational numbers with powers of $p$ as denominators, and $C = Z(p^\infty)$. If $C = \langle c_1, \cdots, c_n, \cdots \rangle$ with $pc_1 = 0$, $pc_{n+1} = c_n$ $(n \geq 1)$, and if we agree to put $p^{-k} c_n = c_{n+k}$ for $k \geq 1$, then a $p$-adic number $\rho = p^k \pi$ [with a $p$-adic unit $\pi$] induces a homomorphism $\eta : Q^{(p)} \to Z(p^\infty)$ such that $p^{-n} \mapsto p^k \pi c_n$. [Notice that the images of $p^{-n} \in Q^{(p)}$ $(n = 1, 2, \cdots)$ in $Z(p^\infty)$ completely determine $\eta$.] It is obvious from preceding example that different $p$-adic numbers $\rho$ give rise to different the homomorphisms, and it is straightforward to check that every element of $\mathrm{Hom}(Q^{(p)}, Z(p^\infty))$ arises in this way. Consequently, $\mathrm{Hom}(Q^{(p)}, Z(p^\infty))$ is isomorphic to the additive group of all $p$-adic numbers, i.e.,

$$\mathrm{Hom}(Q^{(p)}, Z(p^\infty)) \cong \bigoplus_{\mathfrak{c}} Q \qquad \text{with} \quad \mathfrak{c} = 2^{\aleph_0}.$$

*Example 5.* If $A = C = J_p$, the group of $p$-adic integers, then it is evident that multiplication by a fixed $p$-adic integer $\pi$ is an endomorphism $\eta(\pi)$ of $J_p$, and different $p$-adic integers yield different endomorphisms, since they have different effects on $1 \in J_p$. If $\xi$ is any endomorphism of $J_p$ and if $\xi 1 = \pi \in J_p$, then we must have $\xi = \eta(\pi)$. For, then $(\xi - \eta(\pi))1 = 0$ implies $(\xi - \eta(\pi))n = 0$ for every rational integer $n \in J_p$, i.e., $Z$ [as a subgroup of $J_p$] is contained in $\mathrm{Ker}(\xi - \eta(\pi))$. Because of the divisibility of $J_p/Z$, and the reducedness of $J_p$, $J_p/\mathrm{Ker}(\xi - \eta(\pi)) = 0$ whence $\xi = \eta(\pi)$, indeed. We conclude:

$$\mathrm{End}\, J_p \cong J_p.$$

Next we list some simple facts on homomorphism groups.

(A) $\mathrm{Hom}(A, C) = 0$ in the following cases: (i) $A$ is torsion and $C$ is torsion-free; (ii) $A$ is a $p$-group and $C$ is a $q$-group, where $p, q$ are distinct primes; (iii) $A$ is divisible and $C$ is reduced.

(B) *If $C[n] = 0$ for some integer $n$, then $\mathrm{Hom}(A, C)[n] = 0$ for every group $A$.*

Let $\alpha \in \mathrm{Hom}(A, C)$ and $n\alpha = 0$. For every $a \in A$ we have $n(\alpha a) = (n\alpha)a = 0$, whence $C[n] = 0$ implies $\alpha a = 0$, i.e., $\alpha = 0$.

(C) $\mathrm{Hom}(A, C)$ *is torsion-free whenever $C$ is torsion-free.*

(D) *If $C$ is torsion-free and divisible, then $\mathrm{Hom}(A, C)$ is torsion-free and divisible.*

In order to show $\mathrm{Hom}(A, C)$ divisible, let $\alpha \in \mathrm{Hom}(A, C)$ and $n$ a positive integer. To every $a \in A$, there exists a unique $c \in C$ such that $nc = \alpha a$, and thus we may define a mapping $\beta$ as $a \mapsto c$. It follows readily that $\beta$ is a homomorphism satisfying $n\beta = \alpha$.

It follows in the same way: *if $C$ is a group in which division by a prime $p$ is uniquely performable, then the same holds for $\mathrm{Hom}(A, C)$.*

(E) *If for some integer $n > 0$, $nA = A$, then $\mathrm{Hom}(A, C)[n] = 0$.*

Let $\alpha \in \mathrm{Hom}(A, C)$ and $n\alpha = 0$. Write $a \in A$ as $nb = a$ for some $b \in A$. Then $\alpha a = \alpha(nb) = (n\alpha)b = 0$ shows $\alpha = 0$.

(F) *If $A$ is divisible, then $\mathrm{Hom}(A, C)$ is torsion-free.*

(G) *If $A$ is torsion-free and divisible, then the same holds for $\mathrm{Hom}(A, C)$.*

To prove divisibility, let $\alpha \in \mathrm{Hom}(A, C)$ and $n$ an integer $> 0$. Given $a \in A$, there is a unique $b \in A$ such that $nb = a$. The mapping $\beta: a \mapsto \alpha b$ turns out to be a homomorphism $A \to C$ which plainly satisfies $n\beta = \alpha$.

Our next concern is the behavior of $\mathrm{Hom}(A, C)$ toward direct sums and products.

**Theorem 43.1.** *There is a natural isomorphism*

$$\mathrm{Hom}\left(\bigoplus_{i \in I} A_i, C\right) \cong \prod_{i \in I} \mathrm{Hom}(A_i, C).$$

The restriction of $\alpha: \bigoplus A_i \to C$ to $A_i$ is a homomorphism $\alpha_i : A_i \to C$. In this way, we obtain a correspondence $\alpha \mapsto (\cdots, \alpha_i, \cdots)$ of $\mathrm{Hom}(\bigoplus A_i, C)$ into $\prod \mathrm{Hom}(A_i, C)$ which is manifestly a homomorphism $\phi$. Clearly, $\phi$ maps only $\alpha = 0$ on $(0, \cdots, 0, \cdots)$, thus $\phi$ is monic. Since any set $\{\alpha_i\}_{i \in I}$ with $\alpha_i \in \mathrm{Hom}(A_i, C)$ defines an $\alpha \in \mathrm{Hom}(\bigoplus A_i, C)$ such that $\alpha_i = \alpha \,|\, A_i$, $\phi$ is epic too.☐

**Theorem 43.2.** *There exists a natural isomorphism*

$$\mathrm{Hom}\left(A, \prod_{i \in I} C_i\right) \cong \prod_{i \in I} \mathrm{Hom}(A, C_i).$$

If $\pi_i$ denotes the $i$th coordinate projection $\prod C_i \to C_i$, then each $\alpha \in \mathrm{Hom}(A, \prod C_i)$ defines homomorphisms $\pi_i \alpha \in \mathrm{Hom}(A, C_i)$. As in the preceding proof, one concludes that $\alpha \mapsto (\cdots, \pi_i \alpha, \cdots)$ is an isomorphism of $\mathrm{Hom}(A, \prod C_i)$ onto $\prod \mathrm{Hom}(A, C_i)$.$\square$

From these theorems, we are led at once to the following corollaries.

**Corollary 43.3.** *Let $A$ be a direct sum of cyclic groups, and let $\mathfrak{m}$ and $\mathfrak{m}_{p,k}$ denote the cardinal numbers of the set of direct summands isomorphic to $Z$ and to $Z(p^k)$, respectively, in a decomposition of $A$. Then*

$$\mathrm{Hom}(A, C) \cong \prod_{\mathfrak{m}} C \oplus \prod_p \prod_{k=1}^{\infty} \prod_{\mathfrak{m}_{p,k}} C[p^k].\square$$

**Corollary 43.4.** *If $A$ is a torsion group with $p$-components $A_p$, and if $C_p$ denote the $p$-components of $C$, then*

$$\mathrm{Hom}(A, C) \cong \prod_p \mathrm{Hom}(A_p, C_p).\square$$

Observe that the second summand in (43.3) can be determined more explicitly with the aid of (33.3) and example 2 in **40**.

*Example 6.* For any group $A$,

$$\mathrm{Hom}(A, Q) \cong \prod_{\mathfrak{n}} Q \qquad \text{with} \quad \mathfrak{n} = r_0(A).$$

By (D), the structure of $\mathrm{Hom}(A, Q)$ is given by a simple set-theoretic calculation. If $F$ is generated by a maximal independent system of elements of infinite order in $A$, then the elements of $\mathrm{Hom}(A, Q)$ and $\mathrm{Hom}(F, Q)$ are in an obvious one-to-one correspondence. The latter group is evaluated by (43.1). A similar inference leads us to the more general isomorphism

$$\mathrm{Hom}\left(A, \bigoplus_{\mathfrak{m}} Q\right) \cong \prod_{\mathfrak{n}} \left[\bigoplus_{\mathfrak{m}} Q\right] \qquad \text{with} \quad \mathfrak{n} = r_0(A).$$

EXERCISES

1.  (a) Prove (B) and (C) by using the fact that $\mathrm{Hom}(A, C) \leq C^A$.
    (b) Prove that
    $$|\mathrm{Hom}(A, C)| \leq |C|^{|A|}.$$

2.  Show that $\mathrm{Hom}(A, C) \cong \mathrm{Hom}(C, A)$ and $\mathrm{Hom}(A, Q/Z) \cong A$, if both $A$ and $C$ are finite.
3.  If $A$ is torsion-free and $C$ is divisible, then $\mathrm{Hom}(A, C)$ is divisible.
4.  Describe the group $\mathrm{Hom}(A, Z(m))$.
5.  (a) Prove that $\mathrm{Hom}(Q, C)$ is, for a torsion-free $C$, isomorphic to the maximal divisible subgroup of $C$.
    (b) Find $\mathrm{Hom}(Q, C)$ for arbitrary $C$.

6.  Give the structures of $\text{Hom}(Q/Z, Q/Z)$ and $\text{Hom}(J_p, Z(p^\infty))$.
7.  (a) If $A$ and $C$ are nontrivial $p$-groups, then either $\text{Hom}(A, C) \neq 0$ or $\text{Hom}(C, A) \neq 0$.
    (b) Give an example for torsion-free $A$, $C\ (\neq 0)$ such that $\text{Hom}(A, C) = 0 = \text{Hom}(A, C)$,
8.  Let $A$ and $C$ be reduced algebraically compact groups and $A_p$, $C_p$ their $p$-adic components. Prove the isomorphism

$$\text{Hom}(A, C) \cong \prod_p \text{Hom}(A_p, C_p).$$

9.  (Lewis [1]) If the torsion group $A$ satisfies $\text{Hom}(A, C) \cong A$ for some $C$, then $A$ is finite. [*Hint*: reduced; basic subgroup must be finite.]
10.  (Lewis [1]) The rank of $\text{Hom}(A, Z)$ is at least $n$ if and only if $A$ has a direct summand which is a free group of rank $n$. [*Hint*: induction on $n$.]
11.  (a) If $A$ is a torsion group, then the set union $\bigcup \text{Im } \alpha$ for all $\alpha \in \text{Hom}(A, C)$ is a subgroup of $C$.
    (b)* The same is not necessarily true if $A$ is torsion-free. [*Hint*: $A$ of rank $\geq 2$ whose endomorphisms are integers and $C = A \oplus A$.]

## 44.  EXACT SEQUENCES FOR HOM

It is the purpose of this section to investigate the functorial behavior of Hom. This is based on the notion of induced homomorphisms between groups of homomorphisms, so first of all we introduce this notion.

Let $\alpha : A' \to A$ and $\gamma : C \to C'$ be fixed homomorphisms. Every $\eta \in \text{Hom}(A, C)$ gives rise to a homomorphism $A' \to C'$ which is the composite

$$A' \xrightarrow{\alpha} A \xrightarrow{\eta} C \xrightarrow{\gamma} C'.$$

The correspondence $\eta \mapsto \gamma\eta\alpha$ is a homomorphism of $\text{Hom}(A, C)$ into $\text{Hom}(A', C')$ which is denoted as

$$\text{Hom}(\alpha, \gamma) : \text{Hom}(A, C) \to \text{Hom}(A', C')$$

and called the *induced homomorphism*; more precisely, it is induced by $\alpha$ and $\gamma$ as is indicated by our notation $\text{Hom}(\alpha, \gamma)$. From $A'' \xrightarrow{\alpha'} A' \xrightarrow{\alpha} A$ and $C \xrightarrow{\gamma} C' \xrightarrow{\gamma'} C''$ we obtain easily

$$\text{Hom}(\alpha\alpha', \gamma'\gamma) = \text{Hom}(\alpha', \gamma')\text{Hom}(\alpha, \gamma),$$

and manifestly,

$$\text{Hom}(1_A, 1_C) = 1_{\text{Hom}(A,C)}$$

holds. Furthermore, $\text{Hom}(\alpha, \gamma)$ is evidently additive both in $\alpha$ and $\gamma$. Therefore, we conclude:

**Theorem 44.1.** Hom *is an additive bifunctor on* $\mathscr{A} \times \mathscr{A}$ *to* $\mathscr{A}$, *contravariant in the first and covariant in the second variable.*☐

It is convenient at times to use an abbreviated notation for $\text{Hom}(\alpha, 1_C)$ and $\text{Hom}(1_A, \gamma)$; we shall often denote them by $\alpha^*$ and $\gamma_*$, respectively, provided that there is no danger of confusion.

The following result describes the connection of Hom with limits.

**Theorem 44.2** (Cartan and Eilenberg [1]). *Let*

$$A = \{A_i \, (i \in I); \pi_i^j\} \quad and \quad C = \{C_k \, (k \in K), \rho_k^l\}$$

*be direct and inverse systems of groups, respectively, and let* $A = \underline{\lim} \, A_i$, $C = \underline{\lim} \, C_k$ *with the canonical maps* $\pi_i : A_i \to A$ *and* $\rho_k : C \to C_k$. *Then*

$$H = \{\text{Hom}(A_i, C_k) \, ((i, k) \in I \times K); \, \text{Hom}(\pi_i^j, \rho_k^l)\}$$

*is an inverse system of groups whose inverse limit is* $\text{Hom}(A, C)$ *with* $\text{Hom}(\pi_i, \rho_k)$ *as canonical maps.*

It is straightforward to check that $H$ is an inverse system. Let $H$ denote its inverse limit. Owing to the commutativity of

$$(i \leq j; k \leq l),$$

from (12.1) we obtain that there exists a unique $\sigma$ rendering all diagrams

commutative, where the $\sigma_{ik}$ are the canonical maps. In order to show that $\sigma$ is monic, let $\eta \in \text{Ker } \sigma$. Then $\sigma_{ik} \sigma \eta = 0$, that is, $\rho_k \eta \pi_i = \text{Hom}(\pi_i, \rho_k)\eta = 0$ for all $i, k$. Thus $\eta \pi_i : A_i \to C$ is 0, since all the $k$th coordinates are 0, and since $\bigcup \pi_i A_i = A$, $\eta = 0$. Any $\chi \in H$ is of the form

$$\chi = (\cdots, \chi_{ik}, \cdots) \in \prod \text{Hom}(A_i, C_k)$$

· where the coordinates $\chi_{ik}$ satisfy the requisite postulates. Define $\eta : A \to C$ to act in the following way: if $a = \pi_i a_i$, then for this $i \in I$ we put $\eta a = (\cdots, \chi_{ik} a_i, \cdots) \in \prod C_k$. It is straightforward to verify both the independence of $\eta a$ of the chosen $a_i$ and the homomorphism property of $\eta$. Considering that $\sigma_{ik} \chi = \chi_{ik}$ and $\sigma_{ik} \sigma \eta = \rho_k \eta \pi_i = \chi_{ik}$, we must have $\sigma \eta = \chi$ showing that $\sigma$ is epic, and hence an isomorphism.☐

As an application we prove:

**Proposition 44.3.** *For every cardinal* $\mathfrak{m}$,

$$\mathrm{Hom}\!\left(Z(p^\infty),\ \bigoplus_{\mathfrak{m}} Z(p^\infty)\right) \cong p\text{-adic completion of } \bigoplus_{\mathfrak{m}} J_p.$$

Since $Z(p^\infty) = \varprojlim Z(p^n)$, the homomorphism group in question is by (44.2) the inverse limit of $\mathrm{Hom}(Z(p^n),\ \bigoplus_{\mathfrak{m}} Z(p^\infty)) \cong \bigoplus_{\mathfrak{m}} Z(p^n)$. A simple argument shows that the map from the $(n+1)$th onto the $n$th member just sends a component $Z(p^{n+1})$ upon the corresponding $Z(p^n)$. Consequently, the inverse limit must be the same as that for $G/p^n G$, where $G = \bigoplus_{\mathfrak{m}} J_p$ which is just the $p$-adic completion of $G$.□

An extremely important result on Hom states the existence of two exact sequences for Hom, expressing the fact that Hom is a left exact functor.

**Theorem 44.4** (Cartan and Eilenberg [1]). *Let*

(1)                              $0 \to A \xrightarrow{\alpha} B \xrightarrow{\beta} C \to 0$

*be a short exact sequence. Then, for an arbitrary group* $G$, *the induced sequences*

(2)                $0 \to \mathrm{Hom}(C, G) \xrightarrow{\beta^*} \mathrm{Hom}(B, G) \xrightarrow{\alpha^*} \mathrm{Hom}(A, G),$

(3)                $0 \to \mathrm{Hom}(G, A) \xrightarrow{\alpha_*} \mathrm{Hom}(G, B) \xrightarrow{\beta_*} \mathrm{Hom}(G, C)$

*are exact.*

To prove $\beta^*$ a monomorphism, let $\eta : C \to G$ be such that $0 = \beta^* \eta = \eta\beta$. Since $\beta$ is epic, $\eta = 0$. Because of $\alpha^*\beta^* = (\beta\alpha)^* = 0^* = 0$ we need only verify the inclusion $\mathrm{Ker}\ \alpha^* \le \mathrm{Im}\ \beta^*$. Let $\eta \in \mathrm{Ker}\ \alpha^*$, i.e., $\eta\alpha = 0$. By (2.2), there is a $\chi \in \mathrm{Hom}(C, G)$ such that $\chi\beta = \eta$. Therefore, (2) is exact.

If $\eta \in \mathrm{Ker}\ \alpha_*$, i.e., $\alpha\eta = 0$, then $\eta = 0$, $\alpha$ being monic. $\beta_* \alpha_* = (\beta\alpha)_* = 0_* = 0$ shows that it suffices to prove that every $\eta \in \mathrm{Ker}\ \beta_*$ belongs to $\mathrm{Im}\ \alpha_*$. By (2.1), from $0 = \beta_* \eta = \beta\eta$ we infer the existence of a $\chi \in \mathrm{Hom}(G, A)$ such that $\alpha\chi = \eta$, $\eta \in \mathrm{Im}\ \alpha_*$. This establishes the exactness of (3).□

It is a natural thing to ask for conditions on $G$ as to when (2) and (3) can be completed with $\to 0$ at the end to obtain longer exact sequences. It is easy to give a full answer:

**Proposition 44.5.** *Given* $G$, *for every exact sequence* (1), *the sequence* (2) *[resp.* (3)*] with* $\to 0$ *at the end is exact if, and only if,* $G$ *is divisible [free].*

Clearly, $\alpha^*$ is an epimorphism exactly if for every $\xi : A \to G$ there exists an $\eta : B \to G$ such that $\eta\alpha = \xi$; this is just the injectivity of $G$. The assertion on (3) follows dually.□

For our next proposition, we need to say a few words about the effects of endomorphisms of $A$ and $C$ on the group $\mathrm{Hom}(A, C)$. If $\alpha : A \to A$ and $\gamma : C \to C$ are endomorphisms, then the induced maps $\alpha^* : \eta \mapsto \eta\alpha$ and $\gamma_* : \eta \mapsto \gamma\eta$ are evidently endomorphisms of $\mathrm{Hom}(A, C)$. These induced endomorphisms satisfy

$$\alpha^*\gamma_* = \gamma_*\alpha^*$$

as is obvious from the associative law $(\gamma\eta)\alpha = \gamma(\eta\alpha)$. Consequently, the endomorphisms of $A$ and $C$ induce permutable endomorphisms of $\mathrm{Hom}(A, C)$. Of particular interest is the case when they are multiplications by integers:

**Proposition 44.6.** *If $\alpha$ and $\gamma$ are multiplications by the integer $n$, then both $\alpha^*$ and $\gamma_*$ are multiplications by the same integer $n$.*

If $\alpha : a \mapsto na$ $(a \in A)$, then, for every $\eta : A \to C$, $(\alpha^*\eta)a = \eta\alpha a = \eta na = (n\eta)a$ shows $\alpha^*\eta = n\eta$. Similar argument applies for $\gamma_*$.☐

In conclusion, we prove a version of (44.4) for pure-exact sequences.

**Proposition 44.7** (Fuchs [9]). *If the sequence* (1) *is pure-exact, then so are* (2) *and* (3).

To verify the $p$-purity of $\mathrm{Im}\,\beta^*$ in $\mathrm{Hom}(B, G)$, let $\eta \in \mathrm{Hom}(B, G)$ and $\chi \in \mathrm{Hom}(C, G)$ satisfy $p^n\eta = \chi\beta$. Now $\mathrm{Im}\,\alpha \le \mathrm{Ker}\,\chi\beta = \mathrm{Ker}\,p^n\eta$ implies $\mathrm{Im}\,p^n\alpha \le \mathrm{Ker}\,\eta$. By hypothesis and (27.10) there is a direct decomposition

$$B/p^n(\alpha A) = \alpha A/p^n(\alpha A) \oplus B'/p^n(\alpha A)$$

for some $B' \le B$. Denoting by $\pi$ the projection onto the second summand, put $\phi b = \eta'\pi\,(b + p^n\alpha A)$ [with $\eta'(b + p^n\alpha A) = \eta b$] to obtain a homomorphism $\phi : B \to G$ satisfying $p^n\phi = p^n\eta$. In view of $\alpha A \le \mathrm{Ker}\,\phi$, there is a $\theta : C \to G$ with $\phi = \theta\beta$, and $p^n(\theta\beta) = p^n\phi = p^n\eta = \chi\beta$ establishes the $p$-pure exactness of (2).

Turning to (3), let $p^n\eta = \alpha\chi$ with $\eta \in \mathrm{Hom}(G, B)$ and $\chi \in \mathrm{Hom}(G, A)$. Thus $p^n\eta$ maps $G$ into $\alpha A$ and $\eta$ maps $G$ into $p^{-n}\alpha A$. Owing to (28.4), $\alpha A$ is a direct summand of $p^{-n}\alpha A$,

$$p^{-n}\alpha A = \alpha A \oplus B'$$

for some $B' \le B$. If $\pi$ is the projection onto the first summand, then $\phi = \alpha^{-1}\pi\eta \in \mathrm{Hom}(G, A)$ satisfies $p^n\alpha\phi = \alpha\chi$, because $p^n B' = 0$. Hence (3) is $p$-pure-exact.☐

The exact analog of (44.5) is our next result.

**Proposition 44.8.** *For a fixed $G$, the sequence* (2) [*the sequence* (3)] *stays exact with $\to 0$ at the end for every pure-exact sequence* (1) *if and only if $G$ is algebraically compact* [*direct sum of cyclic groups*].

As in the proof of (44.5) one concludes that the stated condition is equivalent to pure-injectivity [pure-projectivity]. $\square$

EXERCISES

1. Prove End $J_p \cong J_p$ *via* the isomorphism

$$\text{End } J_p \cong \varprojlim_n \text{Hom}(J_p, Z(p^n)).$$

2. For a cardinal $\mathfrak{m}$, find the structures of $\text{Hom}(Q/Z, \oplus_\mathfrak{m} Q/Z)$ and $\text{Hom}(Q/Z, \prod_\mathfrak{m} Q/Z)$.
3. Describe the structures of the groups $\text{Hom}(\oplus_\mathfrak{m} Q, \oplus_\mathfrak{m} Q)$ and $\text{Hom}(\oplus_\mathfrak{m} J_p, \oplus_\mathfrak{m} J_p)$ for any cardinal $\mathfrak{m}$.
4. Find $\text{Hom}(A, J_p)$ for arbitrary $A$.
5. If either $A$ or $C$ is a $p$-group, then $\text{Hom}(A, C)$ is a $p$-adic module.
6. (a) If $\alpha$ [$\gamma$] is an automorphism of $A$ [$C$], then it induces an automorphism of $\text{Hom}(A, C)$.
   (b) Derive **43** (G) and (D) from (a).
7. Let $\alpha : A \to B$ be monic such that, for any $G$, the induced map $\alpha^* : \text{Hom}(B, G) \to \text{Hom}(A, G)$ is epic. Show that $\alpha A$ is a direct summand of $B$.
8. Let $\beta : B \to C$ be an epimorphism. If for any $G$, the induced map $\beta_* : \text{Hom}(G, B) \to \text{Hom}(G, C)$ is epic, then the sequence $0 \to \text{Ker } \beta \to B \to C \to 0$ splits.
9. If (1) is an exact sequence such that (2) [or (3)] is pure-exact whatever $G$ is, then (1) is pure-exact. [*Hint*: choose $G = B/p^n \alpha A$ or $G = Z$; in the first case $\text{Hom}(C, *)$ is pure in $\text{Hom}(B/p^n \alpha A, *)$ with bounded quotient.]
10. (1) is pure-exact if, and only if, for every $G = Z(m)$, $\alpha^*$ in (2) [$\beta_*$ in (3)] is epic.
11. Let $0 \to A \to B \to C \to 0$ be an exact sequence. The induced sequence

$$0 \to \text{Hom}(T, A) \to \text{Hom}(T, B) \to \text{Hom}(T, C) \to 0$$

is exact for all torsion $T$ if, and only if, the sequence of torsion parts

$$0 \to T(A) \to T(B) \to T(C) \to 0$$

is splitting exact. [*Hint*: choose $T = T(C)$.]

## 45.* CERTAIN SUBGROUPS OF HOM

In various investigations, some improvements of (44.4) have turned out to be useful in handling certain situations. As in (44.7), if one assumes more than the mere exactness of the sequence (1), then under some circumstances more can be said about the induced sequences (2), (3) in **44**. Our intention is to present a rather general method of deriving statements finer than (44.4). Our discussion follows the author's paper (Fuchs [12]).

Our setting is the category $\mathscr{A}$ of abelian groups together with one of the following two sorts of categories derived from $\mathscr{A}$.

First, we select for each object $A \in \mathscr{A}$ an ideal $\mathbf{I}_A$ of the lattice $\mathbf{L}(A)$ of all subgroups of $A$ [it is *not* assumed that this selection is invariant under isomorphisms]. Define the *category* $(\mathscr{A}, \mathbf{I})$ to consist of the *objects* $(A, \mathbf{I}_A)$ for all $A \in \mathscr{A}$ and of the *morphisms*

(1) $$\alpha : (A, \mathbf{I}_A) \to (B, \mathbf{I}_B)$$

where $\alpha : A \to B$ is a group homomorphism such that

$$\alpha \mathbf{I}_A = \{\alpha A' \mid A' \in \mathbf{I}_A\} \subseteq \mathbf{I}_B \, ;$$

that is to say, subgroups in $\mathbf{I}_A$ are mapped by $\alpha$ onto subgroups in $\mathbf{I}_B$. It is readily checked that $(\mathscr{A}, \mathbf{I})$ is in fact a category.

In contrast to the usual definition in category theory, we shall mean by a *monomorphism* a map (1) which is a group monomorphism and satisfies

$$\alpha \mathbf{I}_A = \{\alpha A \cap B' \mid B' \in \mathbf{I}_B\}.$$

An *epimorphism* is a map (1) which is right-cancellable. It is readily seen that this simply means a morphism in $(\mathscr{A}, \mathbf{I})$ that is epic in the usual group theo-retic sense. Accordingly, we say that the sequence

$$0 \to (A, \mathbf{I}_A) \xrightarrow{\alpha} (B, \mathbf{I}_B) \xrightarrow{\beta} (C, \mathbf{I}_C) \to 0$$

is *exact*, or equivalently, the sequence $0 \to A \xrightarrow{\alpha} B \xrightarrow{\beta} C \to 0$ is **I**-*exact* if $\alpha$ and $\beta$ are mono- and epimorphisms, respectively, in $(\mathscr{A}, \mathbf{I})$.

We mention the following examples. In regard to the definition of mor-phisms, it is enough to refer to the objects, and these are characterized by the ideal $\mathbf{I}_A$. Now $\mathbf{I}_A$ can be chosen to consist of: 1) all subgroups of $A$; 2) finite or finitely (co)generated subgroups of $A$; 3) bounded or torsion subgroups of $A$; 4) bounded or arbitrary $p$-subgroups of $A$; 5) subgroups of cardinality less than an infinite cardinal $\mathfrak{m}$; 6) subgroups which are epimorphic images of groups in a class closed under formation of subgroups and finite direct sums; finally, 7) the objects can be chosen to consist of all pairs $(A, \mathbf{I}_A)$ with $A \in \mathscr{A}$ and $\mathbf{I}_A$ running over all ideals of $\mathbf{L}(A)$. It is readily seen that in 1)–6) **I**-exactness is equivalent to exactness.

Next we choose a dual ideal $\mathbf{D}_A$ in the lattice $\mathbf{L}(A)$, for each $A \in \mathscr{A}$, and let the *category* $(\mathscr{A}, \mathbf{D})$ consist of the *objects* $(A, \mathbf{D}_A)$ for all $A \in \mathscr{A}$ and of the *morphisms*

(2) $$\phi : (A, \mathbf{D}_A) \to (B, \mathbf{D}_B)$$

where $\phi : A \to B$ is a group homomorphism such that to every $B' \in \mathbf{D}_B$ there exists an $A' \in \mathbf{D}_A$ satisfying $\phi A' \leq B'$. In other words, if groups are regarded as being equipped with the $\mathbf{D}$-topology, then the morphisms of $(\mathscr{A}, \mathbf{D})$ are nothing else than the continuous homomorphisms. It follows immediately that $(\mathscr{A}, \mathbf{D})$ is again a category.

Now we use, dually, left-cancellability to define *monomorphisms* in $(\mathscr{A}, \mathbf{D})$; thus (2) is a monomorphism if it is one in the group-theoretical sense. $\phi$ is an *epimorphism* of $(\mathscr{A}, \mathbf{D})$ if it is a group epimorphism satisfying $\phi \mathbf{D}_A = \mathbf{D}_B$. The definition of the *exactness* of

$$0 \to (A, \mathbf{D}_A) \xrightarrow{\alpha} (B, \mathbf{D}_B) \xrightarrow{\beta} (C, \mathbf{D}_C) \to 0$$

and the $\mathbf{D}$-*exactness* of $0 \to A \xrightarrow{\alpha} B \xrightarrow{\beta} C \to 0$ is evident.

Examples for categories of type $(\mathscr{A}, \mathbf{D})$ are abundant. Each of examples 1)–6) above yields one for $(\mathscr{A}, \mathbf{D})$ if the requirement on the subgroups is replaced by the same on the corresponding quotient groups. The analog of 7) is our next example. Finally, $\mathbf{D}_A$ can be chosen to be the dual ideal of all essential subgroups of $A$. Again, in examples 1)–6), $\mathbf{D}$-exactness means nothing else than exactness, while in the last example it means neat-exactness.

Instead of entering into the discussion of the groups of morphisms in category $(\mathscr{A}, \mathbf{I})$ or $(\mathscr{A}, \mathbf{D})$, we turn our attention to more important subgroups of Hom.

Let $A \in \mathscr{A}$ and $(C, \mathbf{I}_C) \in (\mathscr{A}, \mathbf{I})$, and consider all homomorphisms $\eta : A \to C$ (in $\mathscr{A}$) such that

(3)                              $\text{Im } \eta \in \mathbf{I}_C .$

If $\eta_1 : A \to C$ with $\text{Im } \eta_1 \in \mathbf{I}_C$, then $\text{Im}(\eta_1 - \eta) \leq \text{Im } \eta_1 + \text{Im } \eta$ implies that the set of all homomorphisms with (3) is a subgroup of $\text{Hom}(A, C)$. It will be denoted by $\text{Hom}(A, C \,|\, \mathbf{I})$.

Analogously, for $(A, \mathbf{D}_A) \in (\mathscr{A}, \mathbf{D})$ and $C \in \mathscr{A}$, the set of all homomorphisms $\chi : A \to C$ (in $(\mathscr{A})$ subject to the condition

(4)                              $\text{Ker } \chi \in \mathbf{D}_A$

is a subgroup $\text{Hom}(A \,|\, \mathbf{D}, C)$ of $\text{Hom}(A, C)$. In fact, if also $\chi_1 : A \to C$ satisfies $\text{Ker } \chi_1 \in \mathbf{D}_A$, then $\text{Ker } \chi \cap \text{Ker } \chi_1 \leq \text{Ker}(\chi - \chi_1)$ implies that together with $\chi, \chi_1$ also $\chi - \chi_1$ has property (4).

Finally,

$$\text{Hom}(A \,|\, \mathbf{D}, C \,|\, \mathbf{I}) = \text{Hom}(A, C \,|\, \mathbf{I}) \cap \text{Hom}(A \,|\, \mathbf{D}, C)$$

is clearly the set of all homomorphisms $A \to C$ satisfying both (3) and (4).

The morphisms $\alpha : (B, \mathbf{D}_B) \to (A, \mathbf{D}_A)$ and $\phi : (C, \mathbf{I}_C) \to (D, \mathbf{I}_D)$ induce a homomorphism

$$\text{Hom}(\alpha, \phi) : \text{Hom}(A \,|\, \mathbf{D}, C \,|\, \mathbf{I}) \to \text{Hom}(B \,|\, \mathbf{D}, D \,|\, \mathbf{I}),$$

namely, $\eta \mapsto \phi\eta\alpha$. In fact, Im $\phi\eta\alpha \leq$ Im $\phi\eta \in \phi\mathbf{I}_C \subseteq \mathbf{I}_D$, and if $B' \in \mathbf{D}_B$ satisfies $\alpha B' \leq$ Ker $\eta \in \mathbf{D}_A$, then $B' \leq$ Ker $\eta\alpha \leq$ Ker $\phi\eta\alpha$. On account of this, we conclude that Hom($A \mid \mathbf{D}, C \mid \mathbf{I}$) *is an additive bifunctor on* $(\mathscr{A}, \mathbf{D}) \times (\mathscr{A}, \mathbf{I})$ *to* $\mathscr{A}$, *contravariant in the first and covariant in the second variable.*

In the following theorems, for the sake of simplicity, we shall write Hom($\alpha$, 1) = $\alpha^*$ and Hom(1, $\phi$) = $\phi_*$.

**Theorem 45.1** (Fuchs [12]). *If* $0 \to A \xrightarrow{\alpha} B \xrightarrow{\beta} C \to 0$ *is a* **D**-*exact sequence, then for any group G and any category* $(\mathscr{A}, \mathbf{I})$, *the induced sequence*

$$0 \to \mathrm{Hom}(C \mid \mathbf{D}, G \mid \mathbf{I}) \xrightarrow{\beta^*} \mathrm{Hom}(B \mid \mathbf{D}, G \mid \mathbf{I}) \xrightarrow{\alpha^*} \mathrm{Hom}(A \mid \mathbf{D}, G \mid \mathbf{I})$$

*is exact.*

That $\beta^*$ is monic follows directly from (44.4), and for the same reason, $\alpha^*\beta^* = 0$. Thus it suffices to verify the inclusion Ker $\alpha^* \leq$ Im $\beta^*$. Let $\eta \in$ Ker $\alpha^*$; then by (44.4), there is a $\chi \in$ Hom($C$, $G$) such that $\chi\beta = \eta$. Obviously, Im $\chi =$ Im $\eta \in \mathbf{I}_G$, and Ker $\chi \geq \beta$ Ker $\eta$ together with Ker $\eta \in \mathbf{D}_B$ implies Ker $\chi \in \mathbf{D}_C$; therefore, $\chi \in$ Hom($C \mid \mathbf{D}, G \mid \mathbf{I}$). $\square$

**Theorem 45.2** (Fuchs [12]). *If* $0 \to A \xrightarrow{\alpha} B \xrightarrow{\beta} C \to 0$ *is an* **I**-*exact sequence, then for every group G and category* $(\mathscr{A}, \mathbf{D})$ *one has the exact sequence*

$$0 \to \mathrm{Hom}(G \mid \mathbf{D}, A \mid \mathbf{I}) \xrightarrow{\alpha_*} \mathrm{Hom}(G \mid \mathbf{D}, B \mid \mathbf{I}) \xrightarrow{\beta_*} \mathrm{Hom}(G \mid \mathbf{D}, C \mid \mathbf{I}).$$

As in the proof of (45.1), we may restrict ourselves to verifying the inclusion Ker $\beta_* \leq$ Im $\alpha_*$. Let $\eta \in$ Ker $\beta_*$. By (44.4), some $\chi \in$ Hom($G$, $A$) satisfies $\alpha\chi = \eta$. Since $\alpha$ is monic, Ker $\chi =$ Ker $\eta \in \mathbf{D}_G$. Manifestly, $\alpha$ Im $\chi =$ Im $\eta \in \mathbf{I}_B$ whence Im $\chi \in \mathbf{I}_A$, and so $\chi \in$ Hom($G \mid \mathbf{D}, A \mid \mathbf{I}$). $\square$

Finally, we point out a direct limit representation for Hom($A \mid \mathbf{D}, C \mid \mathbf{I}$).

The ideal $\mathbf{I}_C$ determines a direct system $\{C_i \ (i \in J); \pi_i^j\}$ of subgroups $C_i$ of $C$ in the obvious way : $C_i \leq C_j$ means $i \leq j$, and the maps $\pi_i^j : C_i \to C_j$ are the inclusion maps. Similarly, $\mathbf{D}_A$ defines an inverse system $\{A/A_k \ (k \in K); \rho_k^l\}$ of quotient groups of $A$, where $k \leq l$ exactly if $A_k \geq A_l$ and the maps $\rho_k^l : A/A_l \to A/A_k$ are those induced by the identity map of $A$, i.e., $\rho_k^l : a + A_l \mapsto a + A_k$. It is readily checked that

$$\{\mathrm{Hom}(A/A_k, C_i) \ ((k, i) \in K \times J); \mathrm{Hom}(\rho_k^l, \pi_i^j)\}$$

is a direct system of groups. Moreover,

**Proposition 45.3** (Fuchs [12]). *There is a natural isomorphism*

$$\varinjlim \mathrm{Hom}(A/A_k, C_i) \cong \mathrm{Hom}(A \mid \mathbf{D}, C \mid \mathbf{I}).$$

Let $H$ denote the direct limit, and let $\sigma_{ki} : \mathrm{Hom}(A/A_k, C_i) \to H$ denote the canonical homomorphisms, i.e., $\sigma_{lj} \mathrm{Hom}(\rho_k^l, \pi_i^j) = \sigma_{ki}$ for $i \leq j$ and $k \leq l$.

Since $\pi_i^j$ are monic and $\rho_k^l$ are epic, (44.4) implies that $\mathrm{Hom}(\rho_k^l, \pi_i^j)$ are monic, and hence by **11** (f), $\sigma_{ki}$ are monomorphisms. For the natural homomorphisms $\rho_k : A \to A/A_k$ and the inclusion maps $\pi_i : C_i \to C$,

$$\mathrm{Hom}(\rho_k, \pi_i) : \mathrm{Hom}(A/A_k, C_i) \to \mathrm{Hom}(A \,|\, \mathbf{D}, C \,|\, \mathbf{I})$$

satisfy

$$\mathrm{Hom}(\rho_l, \pi_j)\, \mathrm{Hom}(\rho_k^l, \pi_i^j) = \mathrm{Hom}(\rho_k, \pi_i) \qquad (i \leq j; k \leq l),$$

therefore by (11.1) there exists a unique $\sigma : H \to \mathrm{Hom}(A \,|\, \mathbf{D}, C \,|\, \mathbf{I})$ such that

$$(5) \hspace{5cm} \mathrm{Hom}(\rho_k, \pi_i) = \sigma \sigma_{ik}.$$

Again by (44.4), the left side map is monic. The same must hold for $\sigma$, for if $h \in H$ belongs to Ker $\sigma$, then $h = \sigma_{ki}\eta_{ki}$ [with some $\eta_{ki} \in \mathrm{Hom}(A/A_k, C_i)$] satisfies $\mathrm{Hom}(\rho_k, \pi_i)\, \eta_{ki} = \sigma\sigma_{ki}\eta_{ki} = \sigma h = 0$, thus $\eta_{ki} = 0$ and $h = 0$. To prove $\sigma$ epic, take into account that the groups $\mathrm{Hom}(\rho_k, \pi_i)$ together exhaust $\mathrm{Hom}(A \,|\, \mathbf{D}, C \,|\, \mathbf{I})$ whence (5) guarantees that $\sigma$ is an epimorphism, and consequently, an isomorphism. It is clear from the definition that it is natural. $\square$

EXERCISES

1. Check what was stated for **I**- and **D**-exactness of sequences in the mentioned examples.
2. (Fuchs [12]) For a category $(\mathscr{A}, \mathbf{I})$, let $\mathrm{Hom}_{\mathbf{I}}(A, C)$ be the set of all morphisms $(A, \mathbf{I}_A) \to (C, \mathbf{I}_C)$. Prove that:
   (a) $\mathrm{Hom}_{\mathbf{I}}(A, C)$ is a subgroup of $\mathrm{Hom}(A, C)$;
   (b) it is a left-exact functor in both variables.
3. (Fuchs [12]) For a category $(\mathscr{A}, \mathbf{D})$, let $\mathrm{Hom}_{\mathbf{D}}(A, C)$ mean the set of all morphisms $(A, \mathbf{D}_A) \to (C, \mathbf{D}_C)$. Verify the analogs of (a), (b) in Ex. 2.
4. Let $(\mathscr{A}, \mathbf{I})$ be defined such that $\mathbf{I}_A$ consists of all subgroups of $A$ which belong to a class $\mathfrak{X}$ of groups [closed under taking direct sums, subgroups and epimorphisms], and assume $(\mathscr{A}, \mathbf{D})$ is the category with $\mathbf{D}_A$ consisting of all $A' \leq A$ with $A/A' \in \mathfrak{X}$. Show that $\mathrm{Hom}(A \,|\, \mathbf{D}, C) = \mathrm{Hom}(A, C \,|\, \mathbf{I})$.
5. Give examples for $(\mathscr{A}, \mathbf{I})$ and $(\mathscr{A}, \mathbf{D})$ where $\mathrm{Hom}(A \,|\, \mathbf{D}, C \,|\, \mathbf{I})$ commutes with infinite direct sums in both variables. [*Hint*: subgroups of finite order and of finite index, respectively.]
6. Let $(\mathscr{A}, \mathbf{I})$ be the category with $\mathbf{I}_A$ the bounded subgroups of $A$. If $B$ is a basic subgroup of a $p$-group $A$, then:
   (a) $\mathrm{Hom}(A, C \,|\, \mathbf{I}) \cong \mathrm{Hom}(B, C \,|\, \mathbf{I})$;
   (b) $\mathrm{Hom}(B, C \,|\, \mathbf{I})$ is the torsion part of $\mathrm{Hom}(B, C)$;
   (c) if $A = \oplus A_i$, then $\mathrm{Hom}(A, C \,|\, \mathbf{I})$ is the torsion part of the direct product of the $\mathrm{Hom}(A_i, C \,|\, \mathbf{I})$.

7. Let $(\mathscr{A}, \mathbf{I})$ be the category with $\mathbf{I}_A$ the finite subgroups of $A$. Let $A$ be a torsion or a complete group, and $A_p$ the $p$-component of $A$. Prove the isomorphism

$$\operatorname{Hom}(A, C \,|\, \mathbf{I}) \cong \bigoplus_p \operatorname{Hom}(A_p, C).$$

8*. (Pierce [1]) Let $\operatorname{Hom}_s(A, C)$ denote the group of all small homomorphisms of the $p$-group $A$ into $C$ [for definition, see next section]. Prove that the functor $\operatorname{Hom}_s(*, C)$ carries a pure-exact sequence into an exact sequence.

## 46. HOMOMORPHISM GROUPS OF TORSION GROUPS

Having considered some elementary properties of Hom and the exact sequences for Hom, we may turn our attention to find the structure of $\operatorname{Hom}(A, C)$ in certain cases. Evidently, $\operatorname{Hom}(A, C)$ depends on the given groups $A$ and $C$, and $\operatorname{Hom}(Z, C) \cong C$ shows that every group $C$ occurs as Hom. It is, however, a remarkable fact that, if $A$ is supposed to be a torsion group, then $\operatorname{Hom}(A, C)$ must be algebraically compact, and hence its characterization in terms of invariants of $A$ and $C$ can be hoped for. Because of (43.4), only $p$-groups are to be dealt with.

**Theorem 46.1** (Fuchs [9], Harrison [2]). *If $A$ is a torsion group, then* $\operatorname{Hom}(A, C)$ *is a reduced algebraically compact group, for any $C$.*

*First proof.* We prove that if $A$ is a $p$-group, then $\operatorname{Hom}(A, C)$ is complete in its $p$-adic topology [cf. (40.1)]. To prove $H = \operatorname{Hom}(A, C)$ is Hausdorff, let $\eta \in H$ be divisible by every power of $p$. Given $a \in A$, say of order $p^k$, let $\chi \in H$ satisfy $p^k \chi = \eta$. Then $\eta a = p^k \chi a = \chi p^k a = 0$ implies $\eta = 0$. Next let $\eta_1, \cdots, \eta_n, \cdots$ be a Cauchy sequence in $H$; dropping to a subsequence if necessary, we may assume it neat: $\eta_{n+1} - \eta_n \in p^n H$ for every $n$, that is, $\eta_{n+1} - \eta_n = p^n \chi_n$ for some $\chi_n \in H$. Define

$$\eta = \eta_1 + (\eta_2 - \eta_1) + \cdots + (\eta_{n+1} - \eta_n) + \cdots;$$

this is a homomorphism $A \to C$, since for $a \in A$ of order $p^k$, $(\eta_{n+1} - \eta_n)a = 0$ for all $n \geq k$, showing that $\eta a = \eta_1 a + (\eta_2 - \eta_1)a + \cdots + (\eta_k - \eta_{k-1})a$ is well defined. Furthermore,

$$\eta - \eta_n = (\eta_{n+1} - \eta_n) + (\eta_{n+2} - \eta_{n+1}) + \cdots = p^n(\chi_n + p\chi_{n+1} + \cdots)$$

where, again, $\chi_n + p\chi_{n+1} + \cdots$ belongs to $H$, i.e., $\eta - \eta_n \in p^n H$ and $\eta$ is the limit of the given Cauchy sequence. Consequently, $H$ is complete.

*Second proof.* $A$, as a torsion group, is the direct limit of its finite subgroups $A_i$. By (44.2), $\operatorname{Hom}(A, C)$ is then the inverse limit of the groups

$\text{Hom}(A_i, C)$ which are bounded in view of (43.3). Hence, $\text{Hom}(A, C)$ is the inverse limit of reduced algebraically compact groups, therefore the assertion follows from (39.4).□

In order to find out the exact structure of $\text{Hom}(A, C)$ for torsion $A$, we may restrict ourselves, without loss of generality, to the case when both $A$ and $C$ are $p$-groups. Let

(1)                    $$B = \bigoplus_{n=1}^{\infty} B_n \qquad \text{with} \qquad B_n = \bigoplus_{\mathfrak{m}_n} Z(p^n)$$

be a basic subgroup of $A$, and assume $C$ reduced. The exact sequence $0 \to B \xrightarrow{\alpha} A \to A/B \to 0$ induces the exact sequence

$$0 = \text{Hom}(A/B, C) \to \text{Hom}(A, C) \xrightarrow{\alpha^*} \text{Hom}(B, C)$$

showing that $\text{Hom}(A, C)$ can be viewed as a subgroup of $\text{Hom}(B, C)$. The corresponding quotient group is torsion-free, because if $p^k\eta = \chi\alpha$ with $\chi \in \text{Hom}(A, C), \eta \in \text{Hom}(B, C)$, then define $\theta : A \to C$ as $\theta a = \chi g + \eta b$ if $a \in A$ is of the form $a = p^k g + \alpha b$ $(g \in A, b \in B)$. It is readily seen that $\theta$ is actually a homomorphism such that $\theta\alpha = \eta$, establishing the torsion-freeness of $\text{Hom}(B, C)/\alpha^* \text{Hom}(A, C)$. Because of (46.1), $\text{Hom}(A, C)$ and $\text{Hom}(B, C)$ are algebraically compact, so we have

$$\text{Hom}(B, C) \cong \text{Hom}(A, C) \oplus X,$$

where $X$ must be a $p$-adic algebraically compact torsion-free group.
From (43.3) we infer

$$\text{Hom}(B, C) = \prod_{n=1}^{\infty} \prod_{\mathfrak{m}_n} C[p^n].$$

The groups $C[p^n]$ are direct sums of cyclic groups of orders $\leq p^n$, and the same holds for $\prod_{\mathfrak{m}_n} C[p^n]$. The invariants of these groups can be evaluated by means of cardinal invariants of $C$, thus (33.3) can be applied to describe the basic subgroup $V$ of the torsion part of $\text{Hom}(B, C)$. Therefore, $\text{Hom}(A, C)$ is the direct sum of the $p$-adic completion $\hat{V}$ of $V$ and a $p$-adic algebraically compact torsion-free group which will be known as soon as its basic subgroup $W$ is determined.

In the proof of the key lemma (46.3), the following set-theoretic lemma will be made use of.

**Lemma 46.2** (Pierce [1]). *Let $I$ be a set of infinite cardinality* $\mathfrak{m}$, *and* $\mathfrak{n}$ *a cardinal number such that* $0 < \mathfrak{n} \leq \mathfrak{m}$. *Then there exists a set* $\{I_j\}_{j \in J}$ *of subsets of $I$ such that*

   (i) $|I_j| = \mathfrak{n}$ *for every* $j$;

(ii) $|J| = \mathfrak{m}^{\mathfrak{n}}$;

(iii) *the $I_j$ are independent in the sense that, if $I_0, I_1, \cdots, I_n$ are distinct elements of the set, then $I_0$ is not contained in the union $I_1 \cup \cdots \cup I_n$.*

The following simple proof is due to P. Erdös and A. Hajnal.

Let $K$ be a set of cardinality $\mathfrak{m}$; we can decompose it into $\mathfrak{n}$ disjoint subsets $L_l$, each of cardinality $\mathfrak{m}$. Clearly, if $\{K_j\}_{j \in J}$ is the family of all subsets of $K$ which contain exactly one element from each $L_l$, then this family satisfies (i) and (ii), while (iii) is replaced by the condition that no $K_j$ contains another one. Now define $I_j$ as the set of all finite subsets, of $K_j$. Then the set $\{I_j\}_{j \in J}$ satisfies not only (i) and (ii), but (iii) too, since if $c_i \in I_0 \setminus I_i$ for $i = 1, \cdots, n$, then $\{c_1, \cdots, c_n\} \in I_0$ but $\notin I_1 \cup \cdots \cup I_n$. The proof will be completed by identifying the elements of the given set $I$ with the finite subsets of $K$, under a one-to-one correspondence.□

Now we are ready to prove:

**Lemma 46.3** (Pierce [1]). *If $A$, $C$ are reduced $p$-groups whose basic subgroups $B$, $D$ are of infinite final ranks $\mathfrak{m}$, $\mathfrak{n}$, then the basic subgroup $W$ of $X$ is of rank $\mathfrak{n}^{\mathfrak{m}}$.*

An inference like the one at the beginning of the proof of (36.1) shows that bounded direct summands can be separated from $A$ and $C$ to achieve $r(B) = \mathfrak{m}$ and $r(D) = \mathfrak{n}$. Bounded direct summands have no influence on $W$, so the assumption $r(B) = \mathfrak{m}$, $r(D) = \mathfrak{n}$ means no loss of generality.

From (34.4) we get the estimation $|C| \leq |D|^{\aleph_0}$ whence

$$|W| \leq |\mathrm{Hom}(A, C)| \leq |\mathrm{Hom}(B, C)| \leq |C|^{|B|} \leq |D|^{\aleph_0 |B|} = \mathfrak{n}^{\mathfrak{m}}$$

Consequently, it suffices to prove the existence of a $p$-independent set $S$ of elements of infinite order in $\mathrm{Hom}(A, C)$ such that $|S| = \mathfrak{n}^{\mathfrak{m}}$.

The main idea of the proof is to concentrate on *small* homomorphisms $\phi : A \to C$ which we define by the following condition:

(*) for every $k \geq 0$ there exists an $n_\phi(k)$ such that $e(a) \geq n_\phi(k)$ implies $e(\phi a) \leq e(a) - k$.

These have the advantage that they are completely determined by their restrictions to $B$ and can be chosen arbitrarily on $B$. In fact, if $\phi$ is a small homomorphism of $A$ into $C$, and if $a \in A$ is of order $p^k$, then choose $n = n_\phi(k)$ according to (*) and write $a = p^n g + b$ with $g \in A$, $b \in B$ such that $e(p^n g) \leq e(a)$ [which can be done, $B$ being pure in $A$]. Then $e(\phi g) \leq e(g) - k$ implies $p^n \phi g = 0$, and so $\phi a = \phi b$. Furthermore, if $\phi'$ is a small homomorphism of $B$ into $C$, then putting $\phi a = \phi' b$ we obtain a small homomorphism $\phi : A \to C$ [with $n_\phi(k) = n_{\phi'}(k)$]. Therefore, small homomorphisms of $A$ and $B$ are essentially the same.

It is routine to verify that the small homomorphisms of $A$ into $C$ form a subgroup $H$ of $\operatorname{Hom}(A, C)$. The quotient group $\operatorname{Hom}(A, C)/H$ is torsion-free, for if $\eta \in \operatorname{Hom}(A, C)$ satisfies $p^r\eta \in H$, then together with $p^r\eta$, $\eta$ also satisfies (*) with $n_\eta(k) = n_{p^r\eta}(k + r)$. Hence $H$ is pure in $\operatorname{Hom}(A, C)$, and it is therefore enough to find an $S$ of the desired nature in the subgroup $H$.

We distinguish two cases.

*Case I*: $\mathfrak{n} \leqq \mathfrak{m}$. Choose a direct summand

$$D_0 = \bigoplus_{n=1}^{\infty} \langle d_n \rangle$$

of $D$ such that $0 < e(d_1) < \cdots < e(d_n) < \cdots$, and decompose $B$ in the following way:

$$(2) \qquad\qquad B = \bigoplus_{n=0}^{\infty} G_n \qquad \text{where} \quad G_n = \bigoplus_{i \in I_n} \langle b_{ni} \rangle$$

and fin $r(G_n) = |I_n| = \mathfrak{m}$, $e(b_{ni}) \geqq 2e(d_n) \geqq 2n$ for $n \geqq 1$; the index sets $I_n$ ($n = 0, 1, 2, \cdots$) are of course disjoint. In $I_n$ we select subsets $I_{nj}$ ($j \in J$) such that

$$(3) \qquad |I_{nj}| = \mathfrak{m}, \qquad |J| = \mathfrak{n}^{\mathfrak{m}} = \mathfrak{m}^{\mathfrak{m}}, \qquad \text{and} \qquad I_{nj} \text{ are independent};$$

this can be done by (46.2). There is no loss of generality in choosing the same $J$ for every $n$. Let us define homomorphisms $\phi_j : B \to D_0$ as follows:

$$\phi_j b_{ni} = \begin{cases} d_n & \text{if } i \in I_{nj}, \quad n \geqq 1, \\ 0 & \text{otherwise.} \end{cases}$$

These $\phi_j$ are small homomorphisms, for $e(d_n) \leqq \frac{1}{2}e(b_{ni})$, i.e., $e(b_{ni}) \geqq 2k$ implies $e(\phi_j b_{ni}) \leqq e(b_{ni}) - k$, and hence it is easy to conclude that (*) holds with $n_\phi(k) = 2k$. Every $\phi_j$ is of infinite order, because $d_n \in \operatorname{Im} \phi_j$ for all $n$. The $\phi_j$ are $p$-independent, for if

$$\phi = m_0 \phi_0 + m_1 \phi_1 + \cdots + m_s \phi_s \in p^r H \qquad (m_t \neq 0),$$

then by (3), there is an index $i_0 \in I_{n0}$ that does not belong to the union $I_{n1} \cup \cdots \cup I_{ns}$. For this $i_0$, $\phi_0 b_{ni_0} = d_n$, while $\phi_1 b_{ni_0} = \cdots = \phi_s b_{ni_0} = 0$, thus $\phi b_{ni_0} = m_0 d_n$. This element must belong to $p^r C$, whence the purity of $D_0$ implies $p^r \mid m_0$. Analogously, $p^r \mid m_1, \cdots, m_s$, proving that the set $S = \{\phi_j\}_{j \in J}$ is $p$-independent, of cardinality $\mathfrak{n}^{\mathfrak{m}}$.

*Case II*: $\mathfrak{n} > \mathfrak{m}$. This time we decompose $D$ as follows:

$$(4) \qquad\qquad D = \bigoplus_{n=1}^{\infty} G_n' \qquad \text{where} \quad G_n' = \bigoplus_{k \in K_n} \langle d_{nk} \rangle,$$

where $|K_n| = \mathfrak{n}$ and $e(d_{nk}) \geqq n$, and choose subsets $K_{nl}$ ($l \in L$) in $K_n$ such that

$$(5) \qquad |K_{nl}| = \mathfrak{m}, \qquad |L| = \mathfrak{n}^{\mathfrak{m}}, \qquad \text{and} \qquad K_{nl} \text{ are independent}.$$

Let $f_{nl}$ be a mapping of $K_{nl}$ into $I_n$ of (2), such that each $i \in I_n$ is the image of at most one $k \in K_{nl}$, and if $f_{nl}k = i$, then $2e(d_{nk}) \leq e(b_{ni})$. It is clear that such an $f_{nl}$ does exist for all $n$, $l$. For every $l \in L$ we define a homomorphism $\psi_l$: $B \to D$ by putting

$$\psi_l b_{ni} = \begin{cases} d_{nk} & \text{if } i = f_{nl}k, \\ 0 & \text{otherwise.} \end{cases}$$

As above, it follows that $\psi_l$ is a small homomorphism of infinite order. In order to verify the $p$-independence of the set $S = \{\psi_l\}_{l \in L}$, let

$$\psi = m_0 \psi_0 + m_1 \psi_1 + \cdots + m_s \psi_s \in p^r H \qquad (m_t \neq 0).$$

Some $k_0 \in K_{n0}$ exists such that $k_0 \notin K_{n1} \cup \cdots \cup K_{ns}$. For $i_0 = f_{n0} k_0$ we have $\psi_0 b_{ni_0} = d_{nk_0}$, while $\psi_1 b_{ni_0} = \cdots = \psi_s b_{ni_0} = 0$. From $\psi b_{ni_0} = m_0 d_{nk_0}$ we infer as in *Case I* that $p^r \mid m_0$, and analogously, $p^r \mid m_1, \cdots, m_s$. This establishes the $p$-independence of $S$.☐

Now we are in a position to prove the main result. We shall use the following notations. Let (1) denote a basic subgroup of $A$ and

$$(6) \qquad D = \bigoplus_{n=1}^{\infty} D_n \quad \text{with} \quad D_n = \bigoplus_{\mathfrak{n}_n} Z(p^n)$$

a basic subgroup of $C$. Let us put

$$(7) \qquad \mathfrak{m} = \operatorname{fin} r(B), \qquad \mathfrak{n} = \operatorname{fin} r(D), \qquad \mathfrak{p} = \operatorname{fin} r(C).$$

We define, for cardinal numbers $\mathfrak{u}$, $\mathfrak{v}$,

$$d(\mathfrak{u}, \mathfrak{v}) = \begin{cases} \mathfrak{uv} & \text{if } \mathfrak{u} \text{ is finite,} \\ (2\mathfrak{v})^{\mathfrak{u}} & \text{if } \mathfrak{u} \text{ is infinite,} \end{cases}$$

and notice that

$$\left[ \bigoplus_{\mathfrak{v}} Z(p^n) \right]^{\mathfrak{u}} = \bigoplus_{d(\mathfrak{u},\mathfrak{v})} Z(p^n).$$

**Theorem 46.4**  (Pierce [1]). *Let $A$ and $C$ be reduced $p$-groups. Then $\operatorname{Hom}(A, C)$ is a $p$-adic algebraically compact group whose basic submodule is*

$$\bigoplus_{n=1}^{\infty} \left[ \bigoplus_{\mathfrak{r}_n} Z(p^n) \right] \oplus \left[ \bigoplus_{\mathfrak{r}_0} J_p \right]$$

*where for $n \geq 1$*

$$\mathfrak{r}_n = d\left( \mathfrak{m}_n, \mathfrak{p} + \sum_{k=n}^{\infty} \mathfrak{n}_k \right) + d\left( \sum_{k=n+1}^{\infty} \mathfrak{m}_k, \mathfrak{n}_n \right)$$

*and*

$$\mathfrak{r}_0 = d\, (\mathfrak{m}, \mathfrak{n}).$$

It is clear that $C[p^n] = D_1 \oplus \cdots \oplus D_{n-1} \oplus E_n$, where $E_n$ is a direct sum of $p + \sum_{k=n}^{\infty} \mathfrak{n}_k$ copies of $Z(p^n)$. Since $\mathfrak{r}_n$ is the cardinal number of the set of direct summands $Z(p^n)$ in a basic subgroup of the torsion part of the group $\prod_{n=1}^{\infty} \prod_{\mathfrak{m}_n} C[p^n]$, as is shown by (33.3), we infer

$$\bigoplus_{\mathfrak{r}_n} Z(p^n) \cong E_n^{\mathfrak{m}_n} \oplus \prod_{k=n+1}^{\infty} D_n^{\mathfrak{m}_n} = E_n^{\mathfrak{m}_n} \oplus D_n^{\mathfrak{p}_n}$$

with $\mathfrak{p}_n = \sum_{k=n+1}^{\infty} \mathfrak{m}_n$. This proves the formula for $\mathfrak{r}_n$ $(n \geq 1)$, while that for $\mathfrak{r}_0$ is the content of (46.3), where it is trivial for vanishing $\mathfrak{m}$ or $\mathfrak{n}$. □

In view of this theorem, the algebraic structure of $\mathrm{Hom}(A, C)$ can completely be described if $A$ is a torsion group and $C$ is reduced. If $C$ happens to contain subgroups $Z(p^\infty)$, then (46.4) and **47**, Exs. 8–10 together yield the structure of $\mathrm{Hom}(A, C)$.

EXERCISES

1. (Pierce [1]) Show that condition (*) is equivalent to the following one: for every $k \geq 0$ there exists an $n$ such that $e(a) \leq k$ and $h(a) \geq n$ imply $\phi a = 0$.
2. (a) If $\phi$ is a small homomorphism of $A$, then

$$B + \mathrm{Ker}\ \phi = A$$

   for any basic subgroup $B$ of $A$.
   (b) $A^1 \leq \mathrm{Ker}\ \phi$.
3. Let $A = B \oplus D$ with $B$ bounded and $D$ divisible. Describe the small homomorphisms of $A$ into any group.
4. (Pierce [1]) (a) In each of the following cases there is a homomorphism of $A$ into $C$ which is not small: (i) $A$ is unbounded and $C$ is not reduced; (ii) $A$ is unbounded and $C$ contains a subgroup $\cong A$; (iii) $A$ has an unbounded basic subgroup and is countable, while $C$ is not bounded.
   (b) All homomorphisms of $A$ into $C$ are small if: (i) $A$ has bounded basic subgroup and $C$ is reduced; (ii) $C$ is bounded.
5. (Pierce [1]) Let $B = \bigoplus_{i \in I} \langle a_i \rangle$ be a basic subgroup of $A$, and let $c_i \in C$ $(i \in I)$ satisfy: (i) $e(c_i) \leq e(a_i)$; (ii) for any $k \geq 0$ there is an $n$ such that $e(c_i) \geq n$ implies $e(c_i) \leq e(a_i) - k$. Then there is a unique small homomorphism $\phi$ of $A$ into $C$ such that $\phi(a_i) = c_i$ $(i \in I)$.
6. (Pierce [1]) Let $G$ be a pure subgroup of the $p$-group $A$. Every small homomorphism of $G$ into $C$ can be extended to one of $A$ into $C$. [*Hint:* Ex. 5.]
7. (Pierce [1]) (a) $\phi$ is a small homomorphism if $p^m \phi$ is small for some $m$.
   (b) A small homomorphism of $p^m A$ into $p^m C$ can be extended to a small homomorphism of $A$ into $C$. [*Hint:* Ex. 5.]

8. Let $A$, $C$ be as in (46.4). Determine the final ranks of $\mathrm{Hom}(A, C)$ and its basic subgroup.
9. Let $C$ be a reduced $p$-adic module, and let $A$ have torsion $p$-basic subgroup. Then $\mathrm{Hom}(A, C)$ is algebraically compact.

## 47. CHARACTER GROUPS

In the preceding section we examined $\mathrm{Hom}(A, C)$ for torsion groups $A$ and reduced groups $C$. Our next aim is to investigate the case when $A$ is arbitrary and $C$ is the additive group $K$ of the real numbers mod 1, in which case $\mathrm{Hom}(A, K)$—equipped with a suitable topology—is known as the *character group* Char $A$ of $A$. In this section, we concentrate on the algebraic structure of Char $A$; the fact that it carries a topology will be irrelevant in most of our results in **47**. Our results will give complete information about the structure of character groups.

Algebraically, $K$ is nothing else than the direct product of quasicyclic groups, one for each prime $p$; hence

$$\mathrm{Char}\ A \cong \prod_{p} \mathrm{Hom}(A, Z(p^{\infty})).$$

Consequently, it suffices to deal with $\mathrm{Hom}(A, Z(p^{\infty}))$.

Let

$$B = \bigoplus_{n=0}^{\infty} B_n \quad \text{where} \quad B_0 = \bigoplus_{\mathfrak{m}_0} Z, \ B_n = \bigoplus_{\mathfrak{m}_n} Z(p^n) \quad \text{for} \quad n \geq 1,$$

be a $p$-basic subgroup of $A$. The $p$-component of $A/B$ is of the form $\bigoplus_{\mathfrak{m}} Z(p^{\infty})$; the fact that $\mathfrak{m}$ is not an invariant is inessential, but it can be made unique by choosing a lower basic subgroup in the $p$-component of $A$. Finally, let $\mathfrak{f}$ denote the torsion-free rank of $A/B$. [These cardinals will be denoted by $\mathfrak{m}_0(p)$, $\cdots$, $\mathfrak{f}(p)$, if their dependence on $p$ cannot be suppressed.] The following theorem gives complete information about the homomorphism groups $\mathrm{Hom}(A, Z(p^{\infty}))$.

**Theorem 47.1** (Fuchs [8]). *For any group $A$,*

$$(1) \qquad \mathrm{Hom}(A, Z(p^{\infty})) \cong \prod_{\mathfrak{m}_0} Z(p^{\infty}) \oplus \prod_{n=1}^{\infty} \prod_{\mathfrak{m}_n} Z(p^n) \oplus \prod_{\mathfrak{m}} J_p \oplus \prod_{\aleph_0} Q.$$

Owing to (44.5) and (44.7), the $p$-pure exact sequence $0 \to B \to A \to A/B \to 0$ implies the $p$-pure exactness of the induced sequence $0 \to \mathrm{Hom}(A/B, Z(p^{\infty})) \to \mathrm{Hom}(A, Z(p^{\infty})) \to \mathrm{Hom}(B, Z(p^{\infty})) \to 0$. By virtue of (43.1)

$$\mathrm{Hom}(B, Z(p^{\infty})) = \prod_{n=0}^{\infty} \mathrm{Hom}(B_n, Z(p^{\infty})) \cong \prod_{\mathfrak{m}_0} Z(p^{\infty}) \oplus \prod_{n=1}^{\infty} \prod_{\mathfrak{m}_n} Z(p^n).$$

If we write $A/B = \bigoplus_m Z(p^\infty) \oplus G$, with $G[p] = 0$, then because of

$$\text{Hom}\left(\bigoplus_m Z(p^\infty), Z(p^\infty)\right) = \prod_m \text{Hom}(Z(p^\infty), Z(p^\infty)) = \prod_m J_p,$$

it remains to evaluate $\text{Hom}(G, Z(p^\infty))$. Let $H$ be the $p$-pure subgroup in $G$ generated by a maximal independent system of elements of infinite order in $G$; this $H$ is well defined, since in $G$ division by $p$ is unique, and evidently, $H \cong \bigoplus_{\mathfrak{f}} Q^{(p)}$ such that $G/H$ is a torsion group with 0 $p$-component. The exactness of $0 \to H \to G \to G/H \to 0$ implies the exactness of the sequence $0 = \text{Hom}(G/H, Z(p^\infty)) \to \text{Hom}(G, Z(p^\infty)) \to \text{Hom}(H, Z(p^\infty)) \to 0$, thus

$$\text{Hom}(G, Z(p^\infty)) \cong \text{Hom}(H, Z(p^\infty)) \cong \prod_{\mathfrak{f}} \text{Hom}(Q^{(p)}, Z(p^\infty)) \cong \prod_{\mathfrak{f}} \left(\bigoplus_{2^{\aleph_0}} Q\right),$$

where we have used example 4 in **43**. We see that $\text{Hom}(A/B, Z(p^\infty))$ is the direct sum of a divisible group and a $p$-adic algebraically compact group, so its $p$-purity in $\text{Hom}(A, Z(p^\infty))$ implies that it is a direct summand of $\text{Hom}(A, Z(p^\infty))$. $\square$

Notice that the group in (1) may be given a more explicit form by making use of examples in **23** and **40**. If we determine the cardinal numbers $\mathfrak{m}_0$, $\mathfrak{m}_n$, $\mathfrak{m}$, $\mathfrak{f}$ for every prime $p$, then Char $A$ can be determined as the direct product of the groups (1), with $p$ ranging over all primes $p$.

Notice that the first and fourth summands in (1) arise from elements of infinite order, while the two middle summands from elements of $p$-power orders. Hence

**Corollary 47.2.** Char $A$ *is reduced if and only if* $A$ *is a torsion group, and is divisible if and only if* $A$ *is torsion-free.* $\square$

A glance at the group (1) shows that it is algebraically compact. This yields the known fact that the character groups are algebraically compact, without making use of the deep theorem that the character groups of discrete abelian groups [with the suitable topology] are just the compact abelian groups.

It is natural to inquire about conditions under which an algebraically compact group can carry a compact topology, that is to say, is a character group. The conditions are not difficult to state: they are inequalities between cardinal numbers [cf. Ex. 5]. We prefer to formulate the relevant conditions in the simpler forms given in the two subsequent corollaries.

Clearly, the cardinal numbers $\mathfrak{m}_n$ and $\mathfrak{m}$ can be chosen arbitrarily and independently for every $p$; we thus have:

**Corollary 47.3** (Hulanicki [2], Harrison [1]). *A reduced group is the character group of some (torsion) group exactly if it is a direct product of cyclic p-adic modules* [p *need not be fixed*]. $\square$

For divisible groups, a simple inequality must be satisfied:

**Corollary 47.4** (T. Harrison [1]). *A divisible group $\neq 0$ is the character group of some (torsion-free) group if, and only if, it is of the form*

$$\prod_p \prod_{\mathfrak{r}_p} Z(p^\infty) \oplus \prod_{\mathfrak{r}} Q \qquad \text{where} \quad \mathfrak{r} \geq \aleph_0 .$$

If $A$ is torsion-free, then its rank is, in the above notation, $\mathfrak{m}_0(p) + \mathfrak{f}(p)$, showing that Char $A$ will have the indicated form with $\mathfrak{r}_p = \mathfrak{m}_0(p)$, unless $\mathfrak{f}(p) = 0$ for every $p$, in which case, however, the direct sum with $\prod_{\aleph_0} Q$ does not change the structure of the first direct product. Conversely, given a divisible group of the above form, a simple calculation shows that the group structure does not change if $\mathfrak{r}$ is replaced by $\mathfrak{r} + \sum_p \mathfrak{r}_p$, in other words, $\mathfrak{r} \geq \mathfrak{r}_p$ may be assumed. If $A$ is defined as the direct sum of $\mathfrak{r}$ rational groups $G_i$ such that exactly $\mathfrak{r}_p$ of them satisfy $pG_i \neq G_i$ and $\mathfrak{r}$ of them satisfy $pG_i = G_i$, then it results easily that Char $A$ is as desired.☐

Another consequence which is sufficiently important to deserve a separate statement is due to S. Kakutani.

**Corollary 47.5.** *The character group of a group of infinite cardinality $\mathfrak{n}$ is of power $2^\mathfrak{n}$.*

It is straightforward to prove that $\mathfrak{n}$ is just the sum of all cardinals $\mathfrak{m}_0(p), \cdots, \mathfrak{f}(p)$, taken for every $p$, and thus the direct product of the groups (1) must be of cardinality $2^\mathfrak{n}$.☐

Since the cardinality of the set of non-isomorphic groups of cardinality $\leq \mathfrak{n}$ is at most $2^\mathfrak{n}$ ($\mathfrak{n} \geq \aleph_0$), it is clear from (47.5) that the set of nonisomorphic compact (abelian) groups of cardinality $\leq 2^\mathfrak{n}$ is at most of cardinality $2^\mathfrak{n}$. Moreover, we have the rather surprising fact [which also shows that—though compact topology restricts the group structure considerably—the group structure has practically no effect at all on the compact topologies on the group]:

**Theorem 47.6** (Fuchs [8]). *For every infinite cardinal $\mathfrak{n}$, there exist $2^\mathfrak{n}$ non-isomorphic compact groups of power $2^\mathfrak{n}$ which are algebraically all isomorphic.*

In the proof, we need a result from Chapter XII. Namely, there exist $2^\mathfrak{n}$ pairwise nonisomorphic $p$-groups of cardinality $\mathfrak{n}$, moreover, they can be chosen so as to have isomorphic basic subgroups $\bigoplus_{n=1}^\infty \bigoplus_\mathfrak{n} Z(p^n)$ and the same final rank $\mathfrak{n}$. By (47.1), their character groups are isomorphic to

$$\prod_{n=1}^\infty \prod_\mathfrak{n} Z(p^n) \oplus \prod_\mathfrak{n} J_p .$$

By the Pontryagin duality theory, they are not isomorphic as topological groups [cf. next section].☐

The last theorem should be compared with the other extreme case: the group $J_p^m$ admits a single compact topology [which is the finite index topology]. In fact, (47.1) shows that the only discrete group whose character group is algebraically isomorphic to $J_p^m$ is the group $\bigoplus_m Z(p^\infty)$. [We have assumed the generalized continuum hypothesis to conclude that $m$ is uniquely determined.]

The methods of this section enable us to describe the groups of homomorphisms into algebraically compact groups. In this connection the basic result is:

**Theorem 47.7** (Fuchs [9]). *If $C$ is algebraically compact, then for every group $A$, $\mathrm{Hom}(A, C)$ is algebraically compact.*

From (38.1) we know that $C$ is a direct summand of a direct product of cocyclic groups. Hence $\mathrm{Hom}(A, C)$ is a direct summand of a group of type $\prod \mathrm{Hom}(A, C_i)$ with cocyclic groups $C_i$ [cf. (43.2)]. If $C_i$ is finite cyclic, say of order $p^n$, then $\mathrm{Hom}(A, C_i)$ is $p^n$-bounded, and hence algebraically compact. If $C_i$ is quasicyclic, then by (47.1), $\mathrm{Hom}(A, C_i)$ is likewise algebraically compact. Consequently, $\prod \mathrm{Hom}(A, C_i)$ is algebraically compact, and hence the result.□

As we know, the algebraically compact groups can be characterized by complete systems of cardinal invariants, so by (47.7), the same can be hoped for $\mathrm{Hom}(A, C)$ for algebraically compact $C$. As a matter of fact, the invariants of $\mathrm{Hom}(A, C)$ can be computed by means of the invariants of $C$ and certain ones of $A$. For details, we refer to the Exercises.

EXERCISES

1.  (a) Let $C$ be an algebraically compact group and $G$ a pure subgroup of $A$. Verify the isomorphism $\mathrm{Hom}(A, C) \cong \mathrm{Hom}(G, C) \oplus \mathrm{Hom}(A/G, C)$.
    (b) Char $A \cong$ Char $G \oplus$ Char $A/G$.
2.  Let $A, B$ be reduced $p$-groups. Give a necessary and sufficient condition for the (algebraic) isomorphism Char $A \cong$ Char $B$.
3.  (a) The additive group of reals admits infinitely many compact topologies under which it is a topological group. [*Hint*: Char($\bigoplus_n Q$).]
    (b) For which $A$ is Char $A$ torsion-free and divisible? Verify the statement in (a) for such a Char $A$.
4.  If $A$ is the quotient group of $Z^{\aleph_0}$ mod the direct sum $\bigoplus Z$, then

$$A = \mathrm{Char} \bigoplus_{\aleph_0} (Q \oplus Q/Z).$$

[*Hint*: **42**, Ex. 7.]
5.  (Hulanicki [1]) A nonzero divisible group

$$D = \bigoplus_m Q \oplus \bigoplus_p \bigoplus_{m_p} Z(p^\infty)$$

admits a compact topology exactly if the following conditions are satisfied:

(i) $\mathfrak{m}$ is of the form $2^{\mathfrak{n}}$ with infinite $\mathfrak{n}$;

(ii) $\mathfrak{m}_p$ is finite or of the form $2^{\mathfrak{n}_p}$, $\mathfrak{n}_p$ infinite;

(iii) $\mathfrak{m} \geqq \mathfrak{m}_p$ for every $p$.

6.   If $C$ is a complete group, then $\mathrm{Hom}(A, C)$ is the inverse limit of bounded groups.

7.   If $C$ is $p$-adic algebraically compact, and if $B$ is a $p$-basic subgroup of $A$, then $\mathrm{Hom}(B, C) \cong \mathrm{Hom}(A, C)$.

8.   (Pierce [1]) If $A \cong \bigoplus_{\mathfrak{m}} Z(p^\infty)$ and $C \cong \bigoplus_{\mathfrak{n}} Z(p^\infty)$, then $\mathrm{Hom}(A, C) \cong$ $p$-adic completion of the direct sum of $d(\mathfrak{m}, \mathfrak{n})$ copies of $J_p$. [*Hint*: apply (44.4) to $0 \to A[p] \to A \xrightarrow{p} A \to 0$ and $C$.]

9.   Determine the invariants of the algebraically compact group $\mathrm{Hom}(B, C)$, if $B$ is a direct sum of cyclic $p$-groups and $C = \bigoplus_{\mathfrak{n}} Z(p^\infty)$.

10.  Let $A$ be torsion-free and $C = \bigoplus_{\mathfrak{n}} Z(p^\infty)$. Determine the structure of $\mathrm{Hom}(A, C)$. [*Hint*: take $p$-basic in $A$.]

11.  By making use of Exs. 7–10, determine the invariants of $\mathrm{Hom}(A, C)$ for algebraically compact $C$.

12.  (Pierce [1]) Find the invariants of $\mathrm{Hom}(A, C)$ if $A$ is an arbitrary $p$-group. [*Hint*: (46.4), Exs. 8 and 9.]

## 48.* DUALITY BETWEEN DISCRETE TORSION AND 0-DIMENSIONAL COMPACT GROUPS

In this section we continue the study of character groups of abelian groups, but we no longer disregard the topology. As a matter of fact, the topology carried by character groups is very essential in the surprising duality between discrete and compact abelian groups.

This duality theory for the general case of locally compact abelian groups is due to L. S. Pontryagin and E. R. van Kampen, and is based on a deep theorem which guarantees the existence of sufficiently many characters for a compact group. A closer examination of the proof of the duality shows, however, that this result is not needed if we restrict ourselves to 0-dimensional compact groups and their duals; in fact, in this case rather routine topological arguments suffice to establish duality. Our aim is to prove duality in this special case, i.e., the duality between discrete torsion groups on one hand and totally disconnected [that is, 0-dimensional] compact groups on the other hand. All topological groups in this section are assumed to be Hausdorff.

To avoid tedious repetition of notation, we agree in denoting by $K$ the real numbers mod 1, i.e., the circle group equipped with the usual topology, and by $A^*$ the group of all continuous homomorphisms of the topological group $A$ into $K$, i.e., the *character group* of $A$. $A^*$ is awarded the *compact-open topology*, i.e., a fundamental system of neighborhoods about 0 is formed by all sets of the form

$$U(C, \varepsilon) = \{\chi \in A^* \mid \chi C \subset K_\varepsilon\}$$

where $K_\varepsilon$ is the $\varepsilon$-neighborhood of 0 in $K$, and $C$ is a compact subset of $A$. Moreover, we shall suppose that $\varepsilon$ is so small that $K_\varepsilon$ contains no subgroup $\neq 0$ of $K$.

We start with a few lemmas which we shall neither state nor prove in full generality, only in the case where we need them. In the proofs, standard results on topological groups will be taken for granted.

(a) *If $A$ is discrete torsion, then $A^*$ is a 0-dimensional compact group.*

The functions $f$ on $A$ to $K$ form a group $K^A$; this is compact in the product topology, as $K$ is compact. For fixed $a, b \in A$, define

$$H(a, b) = \{f \in K^A \mid f(a + b) = f(a) + f(b)\}$$

which is a closed subset of $K^A$ [being defined in terms of an equation]. Clearly, $A^*$ is the intersection of all $H(a, b)$ if we let $a, b$ run over all elements of $A$; hence $A^*$ is closed in $K^A$ and thus compact in the induced topology. The proof of compactness will be completed when we have shown that this topology is the same as the compact-open topology on $A^*$. Since $A$ is discrete, compactness means finiteness: $C = \{a_1, \cdots, a_n\}$. If $V(C, \varepsilon)$ denotes the neighborhood of 0 in $K^A$ where the coordinates in the components corresponding to the elements of $C$ belong to $K_\varepsilon$, then $A^* \cap V(C, \varepsilon) = U(C, \varepsilon)$. Since the $V(C, \varepsilon)$ form a fundamental system of neighborhoods of 0 in $K^A$, the compactness follows. In order to establish 0-dimensionality, notice that no generality is lost if $C$ is assumed to be a subgroup of $A$, $A$ being torsion. If $\varepsilon$ is chosen as agreed upon, then the condition $\chi C \subset K_\varepsilon$ amounts to $\chi C = 0$, i.e., $U(C, \varepsilon) = $ Ann $C$, the *annihilator* of $C$, defined as

$$\text{Ann } C = \{\chi \in A^* \mid \chi C = 0\}.$$

Thus $A^*$ has a fundamental system of neighborhoods about 0 consisting of subgroups $U(C, \varepsilon)$; these are then open and closed, and hence $A^*$ is 0-dimensional.

(b) *If $A$ is discrete torsion and $a \neq 0$ in $A$, then there exists a character $\chi$ of $A$ such that $\chi a \neq 0$.*

If $c \in K$ is of the same order as $a$, then $a \mapsto c$ extends to a homomorphism $\langle a \rangle \to K$. The latter group is divisible, hence this homomorphism can be extended to a $\chi: A \to K$.

(c) *If $A$ is discrete torsion and $C$ is a finite subgroup of $A$, then*

$$C^* \cong A^*/\text{Ann } C.$$

Because of the divisibility of $K$, every character of $C$ can be extended to a character of $A$, that is, the map $A^* \to C^*$ induced by the injection $C \to A$ is epic. Ann $C$ consists of all $\chi \in A^*$ that induce the $0$ character on $C$, i.e., Ann $C$ is the kernel of $A^* \to C^*$.

(d) *If $G$ is a $0$-dimensional compact group, then its topology is linear.*

Let $U$ be an open-closed neighborhood of $0$. For every $u \in U$ there is a neighborhood $W_u$ of $0$ such that $u + W_u \subset U$, and there is one $V_u$ such that $V_u + V_u \subset W_u$. Clearly, $U$ is covered by $\bigcup_u (u + V_u)$, hence by the compactness of $U$, there is a finite set $\{u_1, \cdots, u_n\}$ such that $\bigcup_{i=1}^n (u_i + V_{u_i})$ contains $U$. If $V = \bigcap_i V_{u_i}$, then

$$U + V \subseteq \bigcup_i (u_i + V_{u_i} + V) \subseteq \bigcup_i (u_i + W_{u_i}) \subseteq U.$$

Let $W$ be a neighborhood of $0$ satisfying $W = -W \subseteq U \cap V$. Then also $W + W \subseteq U + V \subseteq U$, and by a simple induction, $W + W + \cdots + W \subseteq U$. This shows that $\langle W \rangle \subseteq U$. Since $\langle W \rangle$ is the union of the open sets $W + \cdots + W$, it is an open subgroup, and the assertion follows.

(e) *If $G$ is a group with a linear topology, then a homomorphism $\gamma: G \to K$ is continuous if and only if* Ker $\gamma$ *is open.*

If Ker $\gamma$ is open, then $G/\mathrm{Ker}\, \gamma$ is discrete and so $\gamma$ is obviously continuous. Conversely, if $\gamma$ is continuous, then $\gamma^{-1} K_\varepsilon$ is open in $G$, whence there is an open subgroup $H$ of $G$ such that Ker $\gamma \supseteq H$. Hence Ker $\gamma$ is open.

(f) *If $G$ is a $0$-dimensional compact group, then $G^*$ is a discrete torsion group.*

If $\gamma \in G^*$, then from (d) and (e) we see that Ker $\gamma$ is open, and therefore $G/\mathrm{Ker}\, \gamma$ is discrete. It is compact, too, hence finite. Thus Im $\gamma$ is a finite subgroup of $K$, and so $n\gamma = 0$ for some integer $n > 0$. This proves $G^*$ torsion. Now $U(G, \varepsilon) = 0$; thus $G^*$ is discrete.

(g) *If $G$ is a $0$-dimensional compact group and $g \neq 0$ in $G$, then there is a $\gamma \in G^*$ satisfying $\gamma g \neq 0$.*

By (d), there exists an open subgroup $H$ of $G$ excluding $g$. Now $G/H$ is a discrete compact group, hence finite. (b) completes the proof.

The assertions of (b) and (g) express the fact: "the groups $A$ and $G$ have sufficiently many characters." This is fundamental in the duality.

The *second character group* $A^{**} = (A^*)^*$ of a group $A$ contains characters $a^{**}$ of $A^*$ induced by elements $a \in A$ in the following fashion:

$$a^{**}(\chi) = \chi a \qquad \text{for} \quad \chi \in A^*.$$

In view of the definition of addition of characters, $a^{**}$ is in fact a character of $A^*$ and the canonical map $\phi: a \mapsto a^{**}$ of $A$ into $A^{**}$ is plainly a homomorphism.

(h) *If $A$ is a discrete torsion group, then the canonical map $\phi: A \to A^{**}$ is an isomorphism.*

First we prove $\phi$ monic. If $a^{**} = 0$, then $\chi a = 0$ for all $\chi \in A^*$. In view of (b), this can happen only if $a = 0$; thus $\phi$ is monic.

If $A$ is finite, then it is a direct sum of finite cyclic groups, and since $Z(n)^* \cong Z(n)$, we have $A \cong A^*$, and a repeated application gives $A \cong A^{**}$. Thus $\phi$ is an isomorphism for finite $A$.

If $A$ is any discrete torsion group, then, for a character $\gamma: A^* \to K$, Ker $\gamma$ must be open because of (a), (d) and (e). Consequently, there is a finite subgroup $C$ of $A$ such that Ann $C \leqq$ Ker $\gamma$ [see the proof of (a)]. Now $\gamma$ induces a character $\bar{\gamma}$ of $A^*/\text{Ann } C \cong C^*$ [cf. (c)], and $C^*$ being finite, we know from what has already been proved that $\bar{\gamma}$ is induced by some element $c \in C$, i.e., $\bar{\gamma}(\bar{\chi}) = \bar{\chi}c$ for all $\bar{\chi} \in C^*$. It is now readily checked that $\gamma(\chi) = \chi c$ for all $\chi \in A^*$ which proves $\gamma$ is of the form $\gamma = c^{**}$. Thus $\phi$ is epic, and hence an isomorphism.

(j) *If $G$ is a 0-dimensional compact group, then the canonical map $\psi: G \to G^{**}$ is a topological isomorphism.*

That $\psi$ is monic follows in the same way as in (h), except that the reference to (b) must be replaced by that to (g). To prove $\psi$ continuous, we show that if $U$ is a fundamental neighborhood of $G^{**}$, then $\psi$ maps some open subgroup $V$ of $G$ into $U$. Now $G^*$ is discrete torsion by (f), so by the proof of (a), we may write $U = \text{Ann } C$ for some finite subgroup $C = \{\gamma_1, \cdots, \gamma_n\}$ of $G^*$. Put $V = \bigcap_{j=1}^{n} \text{Ker } \gamma_j$. By (d) and (e), Ker $\gamma_j$ is open, and so is $V$. Now for $g \in V$ we find $g^{**}(\gamma_j) = \gamma_j g = 0$ for $j = 1, \cdots, n$, that is, $g^{**} \in \text{Ann } C$. This proves that $\psi$ maps $V$ into $U$, i.e., $\psi$ is continuous. Hence $\psi$ is a topological isomorphism between the compact groups $G$ and $\psi G$.

Since $\psi G$ is compact, it is closed in the group $G^{**}$. Now $G^{**}/\psi G$ is again 0-dimensional compact. If it is not 0, then application of (g) guarantees the existence of a nonzero character $\bar{\tau}: G^{**}/\psi G \to K$. This is induced by a character $\tau: G^{**} \to K$ such that $\tau \psi G = 0$. By the discrete case (h), there is a $\chi: G^* \to K$ such that $\tau = \chi^{**}$, in other words, $\tau(\gamma) = \gamma(\chi)$ for all $\gamma \in G^{**}$. Choose $\gamma = g^{**}$ (with $g \in G$) to obtain $\chi g = g^{**}(\chi) = \tau(g^{**}) = 0$. It follows that $\chi = 0$ and $\tau = 0$, in contradiction to $\bar{\tau} \neq 0$. Therefore $\psi G = G^{**}$.

To sum up, we have proved:

**Theorem 48.1** (Pontryagin [1]). *Let $A$ be a discrete torsion [0-dimensional compact] group. Then the group $A^*$ of its continuous characters is a 0-dimensional compact [discrete torsion] group, and the correspondence $a \mapsto a^{**}$ of $A$ into the second character group $A^{**}$ is a topological isomorphism.*□

EXERCISES

1. (48.1) continues to hold if characters are taken into the discrete group $Q/Z$.
2. Let $A_i$ $(i \in I)$ be discrete torsion groups, and let $\bigoplus A_i$ have the discrete topology. Then $(\bigoplus A_i)^*$ is topologically isomorphic to $\prod A_i^*$ equipped with the product topology.
3. If $G$ is a compact topological group and $H$ is a closed subgroup, then

$$(G/H)^* \cong \text{Ann } H.$$

4. Let $G$ be a 0-dimensional compact group.
   (a) The $Z$-adic topology of $G$ is finer than the given topology.
   (b) For every $n \in Z$, $n > 0$, the subgroups $nG$ and $G[n]$ are closed.
5. A group $A$ is compact in its $Z$-adic topology if and only if it is complete in the $Z$-adic topology and $A/pA$ is finite for every prime $p$.
6. If a group is linearly compact in its $Z$-adic topology, then it is compact in this topology. [Hint: Ex. 5.]
7. A group $A$ is locally compact in the $Z$-adic topology if and only if for some $n > 0$, $nA$ is compact in the $Z$-adic topology.
8. If $A$ is discrete or compact, then

$$\text{Ann } nA \cong A^*[n] \quad \text{and} \quad \text{Ann } A[n] \cong nA^*.$$

9. For any compact topological group $C$, $\text{Hom}(A, C)$ can be made into a compact topological group.

## NOTES

It has long been known that the homomorphisms of an abelian group into another form a group. The importance of Hom was recognized by Eilenberg and Mac Lane [1], and Hom as a fundamental functor was fully developed by Cartan and Eilenberg [1].

The idea of stating in (44.4) more than exactness for (2) and (3) whenever more than mere exactness is assumed for (1), seems to appear first in the author's papers, Fuchs [8] and [9]. Various other generalizations of (44.4) and (51.3) were given by Harrison et al. [1], Irwin et al. [1], Pierce [1], and others.

The algebraic structure of Hom was known in some special cases. A major contribution was made by Pierce [1], who computed the invariants of $\text{Hom}(A, C)$ as an algebraically compact group for torsion groups $A$. The same for Char $A$ was done earlier by the author (Fuchs [8]), slightly improving the algebraic description of compact groups due to Hulanicki [1], [2] and Harrison [1]. [An excellent presentation of compact and locally compact abelian groups may be found in E. Hewitt and K. A. Ross, "Abstract Harmonic Analysis," Vol. I, where the reader will also find more about dualities.] A remarkable duality between discrete and linearly compact $p$-adic modules has been discovered by I. Kaplansky [Proc. Amer. Math. Soc. 4 (1953), 213–219] and H. Schöneborn [Math. Z. 59 (1954), 455–473, and 60 (1954), 17–30]. It turns out that every linearly compact abelian group is in a natural way a module over the $Z$-adic completion of the ring of integers [this ring is the direct product of the rings of $p$-adic integers, one ring for every prime $p$]. The structure of linearly compact abelian groups can be completely described; they form a class between algebraically compact and compact groups (see Fuchs [13]).

*Problem 30.* Describe Hom($A$, $C$), in particular, if $C$ is torsion-free [of rank 1 or a direct sum of groups of rank 1].

Torsion-free groups of rank 1 are characterized in Chapter XIII.

*Problem 31.* Find conditions on a group to be of the form End $A$ for some $A$. How many of these $A$ may be nonisomorphic?

Notice that for torsion $A$, (46.4) leads to a solution.

*Problem 32.* Does Hom($A$, $C$) have a "natural" compact topology for torsion groups $A$?

*Problem 33.* Which classes of abelian groups [subclasses of algebraically compact or cotorsion groups] $A$ are closed under the correspondences $A \mapsto \text{Hom}(G, A)$ where $G$ can be arbitrary?

*Problem 34.* Does there exist a set $\mathfrak{X}$ of groups $X$ such that Hom($A$, $X$) $\cong$ Hom($B$, $X$) for every $X \in \mathfrak{X}$ implies $A \cong B$?

*Problem 35.* For which categories $(\mathscr{A}, \mathbf{I})$ and $(\mathscr{A}, \mathbf{D})$ are Hom($A$, $C \mid \mathbf{I}$) and Hom($A \mid \mathbf{D}$, $C$) always algebraically compact whenever $A$ is torsion?

*Problem 36.* Investigate the sets of endomorphisms of Hom($A$, $C$) induced by those of $A$ and $C$, respectively [common elements, centralizers, etc.].

*Problem 37.* Do there exist, for every infinite cardinal $\mathfrak{m}$, $2^{\mathfrak{m}}$ nonisomorphic compact and connected groups of cardinality $\leq 2^{\mathfrak{m}}$ [that are algebraically isomorphic]?

# IX

## GROUPS OF EXTENSIONS

The extension problem for abelian groups [as a special case of the general group theoretical problem formulated by O. Schreier] consists in determining the group from a subgroup and the corresponding quotient group. The classical way of discussing extensions is *via* factor sets. It was a profound discovery of R. Baer's, that the extensions—under a suitable equivalence relation—themselves formed a group Ext, the group of extensions. It is the study of this group which is our main topic in this chapter.

An intimate relationship between groups of extensions and groups of homomorphisms has been pointed out by Eilenberg and Mac Lane [1]; this led to the interpretation of Ext as the socalled derived functor of Hom, and has been exploited extensively by Cartan and Eilenberg [1]. Ext can be discussed in various ways; we shall rely upon the elegant method of Mac Lane [3].

Our main objective is to discover the group theoretical properties of Ext $(C, A)$, its dependence upon the groups $A$ and $C$, and its relation with known constructions. Some of our goals are beyond the limits of our methods, but a good deal of information can be obtained about the general case which settles the problem in a number of special cases. The exact sequences connecting Hom and Ext, their generalizations, and the theory of cotorsion groups are the most relevant results in this chapter.

## 49. GROUP EXTENSIONS

Given the groups $A$ and $C$, the *extension problem* consists in finding groups $B$ such that $B$ contains a subgroup $A'$ isomorphic to $A$ and $B/A' \cong C$. This situation can be expressed in terms of a short exact sequence

$$E : 0 \to A \xrightarrow{\mu} B \xrightarrow{v} C \to 0$$

where $\mu$ stands for the inclusion map and $v$ is an epimorphism with kernel $\mu A$. In this case, one says that $B$ is an *extension of A by C*, and it is our present

209

aim to survey all extensions of $A$ by $C$. This can be done in various ways. In this section we describe the extensions in terms of factor sets, while in the next section we present the discussion based on short exact sequences.

Let $a, b, \cdots$ denote elements of $A$, and $u, v, w, \cdots$ those of $C$. Let $g : C \to B$ be a *representative function*, i.e., $g(u)$ is a representative of the coset $u$, $g(u) \in v^{-1}u$. Every $b \in B$ can be written uniquely in the form $b = g(u) + \mu a$ with $a \in A$. Clearly, $g(u) + g(v)$ and $g(u + v)$ belong to the same coset mod $\mu A$, hence there is an $f(u, v) \in A$ such that

(1) $$g(u) + g(v) = g(u + v) + \mu f(u, v).$$

Thus we have a function

(2) $$f : C \times C \to A$$

which is uniquely determined by the extension $B$ and the choice of the representatives $g(u)$. The commutative and associative laws in $B$ imply the identities

(3) $$f(u, v) = f(v, u),$$

(4) $$f(u, v) + f(u + v, w) = f(u, v + w) + f(v, w)$$

for all $u, v, w \in C$. Choosing $g(0) = 0$, we have in addition

(5) $$f(u, 0) = f(0, v) = 0$$

for all $u, v \in C$. A function (2) satisfying (3)–(5) is said to be a *factor set* [on $C$ to $A$].

Assume, conversely, we are given two groups, $A$ and $C$, together with a factor set (2). We can construct a group $B$ as the set of all pairs $(u, a) \in C \times A$ with the operation

$$(u, a) + (v, b) = (u + v, a + b + f(u, v)).$$

In fact, the commutative and associative laws are consequences of (3) and (4), while $(0, 0)$ is the zero and $(-u, -a - f(-u, u))$ is the inverse to $(u, a)$ in $B$. Manifestly, $\mu : a \mapsto (0, a)$ and $v : (u, a) \mapsto u$ make the sequence

$$0 \to A \xrightarrow{\mu} B \xrightarrow{v} C \to 0$$

exact, thus $B$ is an extension of $A$ by $C$. The choice $g(u) = (u, 0)$ yields the given factor set $f$. Thus *the extension problem can be solved by determining all factor sets*.

The direct sum $C \oplus A$ is one of the extensions of $A$ by $C$; it is also referred to as the *splitting extension*. If $g(u) = u \in C$ is chosen, then $f$ is identically 0. Another choice, say, $g(u) = u + \mu h(u)$ with $h(u) \in A$ yields the factor set

(6) $$f(u, v) = h(u) + h(v) - h(u + v).$$

Conversely, if $h : C \to A$ is any function with $h(0) = 0$, then (6) is a factor set, such that the pairs $(u, -h(u))$ form a complement to $\mu A$ in $B$, i.e., the extension splits. Thus an extension of $A$ by $C$ is splitting exactly if it is defined in terms of a factor set $f$ of the form (6) [where—we emphasize—$h : C \to A$]; such an $f$ is called a *transformation set*.

The dependence of the extension upon the chosen function $g : C \to B$ can be excluded by introducing the following equivalence relation in the set of extensions. If $g_1, g_2$ are both representative functions $C \to B$, then clearly $g_1(u) - g_2(u) = \mu h(u)$ for some function $h : C \to A$. The corresponding factor sets $f_1, f_2$ satisfy

(7)
$$f_1(u, v) - f_2(u, v) = h(u) + h(v) - h(u + v).$$

Accordingly, we define the factor sets $f_1, f_2 : C \times C \to A$ *equivalent*, if (7) holds for some $h : C \to A$. The extensions $B_1$, $B_2$ of $A$ by $C$ corresponding to equivalent $f_1$, $f_2$ are then isomorphic under the correspondence $\beta : (u, a)_1 \mapsto (u, a + h(u))_2$ which makes the diagram

(8)
$$\begin{array}{ccccccccc}
E_1 : & 0 \to & A & \xrightarrow{\mu} & B_1 & \xrightarrow{\nu} & C & \to 0 \\
& & \| & & \downarrow{\scriptstyle\beta} & & \| & \\
E_2 : & 0 \to & A & \xrightarrow{\mu} & B_2 & \xrightarrow{\nu} & C & \to 0
\end{array}$$

commutative. In this case, the extensions $E_1$ and $E_2$ themselves are called *equivalent*. Considering that if (8) is commutative, and if $g_1 : C \to B_1$ is a representative function, then $g_2 = \beta g_1 : C \to B_2$ is also one, and the corresponding factor sets are equivalent, we conclude that *there is a one-to-one correspondence between the equivalence classes of extensions of $A$ by $C$ and the equivalence classes of factor sets $f : C \times C \to A$*. In particular, an extension is equivalent to the splitting extension if, and only if, the corresponding equivalence class of factor sets is the class of transformation sets.

If $f_1, f_2 : C \times C \to A$ are factor sets, then their sum $f_1 + f_2$ defined as

$$(f_1 + f_2)(u, v) = f_1(u, v) + f_2(u, v)$$

is again a factor set, and so is $-f_1$ too. Consequently, the factor sets on $C$ to $A$ form a group Fact$(C, A)$. The transformation sets form obviously a subgroup Trans$(C, A)$ of this group, and what has been shown above can also be formulated by asserting a one-to-one correspondence between the equivalence classes of extensions of $A$ by $C$ and the elements of the quotient group Fact$(C, A)$/Trans$(C, A)$. This quotient group is the *group of extensions of $A$ by $C$*:

$$\mathrm{Ext}(C, A) = \mathrm{Fact}(C, A)/\mathrm{Trans}(C, A).$$

Another method of discusssion is based on producing a presentation of the extension in terms of $A$ and a presentation of $C$. This method will be outlined briefly at the end of **51**.

EXERCISES

1.  If $\text{Hom}(C, A)$ is regarded as a subgroup of $A^{C'}$ [cf. **43**], then
$$A^{C'}/\text{Hom}(C, A) \cong \text{Trans}(C, A) \qquad \text{where} \quad C' = C\backslash 0.$$

2.  A factor set on any group to a divisible group is a transformation set. [*Hint*: (22.1).]

3.  Every factor set on $C$ to $A$ is equivalent to a factor set on $C$ to a subgroup $B$ of $A$, if $A/B$ is divisible.

4.  If $A \cong C \cong Z(p)$, then there are two nonisomorphic and (at least) $p$ nonequivalent extensions of $A$ by $C$.


## 50.  EXTENSIONS AS SHORT EXACT SEQUENCES

In the preceding section we used the method of factor sets to describe the extensions of a group $A$ by another one $C$. Another approach is based upon short exact sequences. This decisive new idea will lead to a number of instructive relations, as we shall now see.

If the extension $B$ of $A$ by $C$ is visualized as an exact sequence
$$0 \to A \xrightarrow{\mu} B \xrightarrow{\nu} C \to 0,$$

then one can try to build up a category in which the objects are just the short exact sequences. An adequate definition of a morphism between two exact sequences is rather clear: it is a triple $(\alpha, \beta, \gamma)$ of group homomorphisms such that the diagram

$$
\begin{array}{ccccccccc}
E: & 0 \to & A & \xrightarrow{\mu} & B & \xrightarrow{\nu} & C & \to 0 \\
   &       & \alpha\downarrow & & \beta\downarrow & & \gamma\downarrow & \\
E': & 0 \to & A' & \xrightarrow{\mu'} & B' & \xrightarrow{\nu'} & C' & \to 0
\end{array}
$$

(1)

has commutative squares. It is straightforward to show that in this way a category $\mathscr{E}$ arises.

In accordance with the previous definition of equivalent extensions, we say that the extensions $E$ and $E'$ with $A = A'$, $C = C'$ are *equivalent*, in sign: $E \equiv E'$, if there is a morphism $(1_A, \beta, 1_C)$ with $\beta : B \to B'$ an isomorphism. Actually, the condition of $\beta$ being an isomorphism can be omitted, since this follows already from (2.3).

First we study extensions with $A$ fixed. If $\gamma : C' \to C$ is any homomorphism, then to the extension $E$ in (1), there is, by (10.1), a pullback square

$$
\begin{array}{ccc}
B' & \xrightarrow{\nu'} & C' \\
\beta\downarrow & & \downarrow\gamma \\
0 \to A \xrightarrow{\mu} B & \xrightarrow{\nu} & C \to 0
\end{array}
$$

with suitable $B'$, $\beta$ and $v'$. From **10** (a) we know that $v'$ is epic [since $v$ is epic], and a glance at (3) in **10** shows that $\operatorname{Ker} v' \cong \operatorname{Ker} v \cong A$, hence there is a monomorphism $\mu' : A \to B'$ [namely, $\mu'a = (\mu a, 0) \in B'$ if $B' \leq B \oplus C'$] such that the diagram

$$E\gamma:\quad 0 \to A \xrightarrow{\mu'} B' \xrightarrow{v'} C' \to 0$$
$$\Big\| \qquad \Big\downarrow \beta \qquad \Big\downarrow \gamma$$
$$E':\quad 0 \to A \xrightarrow{\mu} B \xrightarrow{v} C \to 0$$

with exact rows and pullback right square commutes. The top row is an extension of $A$ by $C'$ which we have denoted by $E\gamma$ to indicate its origin from $E$ and $\gamma$. Notice that $\gamma^* = (1_A, \beta, \gamma)$ is a morphism $E\gamma \to E$ in $\mathcal{E}$.

If the diagram

$$E^\circ:\quad 0 \to A \xrightarrow{\mu^\circ} B^\circ \xrightarrow{v^\circ} C' \to 0$$
$$\Big\| \qquad \Big\downarrow \beta^\circ \qquad \Big\downarrow \gamma$$
$$E:\quad 0 \to A \xrightarrow{\mu} B \xrightarrow{v} C \to 0$$

has exact rows and commutes, then by (10.1) there is a unique $\phi : B^\circ \to B'$ such that $v'\phi = v^\circ$ and $\beta\phi = \beta^\circ$. Since the maps $\phi\mu^\circ$, $\mu' : A \to B'$ are such that $\beta(\phi\mu^\circ) = \beta^\circ \mu^\circ = \mu = \beta\mu'$ and $v'(\phi\mu^\circ) = v^\circ \mu^\circ = 0 = v'\mu'$, the uniqueness assertion in (10.1) implies $\phi\mu^\circ = \mu'$. Hence $(1_A, \phi, 1_{C'})$ is a morphism of $E^\circ$ to $E\gamma$, and so $E^\circ \equiv E\gamma$. This shows that $E\gamma$ is unique up to equivalence, and this yields the equivalences

$$E1_C \equiv E \quad\text{and}\quad E(\gamma\gamma') \equiv (E\gamma)\gamma'$$

for $C'' \xrightarrow{\gamma'} C' \xrightarrow{\gamma} C$. Now the contravariance of $E$ on $C$ is evident.

Next we keep $C$ fixed and let $A$ vary. Given $\alpha : A \to A'$, let $B'$ be defined by the pushout square

$$0 \to A \xrightarrow{\mu} B \xrightarrow{v} C \to 0$$
$$\Big\downarrow \alpha \qquad \Big\downarrow \beta$$
$$A' \xrightarrow{\mu'} B'$$

Here $\mu'$ is a monomorphism, since $\mu$ is one [cf. **10**(b)]. Moreover, if $B'$ is defined as a quotient group of $A' \oplus B$ [as in the proof of (10.2)], then $v'((a', b) + H = vb$ makes the diagram

$$E:\quad 0 \to A \xrightarrow{\mu} B \xrightarrow{v} C \to 0$$
$$\Big\downarrow \alpha \qquad \Big\downarrow \beta \qquad \Big\|$$
$$\alpha E:\quad 0 \to A' \xrightarrow{\mu'} B' \xrightarrow{v'} C \to 0$$

with exact rows commutative. The bottom row is an extension of $A'$ by $C$ which we have denoted by $\alpha E$. Here $\alpha_* = (\alpha, \beta, 1_C)$ is a morphism $E \to \alpha E$ in $\mathscr{E}$.

If

$$E: \quad 0 \to A \xrightarrow{\mu} B \xrightarrow{\nu} C \to 0$$
$$\downarrow{\scriptstyle \alpha} \qquad \downarrow{\scriptstyle \beta_o} \qquad \|$$
$$E_o: \quad 0 \to A' \xrightarrow{\mu_o} B_o \xrightarrow{\nu_o} C \to 0$$

is a commutative diagram with exact rows, then in view of (10.2) there exists a unique $\phi : B' \to B_o$ such that $\phi\beta = \beta_o$ and $\phi\mu' = \mu_o$. From $(\nu_o \phi)\beta = \nu_o \beta_o = \nu = \nu'\beta$, $(\nu_o \phi)\mu' = 0 = \nu'\mu'$ we infer that $\nu_o \phi = \nu'$, thus $(1_{A'}, \phi, 1_C)$ is a morphism $\alpha E \to E_o$. Consequently, $\alpha E \equiv E_o$, i.e., $\alpha E$ is unique up to equivalence. Hence

$$1_A E \equiv E \qquad \text{and} \qquad (\alpha\alpha')E \equiv \alpha(\alpha'E)$$

for $A \xrightarrow{\alpha} A' \xrightarrow{\alpha'} A''$, establishing the covariant dependence of $E$ on $A$.

With $\alpha : A \to A'$ and $\gamma : C' \to C$ we have the important associative law

(2)                    $$\alpha(E\gamma) \equiv (\alpha E)\gamma.$$

Indeed, by making use of the pullback property of $(\alpha E)\gamma$, it is easy to prove the existence of a morphism $(\alpha, \beta', 1) : E\gamma \to (\alpha E)\gamma$ and to show the commutativity of the square

$$\begin{array}{ccc} E\gamma & \xrightarrow{(1, \beta_1, \gamma)} & E \\ {\scriptstyle (\alpha, \beta', 1)}\downarrow & & \downarrow{\scriptstyle (\alpha, \beta_2, 1)} \\ (\alpha E)\gamma & \xrightarrow{(1, \beta, \gamma)} & \alpha E. \end{array}$$

Let us pause for a moment to point out that both $E\gamma$ and $\alpha E$ can be described easily in terms of factor sets, if $E$ is given by a factor set $f: C \times C \to A$. Owing to the definition of $E\gamma$ in terms of pullback, the factor set belonging to $E\gamma$ is the composite function

$$C' \times C' \xrightarrow{\gamma \times \gamma} C \times C \xrightarrow{f} A,$$

as is readily seen from the fact that $B \oplus C'$ is an extension of $A$ by $C \oplus C'$ with the factor set $f(c_1 + c_1', c_2 + c_2') = f(c_1, c_2)$ ($c_i \in C$, $c_i' \in C'$) and restriction to $B'$ means $c_1 = \gamma c_1'$, $c_2 = \gamma c_2'$, i.e., the factor set is $f(\gamma c_1', \gamma c_2')$. Similarly, the factor set describing $\alpha E$ is the composite function

$$C \times C \xrightarrow{f} A \xrightarrow{\alpha} A'.$$

Indeed, in $\alpha E$, $B'$ is a quotient group of the extension $A' \oplus B$ of $A' \oplus A$ by $C$ with the factor set $f(c_1, c_2)$ viewed as $C \times C \to A' \oplus A$; passing to the quotient group, the pushout property shows that $B'$—as an extension of $A'$ by $C$—will have the factor set $\alpha f(c_1, c_2)$.

Returning to short exact sequences, assume we are given two extensions $E_1$ and $E_2$ of $A$ by $C$. As has been shown in the preceding section, the extensions of $A$ by $C$ [more correctly, their equivalence classes] form a group.

In order to describe the group operation in the language of short exact sequences, we make use of the diagonal map $\Delta_G: g \mapsto (g, g)$ and the codiagonal map $\nabla_G: (g_1, g_2) \mapsto g_1 + g_2$ of a group $G$. If we understand by the *direct sum* of two extensions

$$E_i: \quad 0 \to A_i \xrightarrow{\mu_i} B_i \xrightarrow{\nu_i} C_i \to 0 \qquad (i = 1, 2)$$

the extension

$$E_1 \oplus E_2 : 0 \to A_1 \oplus A_2 \xrightarrow{\mu_1 \oplus \mu_2} B_1 \oplus B_2 \xrightarrow{\nu_1 \oplus \nu_2} C_1 \oplus C_2 \to 0,$$

we then have:

**Proposition 50.1** (Mac Lane [3]). *The sum of two extensions $E_1$, $E_2$ of $A$ by C is the extension*

$$(3) \qquad\qquad E_1 + E_2 = \nabla_A(E_1 \oplus E_2) \Delta_C.$$

What we have to verify is that if $f_i : C \times C \to A$ is a factor set belonging to $E_i$ ($i = 1, 2$), then $f_1 + f_2$ belongs to $\nabla_A(E_1 \oplus E_2) \Delta_C$. Clearly, $(f_1(c_1, c_2), f_2(c'_1, c'_2))$ with $c_i, c'_i \in C$ is a factor set belonging to the direct sum $E_1 \oplus E_2$, and $(f_1(c_1, c_2), f_2(c_1, c_2))$ is one corresponding to $(E_1 \oplus E_2)\Delta_C$. An application of $\nabla_A$ yields the factor set $f_1(c_1, c_2) + f_2(c_1, c_2)$. $\square$

It is of course possible to avoid any reference to factor sets and to develop extensions solely *qua* short exact sequences. In doing so, (3) would serve as the definition of the sum of extensions and then (50.1) should be replaced by the assertion that $E_1 + E_2$ is actually an extension of $A$ by $C$ which stays in the same equivalence class if $E_1$ and $E_2$ are replaced by equivalent extensions, and moreover, the equivalence classes of extensions form a group under this operation. [For a proof of this, without using factor sets, we refer to Mac Lane [3].]

From what has been said above about the factor sets belonging to $E\gamma$ and $\alpha E$ it is now evident that for homomorphisms $\alpha : A \to A'$ and $\gamma : C' \to C$, the following equivalences hold true for extensions $E_1, E_2, E$ of $A$ by $C$:

$$(4) \qquad \alpha(E_1 + E_2) \equiv \alpha E_1 + \alpha E_2, \qquad (E_1 + E_2)\gamma \equiv E_1\gamma + E_2\gamma,$$

$$(5) \qquad (\alpha_1 + \alpha_2)E \equiv \alpha_1 E + \alpha_2 E, \qquad E(\gamma_1 + \gamma_2) \equiv E\gamma_1 + E\gamma_2.$$

The equivalences (4) express the fact that $\alpha_* : E \mapsto \alpha E$ and $\gamma^* : E \mapsto E\gamma$ are group homomorphisms

$$\alpha_* : \text{Ext}(C, A) \to \text{Ext}(C, A'), \qquad \gamma^* : \text{Ext}(C, A) \to \text{Ext}(C', A),$$

while (5) assert that $(\alpha_1 + \alpha_2)_* = (\alpha_1)_* + (\alpha_2)_*$ and $(\gamma_1 + \gamma_2)^* = \gamma_1^* + \gamma_2^*$, i.e., the correspondence

$$\text{Ext} : C \times A \mapsto \text{Ext}(C, A), \quad \gamma \times \alpha \mapsto \gamma^* \alpha_* = \alpha_* \gamma^*$$

is an additive bifunctor on $\mathscr{A} \times \mathscr{A}$ to $\mathscr{A}$ [the last equality is just another form of (2)]:

**Theorem 50.2** (Eilenberg and Mac Lane [1]). Ext *is an additive bifunctor on $\mathscr{A} \times \mathscr{A}$ to $\mathscr{A}$ which is contravariant in the first and covariant in the second variable.*☐

In order to be consistent with the functorial notation for homomorphisms, we shall also use the notation

$$\text{Ext}(\gamma, \alpha) : \text{Ext}(C, A) \to \text{Ext}(C', A')$$

instead of $\gamma^* \alpha_* = \alpha_* \gamma^*$; that is, $\text{Ext}(\gamma, \alpha)$ acts as shown by

$$\text{Ext}(\gamma, \alpha) : E \mapsto \alpha E \gamma.$$

Let us keep in mind that if the extension $E$ is given by (1), then for $\gamma : C' \to C$, $E\gamma$ is represented by $0 \to A \xrightarrow{\mu'} B' \xrightarrow{v'} C' \to 0$ where

(6)    $B' = \{(b, c') \mid b \in B, c' \in C', vb = \gamma c'\}, \qquad \mu' a = (\mu a, 0), \qquad v'(b, c') = c',$

and for $\alpha : A \to A'$, $\alpha E$ is represented by $0 \to A' \xrightarrow{\mu'} B' \xrightarrow{v'} C \to 0$ where

(7)
$$B' = \{(a', b) + H \mid a' \in A', b \in B\},$$
$$\mu' a' = (a', 0) + H, \qquad v'((a', b) + H) = vb$$

with $H = \{(\alpha a, -\mu a) \mid a \in A\}$. These formulas for $E\gamma$ and $\alpha E$ are helpful in subsequent computations.

EXERCISES

1. Characterize the mono- and epimorphisms of the category $\mathscr{E}$ [i.e., the morphisms that are left- and right-cancellable].
2. (Mac Lane [3]) If $E$ is as in (1), then both $\mu E$ and $Ev$ split.
3. (Mac Lane [3]) If $(\alpha, \beta, \gamma) : E \to E'$ is a morphism in the category $\mathscr{E}$, then $\alpha E \equiv E' \gamma$.
4. (a) If $\alpha : A \to A'$ is an epimorphism, then $\alpha E$ [with $E$ in (1)] is equivalent to the extension

$$0 \to A/\text{Ker } \alpha \to B/\mu \, \text{Ker } \alpha \to C \to 0$$

with the obvious maps.

(b) If $\gamma : C' \to C$ is a monomorphism, then $E\gamma$ is equivalent to

$$0 \to A \to v^{-1} \operatorname{Im} \gamma \to \operatorname{Im} \gamma \to 0.$$

5.   Let $\alpha$ [$\gamma$] be an automorphism of $A$ [$C$]. When is $\alpha E$ [$E\gamma$] equivalent to $E$ in (1)?

## 51.   EXACT SEQUENCES FOR EXT

As we have seen in the preceding section, Ext is a functor in both of its variables. The main result of this section states that this functor is right exact, moreover, the exact sequences on Hom and Ext can be amalgamated into long exact sequences.

Given an extension

(1)                                    $E : 0 \to A \xrightarrow{\alpha} B \xrightarrow{\beta} C \to 0,$

representing an element of $\operatorname{Ext}(C, A)$, and a homomorphism $\eta : A \to G$, we know from the preceding section that $\eta E$ is an extension of $G$ by $C$, i.e., $\eta E$ represents an element of $\operatorname{Ext}(C, G)$. In this way we get a map

$$E^* : \operatorname{Hom}(A, G) \to \operatorname{Ext}(C, G)$$

defined as

$$E^* : \eta \mapsto \eta E.$$

Analogously, a homomorphism $\xi : G \to C$ yields from $E$ an extension $E\xi$ of $A$ by $G$, and

$$E_* : \operatorname{Hom}(G, C) \to \operatorname{Ext}(G, A)$$

is a map acting as follows:

$$E_* : \xi \mapsto E\xi.$$

From (5) in **50** it results at once that $E^*$ and $E_*$ are homomorphisms. They are natural, for if $\phi : G \to H$ is any homomorphism, then because of $(\phi\eta)E \equiv \phi(\eta E)$ and $E(\xi\phi) \equiv (E\xi)\phi$ the diagrams

$$
\begin{array}{ccc}
\operatorname{Hom}(A, G) \to \operatorname{Ext}(C, G) & \qquad & \operatorname{Hom}(H, C) \to \operatorname{Ext}(H, A) \\
\downarrow \qquad \downarrow & & \downarrow \qquad \downarrow \\
\operatorname{Hom}(A, H) \to \operatorname{Ext}(C, H) & & \operatorname{Hom}(G, C) \to \operatorname{Ext}(G, A)
\end{array}
$$

with the obvious maps commute. $E^*$ and $E_*$ are called the *connecting homomorphisms* for the short exact sequence (1). This terminology is justified in the light of (51.3).

Before stating this theorem, we prove two technical lemmas.

**Lemma 51.1** (Mac Lane [3]). *Given a diagram*

$$E: \quad 0 \to A \xrightarrow{\alpha} B \xrightarrow{\beta} C \to 0$$

$$\downarrow{\eta} \quad \nearrow{\xi}$$

$$G$$

*with exact row, there exists a $\xi : B \to G$ making the triangle commute if and only if $\eta E$ splits.*

If there is such a $\xi$, then the diagram

$$
\begin{array}{ccccccccc}
E: & 0 \to A & \xrightarrow{\alpha} & B & \xrightarrow{\beta} & C \to 0 \\
& & \downarrow{\eta} & & \downarrow{(\xi \oplus \beta)\Delta} & \| \\
& 0 \to G & \xrightarrow{(1_G \oplus 0)\Delta} & G \oplus C & \xrightarrow{\nabla(0 \oplus 1_C)} & C \to 0
\end{array}
$$

commutes, hence the bottom row is $\equiv \eta E$. Conversely, if $\eta E : 0 \to G \to B' \to C \to 0$ splits, then $B \to B'$ followed by the projection $B' \to G$ yields a map $\xi$ with the desired property.☐

The dual of this argument establishes the exact dual of the preceding lemma:

**Lemma 51.2.** *If the diagram*

$$G$$

$$\nearrow{\xi} \quad \downarrow{\eta}$$

$$E: \quad 0 \to A \xrightarrow{\alpha} B \xrightarrow{\beta} C \to 0$$

*has exact row, then there is a $\xi : G \to B$ such that $\beta \xi = \eta$ if, and only if, $E\eta$ splits.*☐

With the aid of these lemmas, the following theorem on the exact sequences for Ext becomes a straightforward, though mildly intricate calculation.

**Theorem 51.3** (Cartan and Eilenberg [1]). *If* (1) *is an exact sequence, then the sequences*

(2)
$$0 \to \mathrm{Hom}(C, G) \to \mathrm{Hom}(B, G) \to \mathrm{Hom}(A, G) \to$$
$$\xrightarrow{E^*} \mathrm{Ext}(C, G) \xrightarrow{\beta^*} \mathrm{Ext}(B, G) \xrightarrow{\alpha^*} \mathrm{Ext}(A, G) \to 0,$$

(3)
$$0 \to \mathrm{Hom}(G, A) \to \mathrm{Hom}(G, B) \to \mathrm{Hom}(G, C)$$
$$\xrightarrow{E_*} \mathrm{Ext}(G, A) \xrightarrow{\alpha_*} \mathrm{Ext}(G, B) \xrightarrow{\beta_*} \mathrm{Ext}(G, C) \to 0$$

*are exact for every group G.*

Owing to (44.4) we may begin the proof of exactness of (2) at $\mathrm{Hom}(A, G)$. We have to show that $\eta : A \to G$ is extendable to $\xi : B \to G$ exactly if $\eta E \in \mathrm{Ext}(C, G)$ is splitting; but this is just the statement of (51.1). The next step is to show the exactness at $\mathrm{Ext}(C, G)$. By (51.2), $E\beta$ splits, thus for $\eta \in \mathrm{Hom}(A, G)$, $\beta^*E^*\eta = \eta E\beta = 0$. Let $E_1 : 0 \to G \xrightarrow{\mu} H \xrightarrow{v} C \to 0 \in$ $\mathrm{Ext}(C, G)$ be such that $E_1\beta$ splits. By (51.2), there is a $\xi : B \to H$ such that $v\xi = \beta$. Since $v\xi\alpha = \beta\alpha = 0$, by (2.1) there is an $\eta : A \to G$ satisfying $\mu\eta = \xi\alpha$, hence $(\eta, \xi, 1_C)$ maps $E$ upon $E_1$, i.e., $E_1 = \eta E$. To show exactness at $\mathrm{Ext}(B, G)$, notice that obviously $\alpha^*\beta^* = (\beta\alpha)^* = 0^* = 0$. Conversely, to prove that the kernel is contained in image, let $E_2 : 0 \to G \xrightarrow{\mu} H \xrightarrow{v} B \to 0 \in \mathrm{Ext}(B, G)$ satisfy $E_2\alpha = 0$. By (51.2), there is a $\xi : A \to H$ such that $v\xi = \alpha$; $\xi$ is monic. Since $\beta v\xi = \beta\alpha = 0$, there is a $\lambda : H/\xi A \to C$ such that $\beta v = \lambda\rho$ with $\rho : H \to H/\xi A$ the canonical map. Consequently, we have a commutative diagram

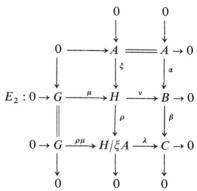

where all the three columns and the first two rows are exact. By the $3 \times 3$-lemma, the bottom row is exact, hence it represents an element of $\mathrm{Ext}(C, G)$ that is mapped by $\beta^*$ upon $E_2$. The exactness of (2) at $\mathrm{Ext}(A, G)$ expresses the fact that every extension of $G$ by $A$ can be prolonged to one of $G$ by $B$: this is true as is shown by (24.6).

Turning to the proof of (3), by (44.4) and (51.2) we may begin the proof at $\mathrm{Ext}(G, A)$. For $\eta \in \mathrm{Hom}(G, C)$, $\alpha_* E_* \eta = \alpha E\eta = 0$, as $\alpha E$ splits because of (51.1). Assume $E_1 : 0 \to A \xrightarrow{\mu} H \xrightarrow{v} G \to 0 \in \mathrm{Ext}(G, A)$ satisfies $\alpha E_1 = 0$; then by (51.1) there is a $\xi : H \to B$ such that $\xi\mu = \alpha$. From $\beta\xi\mu = \beta\alpha = 0$ and (2.2) we infer the existence of an $\eta : G \to C$ such that $\eta v = \beta\xi$, and so $(1_A, \xi, \eta)$ maps $E$ upon $E_1$, i.e., $E_1 = E\eta$. Next we show exactness at $\mathrm{Ext}(G, B)$. By $\beta_* \alpha_* = (\beta\alpha)_* = 0_* = 0$, it suffices to show that kernel is contained in image. Assume $E_2 : 0 \to B \xrightarrow{\mu} H \xrightarrow{v} G \to 0 \in \mathrm{Ext}(G, B)$ satisfies $\beta E_2 = 0$; then by (51.1) there is a $\xi : H \to C$ with $\xi\mu = \beta$. Now $\xi\mu\alpha = 0$ implies the existence of a map $\lambda : A \to \text{Ker } \xi$ with $\rho\lambda = \mu\alpha$, where $\rho : \text{Ker } \xi \to H$ is the injection. Therefore the diagram

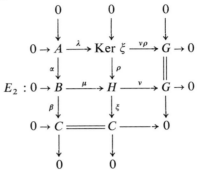

is commutative, has exact columns and the two bottom rows are exact. By the 3 × 3-lemma, the top row is exact, thus it is an element of $\text{Ext}(G, A)$ which is mapped by $\alpha$ upon $E_2$. Finally, the epimorphic character of $\beta_*$ follows again from (24.6).□

The exact sequences (2) and (3) are of cardinal importance in dealing with Hom and Ext. They are extensively made use of in the description of Ext, in particular, in the theory of cotorsion groups. They establish a close connection between Hom and Ext [exploited to a great extent in homological algebra].

It is worthwhile pointing out this connection more closely, since it yields a method of discussing Ext. Given $A$, $C$, let $E_0 : 0 \to H \xrightarrow{\phi} F \xrightarrow{\psi} C \to 0$ be a free resolution of $C$, i.e., both $F$ and $H$ are free. For an $\eta : H \to A$ we can find a $B$ and a $\chi : F \to B$ such that the diagram

$$
\begin{array}{ccccccccc}
E_0 : & 0 \to & H & \xrightarrow{\phi} & F & \xrightarrow{\psi} & C & \to 0 \\
 & & {\scriptstyle \eta}\downarrow & & {\scriptstyle \chi}\downarrow & & \| & \\
\eta E_0 : & 0 \to & A & \xrightarrow{\mu} & B & \xrightarrow{\nu} & C & \to 0
\end{array}
$$

commutes and the bottom row is exact. Now

$$E_0^* : \text{Hom}(H, A) \to \text{Ext}(C, A)$$

is easily seen to be an epimorphism whose kernel consists of all $\eta : H \to A$ that can be extended to an $F \to A$. Notice that if

$$F = \bigoplus_{i \in I} \langle x_i \rangle \qquad \text{and} \qquad H = \bigoplus_{j \in J} \langle y_j \rangle \qquad \text{with} \quad y_j = \sum_i m_{ji} x_i$$

($m_{ji} \in \mathbf{Z}$, almost all $m_{ji}$ with fixed $j$ vanish), then the extension $\eta E_0$ of $A$ by $C$ is the group

$$B = \langle A, x_i \ (i \in I); \ \sum_i m_{ji} x_i = \eta y_j \ (j \in J) \rangle.$$

Two homomorphisms $\eta_1$, $\eta_2 : H \to A$ give rise to equivalent extensions exactly if their difference is extendable to a homomorphism $F \to A$.

EXERCISES

1.  The group $G$ has the property that, for every epimorphism $\beta : B \to C$, the induced mapping $\beta^* : \operatorname{Ext}(C, G) \to \operatorname{Ext}(B, G)$ is monic if and only if $G$ is divisible.
2.  $G$ is free if and only if for every monomorphism $\alpha : A \to B$, the mapping $\alpha_* : \operatorname{Ext}(G, A) \to \operatorname{Ext}(G, B)$ is monic.
3.  Let $\beta : B \to C$ be epic, and assume $\beta^* : \operatorname{Ext}(C, G) \to \operatorname{Ext}(B, G)$ is monic for every $G$. Then $\operatorname{Ker} \beta$ is a direct summand of $B$.
4.  Let $\alpha : A \to B$ be a monomorphism such that $\alpha_* : \operatorname{Ext}(G, A) \to \operatorname{Ext}(G, B)$ is monic for any $G$. Show that $\alpha A$ is a direct summand of $B$.
5.  Let $0 \to H \xrightarrow{\phi} F \xrightarrow{\psi} C \to 0$ be a free resolution of $C$. Prove the isomorphism

$$\operatorname{Ext}(C, A) \cong \operatorname{Hom}(H, A)/\phi^* \operatorname{Hom}(F, A).$$

6.  Let $A$ be a $p$-group and $B$ a basic subgroup of $A$. For any $C$, the groups $\operatorname{Ext}(C, A)$ and $\operatorname{Ext}(C, B)$ are epimorphic images of each other.
7.  Prove that

$$\operatorname{Ext}(Q, Z) \cong Q^{\aleph_0}.$$

    [*Hint*: apply (51.3) to $0 \to Z \to Q \to Q/Z \to 0$.]
8.  Establish the isomorphism

$$\operatorname{Ext}(Z(p^\infty), J_p) \cong J_p.$$

## 52.   ELEMENTARY PROPERTIES OF EXT

Our objective in this section is to record a number of elementary but most useful properties of the groups of extensions. We shall make frequent use of the exact sequences stated in (51.3).

In order not to interrupt our discussion, first we formulate a simple lemma. In accordance with definitions given in **50**, if $E : 0 \to A \xrightarrow{\mu} B \xrightarrow{\nu} C \to 0$ is an extension of $A$ by $C$, and if $\alpha : A \to A$, $\gamma : C \to C$ are endomorphisms of $A$ and $C$, respectively, then $\alpha E$ and $E\gamma$ will again be extensions of $A$ by $C$. The correspondences

$$\alpha_* : E \mapsto \alpha E \qquad \text{and} \qquad \gamma^* : E \mapsto E\gamma$$

are evidently endomorphisms of $\operatorname{Ext}(C, A)$; we call them *induced endomorphisms* of Ext. The formulas $(\alpha_1 + \alpha_2)_* = (\alpha_1)_* + (\alpha_2)_*$ and $(\alpha_1\alpha_2)_* = (\alpha_1)_*(\alpha_2)_*$ show that the endomorphism ring of $A$ acts on $\operatorname{Ext}(C, A)$, and similarly, the dual of the endomorphism ring of $C$ operates on $\operatorname{Ext}(C, A)$. These commute as is shown by $\alpha_* \gamma^* = \gamma^* \alpha_*$; hence $\operatorname{Ext}(C, A)$ is a (unital) bimodule over the endomorphism rings of $A$ and $C$, acting from the left and right, respectively. Now our lemma asserts the following remarkable fact.

**Lemma 52.1.** *Multiplication by the integer n on A or C induces multiplication by n on* $\text{Ext}(C, A)$. *The same holds for p-adic integers.*

If $\alpha_1, \cdots, \alpha_n$ are endomorphisms of $A$, then $(\alpha_1 + \cdots + \alpha_n)_* = (\alpha_1)_* + \cdots + (\alpha_n)_*$. Since $1_A$ obviously induces $1_{\text{Ext}(C,A)}$, the choice $\alpha_1 = \cdots = \alpha_n = 1_A$ leads us to the desired result. The proof for $C$ is analogous.

Multiplication by a *p*-adic integer $\pi$ on $A$ induces an endomorphism $\pi_*$ on $\text{Ext}(C, A)$, and it is clear that in this way the *p*-adic integers act on $\text{Ext}(C, A)$. $\pi$ is the limit of a sequence $n_i \in \mathbf{Z}$ $(i = 1, 2, \cdots)$ in the *p*-adic topology, say $p^i \mid \pi - n_i$ for every $i$. Hence, by what has been proved, it follows that $\pi_*$ is a limit of multiplications by $n_i$ and hence it can be identified with the multiplication by $\pi$.☐

We begin with two rather trivial observations.

(A) *A group C satisfies* $\text{Ext}(C, A) = 0$ *for every A if and only if C is free.*

In other words, every extension of $A$ by $C$ splits, for arbitrary $A$, if and only if $C$ is free. The "if" part is equivalent to (14.4), while the "only if" part follows from (14.6).

(B) *A group A satisfies* $\text{Ext}(C, A) = 0$ *for every C exactly if A is divisible.*
This is a reformulation of the equivalence of (i) and (iii) in (24.5).

(C) Let us turn next to the following theorem.

**Theorem 52.2.** *There exist natural isomorphisms*

(1)
$$\text{Ext}\left(\bigoplus_{i \in I} C_i, A\right) \cong \prod_{i \in I} \text{Ext}(C_i, A),$$

(2)
$$\text{Ext}\left(C, \prod_{j \in J} A_j\right) \cong \prod_{j \in J} \text{Ext}(C, A_j).$$

There are several ways to follow; for instance, we can refer to (51.3). We start with the exact sequences $0 \to G_i \to F_i \to C_i \to 0$, with free $F_i$. These induce the exact sequence $0 \to \bigoplus G_i \to \bigoplus F_i \to \bigoplus C_i \to 0$. Now we get $\text{Hom}(F_i, A) \to \text{Hom}(G_i, A) \to \text{Ext}(C_i, A) \to \text{Ext}(F_i, A) = 0$, the last equality because of (A). In this way, the induced exact sequences arise:

$$\prod \text{Hom}(F_i, A) \to \prod \text{Hom}(G_i, A) \to \prod \text{Ext}(C_i, A) \to 0$$
$$\| \wr \qquad\qquad\qquad \| \wr$$
$$\text{Hom}(\bigoplus F_i, A) \to \text{Hom}(\bigoplus G_i, A) \to \text{Ext}(\bigoplus C_i, A) \to 0$$

where the vertical isomorphisms are natural [cf. (43.1)]. Therefore, the two Ext's are isomorphic in a natural way. The proof for (2) runs dually.☐

(D) *For every group A and for every integer m,*

$$\text{Ext}(Z(m), A) \cong A/mA.$$

The sequence $0 \to Z \xrightarrow{m} Z \to Z/mZ = Z(m) \to 0$ [$m$ stands for the multi-plication by $m$] is exact. Now (51.3) together with $\mathrm{Ext}(Z, A) = 0$ implies the exactness of $\mathrm{Hom}(Z, A) \xrightarrow{m} \mathrm{Hom}(Z, A) \to \mathrm{Ext}(Z(m), A) \to 0$ [cf. (44.6), too]. In view of the natural isomorphism $\mathrm{Hom}(Z, A) \cong A$, $\mathrm{Ext}(Z(m), A)$ must be isomorphic to $A/mA$ [this is, therefore, again a natural isomorphism].

Notice that (52.2) and (D) make it possible to compute $\mathrm{Ext}(C, A)$ for a direct sum $C$ of cyclic groups. Then $\mathrm{Ext}(C, A)$ will be the direct product of groups of the form $A/mA$.

(E) *If $mA = 0$ or $mC = 0$ for some integer $m$, then $m\,\mathrm{Ext}(C, A) = 0$.*
This is an obvious consequence of (52.1).

(F) *For any integer $m$,*
$$\mathrm{Ext}(C, Z(m)) \cong \mathrm{Ext}(C[m], Z(m)).$$

From the exactness of $0 \to C[m] \to C \xrightarrow{m} mC \to 0$ we obtain the induced exact sequence
$$\mathrm{Ext}(mC, Z(m)) \xrightarrow{m} \mathrm{Ext}(C, Z(m)) \to \mathrm{Ext}(C[m], Z(m)) \to 0.$$
The image of the first map here must be 0 owing to (52.1) and (E), whence the desired isomorphism results.

(G) *If $mA = A$ for some $m$, then $m\,\mathrm{Ext}(C, A) = \mathrm{Ext}(C, A)$.*
In fact, the exactness of $A \xrightarrow{m} A \to 0$ implies that of the sequence $\mathrm{Ext}(C, A) \xrightarrow{m} \mathrm{Ext}(C, A) \to 0$.

(H) *An automorphism $\alpha$ of $A$ induces an automorphism $\alpha_*$ of $\mathrm{Ext}(C, A)$.*
If $\bar{\alpha}$ is the inverse to $\alpha$, then clearly $\bar{\alpha}_*$ will be the inverse to $\alpha_*$.—In particu-lar, we conclude that $mA = A$ *and* $A[m] = 0$ *imply* $m\,\mathrm{Ext}(C, A) = \mathrm{Ext}(C, A)$ *and* $\mathrm{Ext}(C, A)[m] = 0$. *Furthermore, if $A$ is torsion-free divisible, then $\mathrm{Ext}(C, A)$, too, is torsion-free divisible.*

(I) $C[m] = 0$ *implies* $m\,\mathrm{Ext}(C, A) = \mathrm{Ext}(C, A)$. *In particular, $\mathrm{Ext}(C, A)$ is divisible if $C$ is torsion-free.*
Assumption guarantees that $0 \to C \xrightarrow{m} C$ is exact. Hence the exactness of $\mathrm{Ext}(C, A) \xrightarrow{m} \mathrm{Ext}(C, A) \to 0$ follows.

(J) *Let $\gamma$ be an automorphism of $C$. Then $\gamma^*$ is an automorphism of $\mathrm{Ext}(C, A)$.*
If $\bar{\gamma}$ is the inverse to $\gamma$, then $\bar{\gamma}^*$ must be the inverse to $\gamma^*$.—Thus *if $mC = C$ and $C[m] = 0$, then $m\,\mathrm{Ext}(C, A) = \mathrm{Ext}(C, A)$ and $\mathrm{Ext}(C, A)[m] = 0$; and if $C$ is torsion-free divisible, then the same holds for $\mathrm{Ext}(C, A)$.*

(K) *If $A$ is $p$-divisible and $C$ is a $p$-group, then $\mathrm{Ext}(C, A) = 0$.*
Let $D$ be the divisible hull of $A$. Then $D/A$ is torsion with 0 $p$-compo-nent, i.e., $\mathrm{Hom}(C, D/A) = 0$. The exactness of $\mathrm{Hom}(C, D/A) \to \mathrm{Ext}(C, A) \to \mathrm{Ext}(C, D) = 0$ implies the assertion.

(L) The following theorem provides us with an essential isomorphism.

**Theorem 52.3** (Eilenberg and Mac Lane [1]). *If A is torsion-free and C is torsion, then*

$$\text{Ext}(C, A) \cong \text{Hom}(C, D/A)$$

*where D is a divisible hull of A. Hence* $\text{Ext}(C, A)$ *is a reduced algebraically compact group.*

We start with the exact sequence $0 \to A \to D \to D/A \to 0$ where $D$ is torsion-free. Hence we obtain the exact sequence

$$0 = \text{Hom}(C, D) \to \text{Hom}(C, D/A) \to \text{Ext}(C, A) \to \text{Ext}(C, D) = 0,$$

establishing the claimed isomorphism. The second assertion now follows at once from (46.1).☐

The choice $A = Z$ leads us to the following interesting isomorphism.

**Corollary 52.4.** *If C is a torsion group, then*

$$\text{Ext}(C, Z) \cong \text{Char } C.$$

In fact, $Q/Z$ being the torsion part of $K$ [reals mod 1], $\text{Hom}(C, Q/Z)$ is just Char $C$.☐

(M) *If A is a torsion-free group whose p-basic subgroup is of rank* $\mathfrak{m}$, *then*

$$\text{Ext}(Z(p^\infty), A) \cong p\text{-adic completion of } \bigoplus_{\mathfrak{m}} J_p.$$

Denote by $B$ a $p$-basic subgroup of $A$. Then $0 \to B \to A \to A/B \to 0$ is exact where $A/B$ is $p$-divisible and has 0 $p$-component, so that both $\text{Hom}(Z(p^\infty), A/B) = 0$ and $\text{Ext}(Z(p^\infty), A/B) = 0$, the latter in view of (K). Hence the sequence $0 \to \text{Ext}(Z(p^\infty), B) \to \text{Ext}(Z(p^\infty), A) \to 0$ is exact, i.e., $\text{Ext}(Z(p^\infty), A) \cong \text{Ext}(Z(p^\infty), B)$. Here $B$ is a free group of rank $\mathfrak{m}$, so its divisible hull is $\bigoplus_{\mathfrak{m}} Q$, and the quotient group is $\bigoplus_{\mathfrak{m}} Q/Z$. Combining (52.3) with (44.3), the claimed isomorphism follows.

(N) *If A is torsion-free,* $\text{Ext}(C, A)$ *is algebraically compact, whatever C is.*

We have an exact sequence $0 \to T \to C \to C/T \to 0$ with $T$ torsion and $C/T$ torsion-free. Hence the induced sequence

$$0 = \text{Hom}(T, A) \to \text{Ext}(C/T, A) \to \text{Ext}(C, A) \to \text{Ext}(T, A) \to 0$$

is exact. By (I), $\text{Ext}(C/T, A)$ is divisible, so the sequence splits:

$$\text{Ext}(C, A) \cong \text{Ext}(C/T, A) \oplus \text{Ext}(T, A).$$

The second summand is, by virtue of (52.3), algebraically compact, and the statement follows.

(O) *If A is algebraically compact, then* Ext(C, A) *is a reduced algebraically compact group.*

Without restricting generality, we may assume $A$ reduced. Then $A$ is a direct summand of a direct product of cyclic $p$-groups [cf. (38.2)]. Since Ext(C, Z($p^n$)) is annihilated by $p^n$ [recall (E)], it is algebraically compact and reduced. Ext(C, A) is a direct summand of a direct product of such groups, as is shown by (52.2), (2). Now the algebraic compactness of Ext(C, A) follows from (38.3).

EXERCISES

1.  If $A$, $C$ are finite groups, then Ext(C, A) $\cong$ Hom(C, A).
2.  Prove that Ext(Q/Z, A) = 0 implies $A$ divisible. [*Hint*: Ext(Q, A) = 0 and every group is a subgroup of a direct sum of copies of $Q/Z$ and $Q$.]
3.  If $C$ is a direct sum of cyclic groups, Ext(C, A) is reduced algebraically compact.
4.  Let the group $A$ and the integer $m > 0$ have the property that $m$ Ext(C, A) = 0 for every $C$. Prove $mA = 0$.
5.  If for some $A$ and $m \in Z$, $m$ Ext(C, A) = Ext(C, A) for every $C$, then $mA = A$.
6.  (a) If for some $C$ and $m \in Z$, $m$ Ext(C, A) = Ext(C, A) for any group $A$, then $C[m] = 0$.
    (b) If $C$ is such that Ext(C, A) is always torsion-free divisible, then $C$ is torsion-free and divisible.
7.  Determine the invariants of Ext(Q/Z, A) as an algebraically compact group with torsion-free $A$, in particular, for $A$ free.
8.  Let $C$ be a $p$-group and $A$ torsion-free. What are the invariants of Ext(C, A) *qua* algebraically compact group? [*Hint*: argue as in (M).]
9.  (Nunke [1]) Show that Hom(C, Z) = Ext(C, Z) = 0 implies $C = 0$. [*Hint*: prove Ext($T(C)$, Z) = Char $T(C)$ = 0 and Ext($C/mC$, Z) = 0; finally, use **51**, Ex. 7.
10. Find nonisomorphic groups $C_1, C_2$ such that Hom($C_1$, Z) $\cong$ Hom($C_2$, Z) and Ext($C_1$, Z) $\cong$ Ext($C_2$, Z).
11. (Nunke [2]) If $A$ is torsion and $C$ torsion-free, then Ext(C, A) is not torsion unless 0. [*Hint*: reduce to $p$-groups $A$, then to direct sums of cyclic groups using (36.1); if $A$ is unbounded, write $A = \oplus_n A_n$ with $A$ an epic image of every $A_n$ and show an epimorphism Ext(C, A)$\rightarrow$ $\prod$ Ext(C, $A_n$) by referring to (42.2).]
12. Prove that if $A$ admits a compact topology, then so does Ext(C, A) for any $C$.

13.  (Baer [4]) $p \operatorname{Ext}(C, A) = \operatorname{Ext}(C, A)$ if and only if either $C[p] = 0$ or $pA = A$.

14.  Prove the isomorphism

$$\operatorname{Ext}(Q_p, Z) \cong Z(p^\infty) \oplus Q^{\aleph_0}.$$

[*Hint*: use $0 \to Z \to Q_p \to \oplus_q Z(q^\infty) \to 0$ to estimate the cardinality and $0 \to Q_p \to Q \to Z(p^\infty) \to 0$ with (J) to find the torsion part.]

15.  Prove that

$$\operatorname{Ext}(Q_p, Q_p) = 0 \qquad \text{and} \qquad \operatorname{Ext}(Q, Q_p) \cong Q^{\aleph_0}.$$

16.  Verify the isomorphisms

$$\operatorname{Ext}(J_p, Z) \cong Z(p^\infty) \oplus Q^{2^{\aleph_0}} \qquad \text{and} \qquad \operatorname{Ext}(J_p, Q_p) \cong \operatorname{Ext}(J_p, Z).$$

17.  Give examples for (a) and (b):
     (a) if $C = \varinjlim C_i$, then $\operatorname{Ext}(C, A)$ is not $\varprojlim \operatorname{Ext}(C_i, A)$. [*Hint*: $C$ is torsion-free, $C_i$ are free];
     (b) if $A = \varprojlim A_i$, then $\operatorname{Ext}(C, A)$ is not $\varprojlim \operatorname{Ext}(C, A_i)$. [*Hint*: **21**, Ex. 6.]

18.  (a) (Nunke [1]) The natural map of Ext into the inverse limit in Ex. 17 (a) is an epimorphism. [*Hint*: an element of the inverse limit defines a direct system of exact sequences; take its direct limit.]
     (b) Prove the same for Ex. 17 (b).

## 53.   THE FUNCTOR PEXT

One of the most astonishing phenomena of group extensions is that the extensions corresponding to pure-exact sequences form a subgroup of Ext which coincides with the first Ulm subgroup of Ext. This leads to a new functor Pext which will be studied in this section.

We begin with two rather general theorems; only very special cases will be needed.

**Theorem 53.1** (Baer [4], Fuchs [7]). *Let the exact sequence*

(1) $$E : 0 \to A \xrightarrow{\mu} B \xrightarrow{\nu} C \to 0$$

*represent an element of* $\operatorname{Ext}(C, A)$, *and let* $\alpha : A \to A$. *Then for the induced endomorphism* $\alpha_*$ *of* $\operatorname{Ext}(C, A)$ *we have*:

   (i) $E \in \operatorname{Im} \alpha_*$ *exactly if* $\operatorname{Im} \mu / \operatorname{Im} \mu\alpha$ *is a direct summand of* $B / \operatorname{Im} \mu\alpha$;
   (ii) *assuming* $\operatorname{Im} \alpha = A$, $E \in \operatorname{Ker} \alpha_*$ *if, and only if,* $\operatorname{Im} \mu / \mu \operatorname{Ker} \alpha$ *is a direct summand of* $B / \mu \operatorname{Ker} \alpha$.

In the proof of (i) (see Tellman [1]), we start with the sequences

$$0 \to \operatorname{Im} \alpha \xrightarrow{\phi} A \xrightarrow{\psi} \operatorname{Im} \mu/\operatorname{Im} \mu\alpha \to 0$$

and

$$0 \to \operatorname{Ker} \alpha \xrightarrow{\chi} A \xrightarrow{\bar{\alpha}} \operatorname{Im} \alpha \to 0,$$

which are manifestly exact if the mappings are the obvious ones. These yield the exact sequences

(2)     $\operatorname{Ext}(C, \operatorname{Im} \alpha) \xrightarrow{\phi_*} \operatorname{Ext}(C, A) \xrightarrow{\psi_*} \operatorname{Ext}(C, \operatorname{Im} \mu/\operatorname{Im} \mu\alpha) \to 0,$

(3)     $\operatorname{Ext}(C, \operatorname{Ker} \alpha) \xrightarrow{\chi_*} \operatorname{Ext}(C, A) \xrightarrow{\bar{\alpha}_*} \operatorname{Ext}(C, \operatorname{Im} \alpha) \to 0.$

Evidently, $\alpha_* = \phi_* \bar{\alpha}_*$ and $\bar{\alpha}_*$ is epic, whence $\operatorname{Im} \alpha_* = \operatorname{Im} \phi_* \bar{\alpha}_* = \operatorname{Im} \phi_* = \operatorname{Ker} \psi_*$. According to the definition of $\psi_*$,

$$\psi_* E = \psi E : 0 \to \operatorname{Im} \mu/\operatorname{Im} \mu\alpha \to B/\operatorname{Im} \mu\alpha \xrightarrow{v'} C \to 0$$

with $v'(b + \operatorname{Im} \mu\alpha) = vb$, whence (i) follows.

Turning to (ii), note that $\operatorname{Im} \alpha = A$ implies $\alpha = \bar{\alpha}$, whence $\operatorname{Ker} \alpha_* = \operatorname{Ker} \bar{\alpha}_* = \operatorname{Im} \chi_*$. From (i) we infer that $E \in \operatorname{Im} \chi_*$ is equivalent to the stated condition. $\square$

**Theorem 53.2** (Baer [4]). *Let $E$ be as in (1) and $\gamma : C \to C$ an endomorphism of $C$; then for the induced endomorphism $\gamma^*$ of $\operatorname{Ext}(C, A)$ the following holds:*

(i) *$E \in \operatorname{Im} \gamma^*$ exactly if $\operatorname{Im} \mu$ is a direct summand of $v^{-1} \operatorname{Ker} \gamma$;*

(ii) *in case $\operatorname{Ker} \gamma = 0$, $E \in \operatorname{Ker} \gamma^*$ if, and only if, $\operatorname{Im} \mu$ is a direct summand of $v^{-1} \operatorname{Im} \gamma$.*

The proof of (i) is analogous to the preceding one. One starts with the exact sequences

$$0 \to \operatorname{Im} \gamma \xrightarrow{\phi} C \xrightarrow{\psi} C/\operatorname{Im} \gamma \to 0 \qquad \text{and} \qquad 0 \to \operatorname{Ker} \gamma \xrightarrow{\chi} C \xrightarrow{\bar{\gamma}} \operatorname{Im} \gamma \to 0$$

to obtain the exact sequences

$$\operatorname{Ext}(C/\operatorname{Im} \gamma, A) \xrightarrow{\psi^*} \operatorname{Ext}(C, A) \xrightarrow{\phi^*} \operatorname{Ext}(\operatorname{Im} \gamma, A) \to 0,$$

$$\operatorname{Ext}(\operatorname{Im} \gamma, A) \xrightarrow{\bar{\gamma}^*} \operatorname{Ext}(C, A) \xrightarrow{\chi^*} \operatorname{Ext}(\operatorname{Ker} \gamma, A) \to 0.$$

Now $\gamma^* = \bar{\gamma}^* \phi^*$, and $\phi^*$ epic implies $\operatorname{Im} \gamma^* = \operatorname{Im} \bar{\gamma}^* = \operatorname{Ker} \chi^*$. Here

$$\chi^* E = E\chi : 0 \to A \xrightarrow{\mu} v^{-1} \operatorname{Ker} \gamma \xrightarrow{v} \operatorname{Ker} \gamma \to 0,$$

whence (i) follows.

If $\operatorname{Ker} \gamma = 0$, then $\bar{\gamma}$ is an isomorphism, and so is $\bar{\gamma}^*$. Hence $\operatorname{Ker} \gamma^* = \operatorname{Ker} \phi^*$. Since $\phi^* E = E\phi : 0 \to A \xrightarrow{\mu} v^{-1} \operatorname{Im} \gamma \to \operatorname{Im} \gamma \to 0$, the assertion of (ii) is evident. $\square$

One of the corollaries we draw is the following theorem.

**Theorem 53.3** (Nunke [1], Fuchs [7]). *E in* (1) *belongs to* $n \, \text{Ext}(C, A)$ *if and only if* $\mu(nA) = \mu A \cap nB$. *It belongs to the first Ulm subgroup of* $\text{Ext}(C, A)$ *if and only if* (1) *is a pure-exact sequence.*

If $\alpha$ is the multiplication by $n$ in $A$, then (52.1) asserts that $\text{Im} \, \alpha_* = n \, \text{Ext}(C, A)$. By (53.1), $E \in n \, \text{Ext}(C, A)$ exactly if $\mu A / n \mu A$ is a direct summand of $B / n \mu A$. If this holds, then $\mu(nA) = \mu A \cap nB$ is obviously true, while the converse follows from (27.5), since the purity of $\mu A / n \mu A$ in $B / n \mu A$ is readily checked. A simple appeal to (27.10) proves the second half of the theorem. $\square$

Consequently, the extensions $E$ of $A$ by $C$ represented by pure-exact sequences form a subgroup of $\text{Ext}(C, A)$ which we call the *group of pure extensions* of $A$ by $C$ and denote by $\text{Pext}(C, A)$ (Harrison [1]). Our theorem asserts:

$$\text{Pext}(C, A) = \text{Ext}(C, A)^1 = \bigcap_n n \, \text{Ext}(C, A).$$

[It is easy to check that an extension equivalent to a pure extension is pure.]
With the aid of Pext, we can reformulate earlier results.

**Proposition 53.4.** *The group $A$ has the property that* $\text{Pext}(C, A) = 0$ *for all $C$ if, and only if, $A$ is algebraically compact. The group $C$ satisfies* $\text{Pext}(C, A) = 0$ *for all $A$ exactly if $C$ is a direct sum of cyclic groups.*

The first assertion follows from the definition of algebraic compactness, the second from (30.2). $\square$

Recall that Ext is a functor and the Ulm subgroups are functorial, thus Pext is again a functor. If $\alpha : A \rightarrow A'$ and $\gamma : C' \rightarrow C$, then the restriction of $\text{Ext}(\gamma, \alpha)$ yields the map

$$\text{Pext}(\gamma, \alpha) : \text{Pext}(C, A) \rightarrow \text{Pext}(C', A').$$

In view of this, it is evident that Pext is an additive bifunctor on $\mathscr{A} \times \mathscr{A}$ to $\mathscr{A}$; it is contravariant in the first and covariant in the second variable. In order to discover its behavior towards short exact sequences, let us first prove the following lemmas.

**Lemma 53.5** (Fuchs [9]). *If* (1) *is a pure-exact sequence, then the induced homomorphisms*

$$\nu^* : \text{Ext}(C, G) \rightarrow \text{Ext}(B, G) \qquad and \qquad \mu_* : \text{Ext}(G, A) \rightarrow \text{Ext}(G, B)$$

*map upon pure subgroups.*

Given the top row, we have in turn the exact rows in the following commutative diagram:

$$0 \to G \to H \to C \longrightarrow 0 \qquad (\in \mathrm{Ext}(C, G))$$

$$0 \to G \to K \to B \longrightarrow 0 \qquad (\in \mathrm{Im}\ v^*)$$

$$0 \to G \to K' \to B[n] \to 0$$

$$0 \to G \to H' \to C[n] \to 0 \qquad (\in (v\iota\rho)^*\ \mathrm{Ext}(C,G))$$

where $\iota$ is the injection map and a $\rho : C[n] \to B[n]$ exists, by (28.4), such that $v\iota\rho$ is the injection $C[n] \to C$. Now assume the second row $\in n\ \mathrm{Ext}(B, G)$; this means, by (53.2) (i), that the third row splits. It follows that the bottom row splits, and hence, again by (53.2) (i), the top row $\in n\ \mathrm{Ext}(C, G)$. We infer that the second row $\in n\ \mathrm{Im}\ v^*$, proving the purity of $\mathrm{Im}\ v^*$ in $\mathrm{Ext}(B, G)$.

The second assertion follows similarly, by making use of the following commutative diagram

$$0 \to \quad A \longrightarrow H \to G \to 0 \qquad (\in \mathrm{Ext}(G, A))$$

$$0 \to \quad B \longrightarrow K \to G \to 0 \qquad (\in \mathrm{Im}\ \mu_*)$$

$$0 \to B/nB \to K' \to G \to 0$$

$$0 \to A/nA \to H' \to G \to 0 \qquad (\in (\rho\pi\mu)\ \mathrm{Ext}(G, A))$$

where the rows are exact, $\pi$ is the natural map and $\rho : B/nB \to A/nA$ exists, due to the assumption on (1), such that $\rho\pi\mu : A \to A/nA$ is the natural map. The argument is the same as above, only the reference to (53.2) (i) must be replaced by that to (53.1) (i). □

**Lemma 53.6.** *Assume that* (1) *represents an element of* $\mathrm{Ext}\ (C, A)^\sigma$, *the $\sigma$th Ulm subgroup of* $\mathrm{Ext}(C, A)$. *Then for every group $G$, the images of the connecting homomorphisms*

$$E^* : \mathrm{Hom}(A, G) \to \mathrm{Ext}(C, G), \qquad E_* : \mathrm{Hom}(G, C) \to \mathrm{Ext}(G, A)$$

*are subgroups of* $\mathrm{Ext}(C, G)^\sigma$ *and* $\mathrm{Ext}(G, A)^\sigma$, *respectively. In particular, if* (1) *is pure-exact, then* $\mathrm{Im}\ E^*$ *and* $\mathrm{Im}\ E_*$ *are contained in the first Ulm subgroups of the respective groups.*

Every $\eta \in \mathrm{Hom}(A, G)$ induces a map $\eta_* : \mathrm{Ext}(C, A) \to \mathrm{Ext}(C, G)$ which carries (1) into the bottom row of the commutative diagram

$$E : 0 \to A \xrightarrow{\mu} B \xrightarrow{\nu} C \to 0$$
$$\downarrow{\eta} \qquad \downarrow \qquad \parallel$$
$$\eta E : 0 \to G \longrightarrow B' \xrightarrow{\nu'} C \to 0.$$

Any homomorphism maps the $\sigma$th Ulm subgroup into the $\sigma$th Ulm subgroup, thus $\eta E \in \mathrm{Ext}(C, G)^\sigma$. But the bottom row is exactly the image of $\eta$ under $E^*$. The dual proof establishes the statement on $E_*$, while the last assertion is a simple corollary, since (53.3) is available. $\square$

Returning to the functor Pext, our immediate goal is to prove the following analog of (51.3).

**Theorem 53.7** (Harrison [1]). *If $E : 0 \to A \xrightarrow{\alpha} B \xrightarrow{\beta} C \to 0$ is pure-exact, then for every group $G$, the following induced sequences are exact:*

(4)
$$0 \to \mathrm{Hom}(C, G) \to \mathrm{Hom}(B, G) \to \mathrm{Hom}(A, G)$$
$$\xrightarrow{E^*} \mathrm{Pext}(C, G) \xrightarrow{\beta^*} \mathrm{Pext}(B, G) \xrightarrow{\alpha^*} \mathrm{Pext}(A, G) \to 0,$$

(5)
$$0 \to \mathrm{Hom}(G, A) \to \mathrm{Hom}(G, B) \to \mathrm{Hom}(G, C)$$
$$\xrightarrow{E_*} \mathrm{Pext}(G, A) \xrightarrow{\alpha_*} \mathrm{Pext}(G, B) \xrightarrow{\beta_*} \mathrm{Pext}(G, C) \to 0.$$

By (53.6) and what has been said above about Pext, it is clear that (4) and (5) make sense. (51.3) and (53.5) imply that

$$0 \to \mathrm{Ext}(C, G)/\mathrm{Im}\ E^* \xrightarrow{\beta^*} \mathrm{Ext}(B, G) \xrightarrow{\alpha^*} \mathrm{Ext}(A, G) \to 0$$

is pure-exact. From (53.6) and (37.1) we conclude that $(\mathrm{Ext}(C, G)/\mathrm{Im}\ E^*)^1 = \mathrm{Ext}(C, G)^1/\mathrm{Im}\ E^*$. Since intersection is an inverse limit, (12.3) implies the exactness of (4) except for the assertion that $\alpha^*$ is epic. Let $E_0 : 0 \to H \to F \to B \to 0$ be a pure-projective resolution of $B$; then $E_0 \alpha : 0 \to H \to F' \to A \to 0$ will be a pure-projective resolution of $A$. From what has already been proved, the exact sequence $\mathrm{Hom}(H, G) \to \mathrm{Pext}(A, G) \to \mathrm{Pext}(F', G) = 0$ arises. Thus the first map $(E_0 \alpha)^* = \alpha^* E_0^*$ is epic, so $\alpha^*$ is epic, too. The exactness of (5) can be verified in a similar way. $\square$

One particular case warrants special mention:

**Corollary 53.8.** *Let $A$ be a group such that $A^1 = 0$. Then*

$$\mathrm{Pext}(Q/Z, A) \cong \mathrm{Hom}(Q/Z, \hat{A}/A)$$

*where $\hat{A}$ is the Z-adic completion of $A$.*

Starting with the pure-exact sequence $0 \to A \to \hat{A} \to \hat{A}/A \to 0$, (5) implies the exactness of

$0 = \text{Hom}(Q/Z, \hat{A}) \to \text{Hom}(Q/Z, \hat{A}/A) \to \text{Pext}(Q/Z, A) \to \text{Pext}(Q/Z, \hat{A}) = 0$;
the last group vanishes by the algebraic compactness of $\hat{A}$.☐

This corollary will be generalized in (57.3).

EXERCISES

1.  (Baer [4]) Let $\alpha : A \to A$ be epic. Then for the induced endomorphism $\alpha_*$ of $\text{Ext}(C, A)$, $\text{Ker } \alpha_*$ is an epimorphic image of $\text{Ext}(C, \text{Ker } \alpha)$.
2.  (Baer [4]) Assume $\alpha$ is an endomorphism of $A$ such that $\text{Ext}(C, \text{Ker } \alpha) = 0$. Then the kernel of the induced endomorphism $\alpha_*$ of $\text{Ext}(C, A)$ is isomorphic to $\text{Hom}(C, A/\alpha A)/\eta_* \text{Hom}(C, A)$, where $\eta_*$ is induced by the canonical map $\eta : A \to A/\alpha A$. [Hint: $\bar{\alpha}_*$ is an isomorphism in (3), extend (2) to the left.]
3.  (Baer [4]) Let $\gamma : C \to C$ be monic. Then the kernel of the induced endomorphism $\gamma^*$ of $\text{Ext}(C, A)$ is an epimorphic image of $\text{Ext}(C/\text{Im } \gamma, A)$.
4.  The extension $E$ in (1) belongs to the Frattini subgroup of $\text{Ext}(C, A)$ if, and only if, $\text{Im } \mu$ is neat in $B$.
5.  $E$ in (1) is $p$-pure-exact exactly if it represents an element of $p^\omega \text{Ext}(C, A)$.
6.  Establish the natural isomorphisms

$$\text{Pext}(\oplus C_i, A) \cong \prod \text{Pext}(C_i, A), \qquad \text{Pext}(C, \prod A_j) \cong \prod \text{Pext}(C, A_j).$$

7.  Let $A$ be a direct sum of cyclic $p$-groups and $C$ a divisible $p$-group. Prove that those extensions of $A$ by $C$ in which $A$ is a basic subgroup form a subgroup of $\text{Ext}(C, A)$. Determine the structure of this subgroup. [Hint: (53.8).]
8.  Let $0 \to A \to B \to C \to 0$ be pure-exact and $G$ algebraically compact. Then

$$0 \to \text{Ext}(C, G) \to \text{Ext}(B, G) \to \text{Ext}(A, G) \to 0$$

is splitting exact. [Hint: (53.6), **52** (O) and (53.5).]
9.  Let $0 \to A \to B \to C \to 0$ be pure-exact and $F$ a direct sum of cyclic groups. Then the sequence $0 \to \text{Ext}(F, A) \to \text{Ext}(F, B) \to \text{Ext}(F, C) \to 0$ is splitting exact.
10. $0 \to A \to B \to C \to 0$ is pure-exact if and only if for every $m > 0$,

$$0 \to \text{Ext}(C, Z(m)) \to \text{Ext}(B, Z(m)) \to \text{Ext}(A, Z(m)) \to 0$$

or

$$0 \to \text{Ext}(Z(m), A) \to \text{Ext}(Z(m), B) \to \text{Ext}(Z(m), C) \to 0$$

is exact (or equivalently, pure-exact).
11. Let $A$ be algebraically compact and $T$ the torsion part of $C$. Prove the natural isomorphism $\text{Ext}(C, A) \cong \text{Ext}(T, A)$.

12.  Let $A$ be torsion-free and $T$ the torsion part of $C$. Then

$$\text{Ext}(C, A) \cong \text{Ext}(T, A) \oplus \text{Ext}(C/T, A).$$

13.  If $A$ is torsion-free and $C$ is a torsion group with basic subgroup $B$, then
$$\text{Ext}(C, A) \cong \text{Ext}(B, A) \oplus \text{Ext}(C/B, A).$$

Show that this enables one to evaluate the invariants of the algebraically compact group $\text{Ext}(C, A)$.

14.  (Baer [4]) (a) Let $A$ be a torsion group and $C$ torsion-free such that $C/pC$ is finite for a prime $p$. Then $\text{Ext}(C, A)[p] = 0$. [*Hint*: Ex. 2; show that every $C \to A/pA$ is induced by a $\xi : C \to A$ with $\text{Im} \, \xi$ a finite subgroup of the basic subgroup of $A$.]
(b) Let $A$ be a torsion group and $C$ a torsion-free group of finite rank. Prove that $\text{Ext}(C, A)$ is torsion-free divisible.
(c) Under the assumptions of (b), $\text{Ext}(C, A) = 0$ or $Q^n$ with $n \geq \aleph_0$. [*Hint*: (52.1).]

15.  If the first Ulm subgroup of $A$ vanishes, then

$$\text{Ext}(Q, A) \cong \text{Hom} \, (Q, \hat{A}/A).$$

## 54.  COTORSION GROUPS

The notion of cotorsion groups is fundamental in the study of Ext. It was discovered by Harrison [1] and found independently by Nunke [1] and Fuchs [9].

A group $G$ is called *cotorsion* if

$$\text{Ext}(J, G) = 0 \qquad \text{for every torsion-free group } J.$$

In other words, $G$ is cotorsion if every extension of $G$ by a torsion-free group splits. Since this means that a cotorsion group is a direct summand in every group in which it is contained with torsion-free quotient group, it is evident that algebraically compact groups are cotorsion. We shall see that the converse is not true.

Every torsion-free group $J$ can be embedded in an exact sequence $0 \to J \to \oplus Q$ with sufficiently many copies of $Q$. This implies by (51.3) the exactness of $\text{Ext}(\oplus Q, G) \cong \prod \text{Ext}(Q, G) \to \text{Ext}(J, G) \to 0$, showing that $\text{Ext}(Q, G) = 0$ already guarantees that $G$ is cotorsion. Thus, the definition of cotorsion groups may also be given as groups $G$ satisfying

$$\text{Ext}(Q, G) = 0.$$

It is clear from (38.5) and (53.4) that $G$ *is algebraically compact if, and only if, it satisfies* $\text{Ext}(Q, G) = 0$ *and* $\text{Pext}(Q/Z, G) = 0$.

It is convenient to list here the following more or less elementary results on cotorsion groups.

(A) *An epimorphic image of a cotorsion group is cotorsion.*
If $G$ is cotorsion and if $G \to H \to 0$ is exact, then the sequence $0 = \mathrm{Ext}(Q, G) \to \mathrm{Ext}(Q, H) \to 0$ is exact. Thus $\mathrm{Ext}(Q, H) = 0$ and $H$ is cotorsion.

(B) *Let $G$ be reduced and cotorsion. For a subgroup $H$ of $G$ to be cotorsion it is necessary and sufficient that $G/H$ is reduced.*
The exact sequence $0 \to H \to G \to G/H \to 0$ yields the exact sequence

$$0 = \mathrm{Hom}(Q, G) \to \mathrm{Hom}(Q, G/H) \to \mathrm{Ext}(Q, H) \to \mathrm{Ext}(Q, G) = 0$$

whence $\mathrm{Ext}(Q, H) \cong \mathrm{Hom}(Q, G/H)$. The latter group vanishes exactly if $G/H$ is reduced.

(C) *If $G$ is reduced and cotorsion, then for every endomorphism $\theta$ of $G$, both $\mathrm{Ker}\,\theta$ and $\mathrm{Im}\,\theta$ are cotorsion.*
This follows from (A) and (B) directly.

(D) *If $H$ is a subgroup of $G$ such that both $H$ and $G/H$ are cotorsion, then $G$ is cotorsion.*
Again from $0 \to H \to G \to G/H \to 0$ we conclude that $0 = \mathrm{Ext}(Q, H) \to \mathrm{Ext}(Q, G) \to \mathrm{Ext}(Q, G/H) = 0$ is exact.

(E) *A direct product $\prod_{i \in I} G_i$ is cotorsion if, and only if, every summand $G_i$ is cotorsion.*
This is an immediate consequence of the isomorphism $\mathrm{Ext}(Q, \prod_i G_i) \cong \prod_i \mathrm{Ext}(Q, G_i)$.

(F) *The inverse limit of reduced cotorsion groups is a reduced cotorsion group.*
The inverse limit $G^*$ of reduced cotorsion groups $G_i$ is a subgroup of $\prod G_i$; the latter is, by (E), reduced and cotorsion. Thus, by (B), we need only verify the reducedness of the quotient group $(\prod G_i)/G^*$. Recall that $G^*$ is the intersection of kernels of certain endomorphisms $\theta$ of $\prod G_i$, and therefore $(\prod G_i)/G^*$ is a subdirect sum of the groups $\mathrm{Im}\,\theta$. The reducedness of $\mathrm{Im}\,\theta$ completes the proof.

(G) *If $G$ is cotorsion, then $\mathrm{Hom}(A, G)$ is cotorsion for any $A$.*
It suffices to prove this, because of (47.7) for reduced $G$. Let $0 \to H \to F \to A \to 0$ be a free resolution of $A$. From (44.4) we infer the exactness of $0 \to \mathrm{Hom}(A, G) \to \mathrm{Hom}(F, G) \to \mathrm{Hom}(H, G)$. Here the middle group is the direct product of copies of $G$, hence it is a reduced cotorsion group. $\mathrm{Hom}(H, G)$ being reduced, (B) implies what was asserted.
The next statement is a fundamental isomorphism.

(H) *For a reduced cotorsion group G, there is a natural isomorphism*

$$\text{Ext}(Q/Z, G) \cong G.$$

Let us start off with the exact sequence $0 \to Z \to Q \to Q/Z \to 0$ which we shall frequently use in the sequel. This induces the exact sequence

$$0 = \text{Hom}(Q, G) \to \text{Hom}(Z, G) \cong G \to \text{Ext}(Q/Z, G) \to \text{Ext}(Q, G) = 0.$$

The connecting homomorphism produces the desired isomorphism.

(I) *A reduced cotorsion group G may be written uniquely in the form*

$$G = \prod_p G_p$$

*where, for each prime p, $G_p$ is a reduced cotorsion group which is a p-adic module.*

By (H), we may write

$$G \cong \text{Ext}(Q/Z, G) = \text{Ext}\Big(\bigoplus_p Z(p^\infty), G\Big) = \prod_p \text{Ext}(Z(p^\infty), G).$$

Since the $p$-adic numbers act on $Z(p^\infty)$, from (52.1) we infer that $G_p = \text{Ext}(Z(p^\infty), G)$ is a $p$-adic module which is by (E) a cotorsion group. The uniqueness of the decomposition follows from the fact that if we iterate transfinitely the process of taking the intersections $\bigcap nG$ with $(n, p) = 1$, then the process terminates at $G_p$.

Now we are ready to derive a number of consequences which enable us to get a deeper insight into the nature of cotorsion groups, especially, to find out the precise relationship of cotorsion to algebraically compact groups.

**Proposition 54.1.** *A group is cotorsion if and only if it is an epimorphic image of an algebraically compact group.*

Because of (A), it suffices to prove that a reduced cotorsion group $G$ is an epimorph of a suitable algebraically compact group $A$. We can embed $G$ in an exact sequence $0 \to G \to D \to D/G \to 0$ with divisible $D$, whence we get an exact sequence

$$\text{Hom}(Q/Z, D/G) \to \text{Ext}(Q/Z, G) \to \text{Ext}(Q/Z, D) = 0.$$

Here $\text{Hom}(Q/Z, D/G)$ is algebraically compact, as follows from (46.1) or (47.7), while the middle group is by (H) isomorphic to $G$, and the assertion follows.☐

**Proposition 54.2** (Fuchs [9]). *A reduced cotorsion group is algebraically compact if and only if its first Ulm subgroup vanishes.*

By one of our introductory remarks, a cotorsion $G$ is algebraically compact exactly if $\text{Pext}(Q/Z, G) = 0$. In view of (53.3), assertion (H) concludes the proof.□

**Theorem 54.3** (Fuchs [9]). *The Ulm subgroups of cotorsion groups are again cotorsion, and the Ulm factors of cotorsion groups are algebraically compact.*

If $G^\sigma$ is the $\sigma$th Ulm subgroup of the cotorsion group $G$, then $G/G^\sigma$ is reduced, and hence the first statement follows from (B). Owing to (A), the 0th Ulm factor $G^\sigma/G^{\sigma+1}$ of $G^\sigma$ is again cotorsion. Its first Ulm subgroup vanishes, so it is, by (54.2), algebraically compact.□

A finite number of applications of (D) leads us to the conclusion [which is a partial converse of (54.3)]: if $G$ is of finite Ulm length and all of its Ulm factors are algebraically compact, then $G$ is cotorsion. If, however, $G$ is of infinite Ulm length, then it need not be cotorsion even if all Ulm factors of $G$ are algebraically compact.

**Corollary 54.4** (Harrison [1], Nunke [1]). *A torsion group is cotorsion if and only if it is a direct sum of a divisible group and a bounded group.*

The first Ulm factor of a torsion cotorsion group $G$ is a reduced torsion algebraically compact group, hence, by (40.3), it is bounded. If $m$ is a bound, then $mG \leq G^1$ implies $mG \leq nmG$ for every $n$, whence $mG = G^1$ is divisible. Consequently, $G$ has the indicated structure. The converse is trivial.□

**Corollary 54.5** (Fuchs [9]). *A necessary and sufficient condition for a torsion-free group to be cotorsion is algebraic compactness.*

Because of the unicity of division in torsion-free groups $G$, $G^1$ must be divisible. Thus any torsion-free group $G$ is the direct sum of a divisible group and a group isomorphic to its 0th Ulm factor $G_0$. By (E), $G$ is cotorsion exactly if $G_0$ is cotorsion. By (54.2), the cotorsion character of $G_0$ is equivalent to its algebraic compactness.□

The following result is a trivial consequence of an isomorphism, proved in homological algebra, stating a kind of associativity for Ext. Our knowledge of cotorsion groups makes it possible to prove without making use of the mentioned isomorphism:

**Theorem 54.6.** $\text{Ext}(C, A)$ *is cotorsion for all* $A$, $C$.

We embed $A$ in an exact sequence $0 \to A \to D \to D/A \to 0$ with divisible $D$ to get the exact sequence

$$\text{Hom}(C, D/A) \to \text{Ext}(C, A) \to \text{Ext}(C, D) = 0.$$

$D/A$ being divisible, from (47.7) we infer that the first group is algebraically compact. Thus $\text{Ext}(C, A)$ is—as an epimorphic image of an algebraically compact group—cotorsion.□

This result yields a device for manufacturing as many cotorsion groups as desired—simply take Ext($C$, $A$) for any two groups $C$ and $A$. Assertion (H) shows that this procedure is the most general in the sense that all reduced cotorsion groups arise in this way. On the other hand, (54.6) also shows that, in order to get more information about the structure of Ext, one has to study more carefully the cotorsion groups in general.

EXERCISES

1.   Let $G$ be a reduced cotorsion group and $H$ a subgroup of $G$. There is a unique minimal cotorsion subgroup of $G$ containing $H$.

2.   Show that assertion (C) fails to hold in general if the reducedness hypothesis is dropped.

3.   If $G$ is a cotorsion group with bounded torsion part, then $G$ is algebraically compact.

4.   A countable cotorsion group is the direct sum of a divisible and a bounded group.

5.   Let $G \neq 0$ be a reduced cotorsion group. Then it has a direct summand which is a cyclic $p$-adic module, for some prime $p$.

6.   (Rotman [1]) A reduced group $G$ is cotorsion if and only if $E/G$ is reduced whenever $E$ is a reduced group containing $G$.

7.*  For a reduced cotorsion $G$, describe the Ulm factors of Hom($A$, $G$). [*Hint*: consider the torsion part $T$ of $A$ and $p$-basic subgroups of $A/T$.]

8.   Prove the analog of (39.9) for reduced cotorsion $A$.

9.   Show that the natural homomorphism $\mu$ of a cotorsion group $G$ into its $Z$-adic completion $\hat{G}$ is an epimorphism.

10.  If $D$ is the divisible hull of the cotorsion group $G$ and $E$ ($\leq D$) is the divisible hull of $G^1$, then $E + G$ is algebraically compact; moreover, it is the pure-injective hull of $G$.

11.  Let $A$ satisfy $A^1 = 0$, and let $\hat{A}$ be its $Z$-adic completion. Show that the 0th Ulm factor of Ext($C$, $A$) is isomorphic to Ext($C$, $\hat{A}$). [*Hint*: $0 \to A \to \hat{A} \to \hat{A}/A \to 0$ is pure-exact and $\hat{A}/A$ is divisible.]

12.  Prove that to every cotorsion group $G$ there exists a cotorsion $A$ such that $A^1 \cong G$. [*Hint*: **37**, Ex. 11, and (24.6).]

13.* (a) For an integer $n$, find a cotorsion group of Ulm length $n$. [*Hint*: Ex. 12.]
     (b) Construct a cotorsion group of Ulm length $\omega$. [*Hint*: inverse limit.]

14.* There exists a group of Ulm length $\omega$ whose Ulm factors are algebraically compact, but which fails to be cotorsion. [*Hint*: if $G$ is cotorsion of length $\omega$, define a subgroup in $\varprojlim G/G^n$ of the desired kind.]

15.  Prove (54.5) by making use of (54.1) and **38**, Ex. 3.

## 55.   THE STRUCTURE OF COTORSION GROUPS

Our next purpose is to expound the theory of cotorsion groups in full. Though there is an apparent analogy between algebraically compact and cotorsion groups [which will be set forth in the next sections], there is no general structure theorem on cotorsion groups. Namely, it turns out that the classifications of torsion and cotorsion groups are essentially the same, and —as will be seen in Chapter XII—a complete structure theorem can be given only for some classes of torsion groups which are not too large.

As a preparation for the main result, we prove two important lemmas.

**Lemma 55.1** (Harrison [1]). *If $T$ is a reduced torsion group, then the torsion part of $\mathrm{Ext}(Q/Z, T)$ is isomorphic to $T$ and the corresponding quotient group is torsion-free and divisible.*

From the exact sequence $0 \to Z \to Q \to Q/Z \to 0$ we derive the exact sequence

$$0 = \mathrm{Hom}(Q, T) \to \mathrm{Hom}(Z, T) \cong T$$

$$\to \mathrm{Ext}(Q/Z, T) \to \mathrm{Ext}(Q, T) \to \mathrm{Ext}(Z, T) = 0.$$

Hence the lemma will be proved as soon as we have proved that $\mathrm{Ext}(Q, T)$ is torsion-free and divisible. This being true because of **52** (J), the proof is finished.☐

**Lemma 55.2.** *If $T$ is the torsion part of the mixed group $A$, then*

$$\mathrm{Ext}(Q/Z, A) \cong \mathrm{Ext}(Q/Z, T) \oplus \mathrm{Ext}(Q/Z, A/T).$$

With the notation $J = A/T$, the exactness of $0 \to T \to A \to J \to 0$ yields the exact sequence

$$0 = \mathrm{Hom}(Q/Z, J) \to \mathrm{Ext}(Q/Z, T) \to \mathrm{Ext}(Q/Z, A) \to \mathrm{Ext}(Q/Z, J) \to 0.$$

From (52.3) we know that the last group is isomorphic to $\mathrm{Hom}(Q/Z, D/J)$ where $D$ is a divisible hull of $J$. Thus $\mathrm{Ext}(Q/Z, J)$ is the direct product of $\mathrm{Hom}(Z(p^\infty), D/J)$, with $p$ ranging over the primes. (44.3) shows that they are torsion-free. Since $\mathrm{Ext}(Q/Z, T)$ is by (54.6) cotorsion, the last exact sequence splits. Hence the claimed isomorphism.☐

**Lemma 55.3.** *For any torsion group $C$, $\mathrm{Ext}(C, A)$ is reduced.*

If $C$ is torsion, there is a pure-exact sequence $0 \to H \to F \to C \to 0$ with $F$ [and hence $H$] a direct sum of finite cyclic groups [cf. (30.1)]. Hence (53.7) yields the exact sequence

$$0 \to \mathrm{Hom}(C, A) \to \mathrm{Hom}(F, A) \to \mathrm{Hom}(H, A) \to \mathrm{Pext}(C, A) \to 0,$$

since $\mathrm{Pext}(F, A) = 0$ by (53.4). From (46.1) we know that here the Homs are reduced algebraically compact, so by (44.7), $\mathrm{Hom}(C, A)$ is a direct summand of $\mathrm{Hom}(F, A)$. Therefore we obtain an exact sequence $0 \to G \to \mathrm{Hom}(H, A) \to \mathrm{Pext}(C, A) \to 0$ with $G$ and $\mathrm{Hom}(H, A)$ reduced algebraically compact. Applying (51.3), the exact sequence

$$0 = \mathrm{Hom}(Q, \mathrm{Hom}(H, A)) \to \mathrm{Hom}(Q, \mathrm{Pext}(C, A)) \to \mathrm{Ext}(Q, G) = 0$$

arises. This shows that $\mathrm{Pext}(C, A)$, and so $\mathrm{Ext}(C, A)$ is reduced.☐

A cotorsion group that is reduced and has no nonzero torsion-free direct summands is called *adjusted*. For $C = Q/Z$ and torsion groups $A$, (55.3) can be improved:

**Lemma 55.4** (Harrison [1]). *If $T$ is a torsion group, $\mathrm{Ext}(Q/Z, T)$ is adjusted cotorsion.*

By (55.1), every torsion-free direct summand of $\mathrm{Ext}(Q/Z, T)$ must be divisible. An application of (55.3) completes the proof.☐

Now we are in a position to verify the two fundamental results on the structure of cotorsion groups.

**Theorem 55.5** (Harrison [1]). *Let $G$ be a reduced cotorsion group and $T$ its torsion part. There is a direct decomposition*

$$(1) \qquad\qquad\qquad G = A \oplus C$$

*where $A$ is torsion-free and algebraically compact, and $C \cong \mathrm{Ext}(Q/Z, T)$ is an adjusted cotorsion group. $C$ is a uniquely determined subgroup of $G$.*

Combining **54** (H) with (55.2), we obtain

$$G \cong \mathrm{Ext}(Q/Z, G) \cong \mathrm{Ext}(Q/Z, T) \oplus \mathrm{Ext}(Q/Z, G/T).$$

By (55.4), the first summand is adjusted cotorsion; let it be denoted by $C$. The second summand, say $A$, is, in view of (52.3) and (44.3), a reduced and torsion-free algebraically compact group. The uniqueness of $C$ follows at once from the observation that $C$ contains $T$, $C/T$ is torsion-free and divisible [cf. (55.1)], and so $C/T$ is precisely the maximal divisible subgroup of $G/T$.☐

Consequently, we are justified to speak of $C$ as *the adjusted part* of the reduced cotorsion group $G$. The last theorem has the important consequence that any cotorsion group $G$ may be decomposed into a direct sum of three groups:

$$G = A \oplus C \oplus D$$

where $D$ is the maximal divisible subgroup, $C$ is adjusted cotorsion, and $A$ is reduced, torsion-free, and algebraically compact. This decomposition of $G$ is unique up to isomorphism, since both $D$ and $C \oplus D$ are uniquely determined

subgroups of $G$. We know that both $A$ and $D$ can be completely characterized by cardinal invariants in a satisfactory manner; consequently, the
structure problem for cotorsion groups is reduced to adjusted cotorsion
groups. The next remarkable theorem makes the structure problem of these
groups equivalent to that of reduced torsion groups.

**Theorem 55.6** (Harrison [1]). *The correspondence*

$$(2) \qquad\qquad T \mapsto \operatorname{Ext}(Q/Z, T) = G$$

*is one-to-one between the class of reduced torsion groups $T$ and the class of ad-
justed cotorsion groups $G$. The inverse of (2) is the formation of the torsion part of $G$.*

Let $T$ be reduced and torsion. By (55.4), (2) yields an adjusted cotorsion
$G$ whose torsion part is, on account of (55.1), isomorphic to $T$. On the other
hand, let $G$ be adjusted cotorsion and $T$ its torsion part. Then (1) holds with
$A = 0$, since $G$ is adjusted. Thus $G = C \cong \operatorname{Ext}(Q/Z, T)$.□

This theorem assigns to $G$ the same invariants as those of $T$, therefore,
it can be thought of, in a wider sense, as the structure theorem for adjusted
cotorsion groups in the cases when $T$ is known, in particular, when $T$ is
countable or a direct sum of countable groups [see chapter XII].

EXERCISES

1.  Let $A$ be cotorsion and $C$ adjusted cotorsion. If $\eta : A \to C$ and $\operatorname{Im} \eta$
    contains the torsion part of $C$, then $\eta$ is epic.
2.  Let $A, C$ be adjusted cotorsion groups and $S, T$ their torsion parts.
    (a) There is a natural isomorphism

    $$\operatorname{Hom}(A, C) \cong \operatorname{Hom}(S, T).$$

    (b) $\operatorname{Hom}(A, C)$ is algebraically compact.
    (c) Every homomorphism of $S$ into $T$ can uniquely be extended to one
    of $A$ into $C$.
3.  Let $G$ be an adjusted cotorsion group and $T$ its torsion part. The corre-
    spondence $\alpha \mapsto \alpha \mid T$ is an isomorphism between the automorphism groups
    of $G$ and $T$.
4.  A reduced cotorsion group is adjusted if and only if for every prime $p$,
    its $p$-basic subgroup is torsion.
5.  If $G$ is adjusted cotorsion with torsion part $T$, then $|G| \leq |T|^{\aleph_0}$.
6.  Let $G$ be an adjusted cotorsion group and $T = T_1 \oplus T_2$ a direct decom-
    position of its torsion part. Then there is a direct decomposition
    $G = G_1 \oplus G_2$ such that $T(G_i) = T_i$ $(i = 1, 2)$.
7.  Prove (55.3) for $C = Q/Z$ and reduced $A$ by starting with a pure-injective
    resolution $0 \to A \to G \to G/A \to 0$ and hence deriving the isomorphism
    $\operatorname{Pext}(Q/Z, A) \cong \operatorname{Hom}(Q/Z, G/A)$.

8.  Let $G$ be adjusted cotorsion and $T$ its torsion part. Show that either $G = T$ or $G/T$ is uncountable. [*Hint*: $G/T \cong \text{Ext}(Q, T)$ has an epimorphism onto $\text{Hom}(Q, \hat{T}/T_0)$ where $\hat{T}$ is the $Z$-adic completion of $T$.]

## 56.  THE ULM FACTORS OF COTORSION GROUPS

Theorem (55.6) asserts that an adjusted cotorsion group $G$ is completely determined by its torsion part $T$, so [both the known and the still undiscovered] invariants of $G$ ought to be expressed in terms of $T$. Recall that, by (54.3), the Ulm factors of $G$ are algebraically compact, and algebraically compact groups can be described by means of cardinal invariants, so it is natural to raise the question as to how these actually depend on $T$. This section is devoted to this question.

In the discussion of the Ulm factors, an important role is played by the following result which is a slightly improved version of a theorem by Irwin *et al.* [1].

**Proposition 56.1.**  *Let $B$ be a subgroup of the $\sigma$th Ulm subgroup $A^\sigma$ of $A$, and $0 \to B \xrightarrow{\alpha} A \xrightarrow{\beta} A/B \to 0$ exact. Then the image of $\alpha_* : \text{Ext}(G, B) \to \text{Ext}(G, A)$ is contained in the $\sigma$th Ulm subgroup $\text{Ext}(G, A)^\sigma$ of $\text{Ext}(G, A)$, and the sequence*

$$0 \to \text{Hom}(G, B) \to \text{Hom}(G, A) \to \text{Hom}(G, A/B)$$

$$\to \text{Ext}(G, B) \xrightarrow{\alpha_*} \text{Ext}(G, A)^\sigma \xrightarrow{\beta_*} \text{Ext}(G, A/B)^\sigma \to 0$$

*is exact for every group $G$.*

Let $0 \to H \to F \to G \to 0$ be a free resolution of $G$. Hence, from (51.3), we obtain the following commutative diagram with exact rows:

$$\begin{array}{ccc}
\text{Hom}(H, B) \xrightarrow{\phi} \text{Ext}(G, B) \to \text{Ext}(F, B) = 0 \\
{\scriptstyle \alpha_*'}\downarrow \qquad\qquad \downarrow{\scriptstyle \alpha_*} \\
\text{Hom}(H, A) \xrightarrow{\psi} \text{Ext}(G, A) \to \text{Ext}(F, A) = 0
\end{array}$$

where $\phi$ and $\psi$ stand for the connecting homomorphisms. Our first claim was $\text{Im } \alpha_* \leqq \text{Ext}(G, A)^\sigma$. To prove this, notice first $\text{Im } \alpha_* = \text{Im } \alpha_* \phi = \text{Im } \psi \alpha_*'$. The group $H$ is free, therefore $\text{Im } \alpha_*'$ must be contained in the $\sigma$th Ulm subgroup of $\text{Hom}(H, A) \cong \prod A$, and this is clearly mapped by $\psi$ into the $\sigma$th Ulm subgroup of $\text{Ext}(G, A)$.

Since $\alpha_* : \text{Ext}(G, A) \to \text{Ext}(G, A/B)$ maps $\sigma$th Ulm subgroup into $\sigma$th Ulm subgroup, we need only show that every $E \in \text{Ext}(G, A/B)^\sigma$ is the image of some $E'$ in $\text{Ext}(G, A)^\sigma$. We do know from (51.3) that an $E' \in \text{Ext}(G, A)$ exists with $\beta_* E' = E$. Since $\text{Ker } \beta_* \leqq \text{Ext}(G, A)^\sigma$, (37.1) shows that no element not in $\text{Ext}(G, A)^\sigma$ can be mapped into the $\sigma$th Ulm subgroup of $\text{Im } \beta_*$ whence $E' \in \text{Ext}(G, A)^\sigma$.☐

**Corollary 56.2.** *Under the hypotheses of* (56.1), *the ρth Ulm factors of* Ext($G$, $A$) *and* Ext($G$, $A/B$) *are isomorphic for* $\rho < \sigma$.

Combining (53.1) and (56.1), it follows at once [from the $3 \times 3$-lemma, for instance] that Ext($G$, $A$)/Ext($G$, $A$)$^\sigma$ is isomorphic to the quotient Ext($G$, $A/B$)/Ext($G$, $A/B$)$^\sigma$, whence the assertion is evident.□

The beginning step in our program is to relate the Ulm length of a torsion group $T$ to that of the corresponding adjusted cotorsion group $G$:

**Proposition 56.3.** *If $T$ is a reduced torsion group of Ulm length $\sigma$, then* Ext($Q/Z$, $T$) $= G$ *has Ulm length $\sigma$ or $\sigma + 1$, $G^\sigma$ is torsion-free, and there is a natural isomorphism*

$$G^\sigma \cong \mathrm{Hom}(Q/Z, G/(G^\sigma + T)).$$

Since $T \leq G$, it is obvious that the Ulm length of $G$ cannot be less than $\sigma$. Clearly, $G^\sigma \cap T = T^\sigma = 0$, hence $G^\sigma$ is torsion-free. By (54.5), it is algebraically compact, thus its first Ulm subgroup vanishes, $G^{\sigma+1} = 0$.

The sequence $0 \to T \to G/G^\sigma \to G/(G^\sigma + T) \to 0$ is exact, and the group $G/(G^\sigma + T)$ is divisible, being an epimorphic image of $G/T$ [cf. (55.1)]. Hence, by (51.3), we obtain the exact sequence

$$0 \to \mathrm{Hom}(Q/Z, G/(G^\sigma + T)) \to \mathrm{Ext}(Q/Z, T)$$
$$\cong G \to \mathrm{Ext}(Q/Z, G/G^\sigma) \cong G/G^\sigma \to 0$$

[the last isomorphism holds because $G/G^\sigma$ is reduced cotorsion]. The map between the two Exts is natural, thus the naturality of the isomorphisms implies that its kernel is $G^\sigma$, and the assertion follows. □

The next step is to give a formula for the Ulm subgroups of $G$, which will be made more explicit in (56.5).

**Proposition 56.4.** *If $T$ is a reduced torsion group, then for every ordinal $\sigma$*

(2) $$\mathrm{Ext}(Q/Z, T)^\sigma \cong \mathrm{Ext}(Q/Z, T^\sigma) \oplus \mathrm{Ext}(Q/Z, T/T^\sigma)^\sigma.$$

From (56.1) we obtain the exact sequence $0 \to \mathrm{Ext}(Q/Z, T^\sigma) \to$ Ext($Q/Z$, $T$)$^\sigma \to \mathrm{Ext}(Q/Z, T/T^\sigma)^\sigma \to 0$. The first group is cotorsion, while the last is, by (56.3), torsion-free, so the sequence splits.□

**Proposition 56.5** (Harrison [3]). *Let $T$ be a torsion group and $\sigma$ an ordinal. Then*

$$\mathrm{Ext}(Q/Z, T/T^\sigma)^\sigma \cong \mathrm{Hom}(Q/Z, H_\sigma)$$

*where $H_\sigma \cong \hat{T}_{\sigma-1}/T_{\sigma-1}$ [with $\hat{T}_{\sigma-1}$ the Z-adic completion of $T_{\sigma-1}$] if $\sigma - 1$ exists and is otherwise the quotient group of the inverse limit $L_\sigma = \varprojlim T/T^\rho$ ($\rho < \sigma$) taken modulo $T/T^\sigma$ [which is a subgroup of $L_\sigma$].*

Apply (56.4) for $\sigma - 1$ and take the Ulm subgroups; then by (56.3), we get $\mathrm{Ext}(Q/Z, T)^\sigma \cong \mathrm{Ext}(Q/Z, T^{\sigma-1})^1$. This latter group is, again by (56.4), isomorphic to $\mathrm{Ext}(Q/Z, T^\sigma) \oplus \mathrm{Ext}(Q/Z, T_{\sigma-1})^1$ where the second direct summand is, in view of (53.8), isomorphic to $\mathrm{Hom}(Q/Z, \hat{T}_{\sigma-1}/T_{\sigma-1})$. Taking into account that $\mathrm{Ext}(Q/Z, T^\sigma)$ is just the adjusted part of $\mathrm{Ext}(Q/Z, T)^\sigma$, the desired isomorphism follows for isolated ordinals $\sigma$.

If $\sigma$ is a limit ordinal, then there are natural embeddings of $T/T^\sigma$ in $\varprojlim T/T^\rho$ and $G/G^\rho$ in $\varprojlim G/G^\rho$ ($\rho < \sigma$) which yield the commutative diagram

$$
\begin{array}{ccccccccc}
0 & \to & T/T^\sigma & \to & \varprojlim T/T^\rho & \to & L_\sigma & \to & 0 \\
& & \downarrow & & \downarrow & & & & \\
0 & \to & G/G^\sigma & \to & \varprojlim G/G^\rho & \to & K_\sigma & \to & 0
\end{array}
$$

where the rows are exact and the vertical maps [the first acts naturally as is shown by $T/T^\sigma \to (G^\sigma + T)/G^\sigma \le G/G^\sigma$] are monic. Since the middle group in the bottom row is a reduced cotorsion group [see **54** (F)], and since the same holds for $G/G^\sigma$, by **54** (B), $K_\sigma$ must also be reduced. Thus if we pass $\mathrm{mod}(G^\sigma + T)/G^\sigma$, then the divisible part of $(\varprojlim G/G^\rho)/[(G^\sigma + T)/G^\sigma]$ is contained in the image of $\varprojlim T/T^\rho$ in this group, and this is clearly isomorphic to $L_\sigma$. Because of $(G/G^\sigma)/[(G^\sigma + T)/G^\sigma] \cong G/(G^\sigma + T)$, (56.3) completes the proof.☐

Now we are in the possession of a fair amount of information about the Ulm subgroups of $\mathrm{Ext}(Q/Z, T)$. The description of the Ulm factors is done *via* the next lemma.

**Lemma 56.6.** *If $T$ is a torsion group with $T^1 = 0$, then*

$$\mathrm{Ext}(Q/Z, T)_0 \cong \hat{T}$$

*where the index of* $\mathrm{Ext}$ *denotes Ulm factor of* $\mathrm{Ext}$.

From (55.1) and (56.3), we infer that the torsion part of $\mathrm{Ext}(Q/Z, T)_0$ contains $T$ as a subgroup, such that the corresponding quotient group is divisible. Since $T$ is pure in $\mathrm{Ext}(Q/Z, T)$, it is readily seen that it stays pure in $\mathrm{Ext}(Q/Z, T)_0$ which is, by (54.3), a complete group. Thus $T$ is pure and dense in $\mathrm{Ext}(Q/Z, T)_0$, and therefore this must be the $Z$-adic completion $\hat{T}$ of $T$.☐

**Theorem 56.7** (Harrison [3]). *Let $T$ be a reduced torsion group and $\sigma$ an ordinal. The $\sigma$th Ulm factor of* $\mathrm{Ext}(Q/Z, T)$ *is given by*

$$\mathrm{Ext}(Q/Z, T)_\sigma \cong \hat{T}_\sigma \oplus \mathrm{Hom}(Q/Z, H_\sigma)$$

*where $H_\sigma$ is defined in (56.5).*

The 0th Ulm factor of $G = \mathrm{Ext}(Q/Z, T)$ is the same as for $\mathrm{Ext}(Q/Z, T_0)$ which is isomorphic to $\hat{T}_0$ [cf. (56.2) and (56.6)]. From (56.4) we infer that the $\sigma$th Ulm factor of $G$ is the direct sum of the 0th Ulm factor of $\mathrm{Ext}(Q/Z, T^\sigma)$ and a group which is, by (56.5), isomorphic to $\mathrm{Hom}(Q/Z, H_\sigma)$. By what has been shown, the 0th Ulm factor of $\mathrm{Ext}(Q/Z, T^\sigma)$ is isomorphic to the $Z$-adic completion of the 0th Ulm factor $T_\sigma$ of $T^\sigma$.☐

EXERCISES

1. Prove the analog of (56.1) by replacing subgroups $X^\sigma$ by $p^\sigma X$ throughout.
2. For a torsion group $T$, the following conditions are equivalent:

   (i) $\mathrm{Ext}(Q/Z, T)$ is algebraically compact;
   (ii) $\mathrm{Ext}(Q/Z, T) \cong \hat{T}$;
   (iii) $T$ is the torsion part of an algebraically compact group.

3. Let $G$ be a reduced cotorsion group and $T$ its torsion part. The first Ulm subgroup $G^1$ of $G$ is adjusted cotorsion exactly if $(G^1 + T)/G^1$ is the torsion part of $G/G^1$.
4. Construct a cotorsion group of length $\omega$ all of whose Ulm subgroups are adjusted.
5. Describe the Ulm factors of $\mathrm{Ext}(Q/Z, A)$ where $A$ is the group of the example in **35**.
6. If $T$ and $T'$ are torsion groups with isomorphic basic subgroups, then the 0th Ulm factors of $\mathrm{Ext}(Q/Z, T)$ and $\mathrm{Ext}(Q/Z, T')$ are isomorphic.
7. Exhibit non-isomorphic reduced cotorsion groups [of length 2] whose corresponding Ulm factors are isomorphic. [*Hint*: let $B$ be the direct sum of $Z(p^n)$, $n = 1, 2, \cdots$, and $T$ a pure subgroup of the torsion part $\bar{B}$ of the of direct product of the $Z(p^n)$ such that $B < T$ and both $|\bar{B} : T|$ and $|T : B|$ are of power $2^{\aleph_0}$; prove that $\mathrm{Ext}(Q/Z, B)$ and $\mathrm{Ext}(Q/Z, T)$ are as stated.]

## 57.* APPLICATIONS TO EXT

Our study of cotorsion groups paves the way for a study of the groups of extensions. Of course, the basic problem consists in finding the structure of $\mathrm{Ext}(C, A)$ in terms of the groups $A$ and $C$. We know that this depends, to a great extent, on determining the torsion part of $\mathrm{Ext}(C, A)$. Our knowledge of torsion groups being essentially limited to the countable case [see Chapter XII], the precise description of Ext is not possible in general.

The first attempt at characterizing $\mathrm{Ext}(C, A)$ in terms of $A$ and $C$ is to find the invariants of its Ulm factors. This is not difficult for the 0th Ulm factor, but for higher Ulm factors these invariants are not known.

The next lemma is crucial.

**Lemma 57.1** (Harrison [3]). *If $E : 0 \to A \to B \to C \to 0$ is a pure-exact sequence, then for any group $G$, the induced sequences of 0th Ulm factors*

(1)                    $0 \to \mathrm{Ext}(C, G)_0 \to \mathrm{Ext}(B, G)_0 \to \mathrm{Ext}(A, G)_0 \to 0$

(2)                    $0 \to \mathrm{Ext}(G, A)_0 \to \mathrm{Ext}(G, B)_0 \to \mathrm{Ext}(G, C)_0 \to 0$

*are splitting exact.*

The exactness of $0 \to \mathrm{Ext}(C, G)/\mathrm{Im}\ E^* \to \mathrm{Ext}(B, G) \to \mathrm{Ext}(A, G) \to 0$ follows from (51.3), that of $0 \to \mathrm{Pext}(C, G)/\mathrm{Im}\ E^* \to \mathrm{Pext}(B, G) \to \mathrm{Pext}(A, G) \to 0$ from (53.7), consequently, the exactness of (1) is a simple result of the $3 \times 3$-lemma. On account of (53.5), the former sequence is pure-exact, and therefore (1), too, is pure-exact. The groups in (1) are algebraically compact, so (1) splits. An analogous proof applies to (2). $\square$

This lemma enables one to compute the 0th Ulm factor of $\mathrm{Ext}(C, A)$ in terms of invariants of $A$ and $C$.

**Theorem 57.2** (Mac Lane [2], Harrison [3]). *Let $A$, $C$ be arbitrary groups, and let $T(A)$, $T(C)$, and $T_b(A)$, $T_b(C)$ denote their torsion parts and basic subgroups of their torsion parts, respectively. Then*

(3)
$$\mathrm{Ext}(C, A)_0 \cong \mathrm{Ext}(T_b(C), T_b(A)) \oplus \mathrm{Ext}(T_b(C), A/T(A))$$
$$\oplus\ \mathrm{Ext}(T(C)/T_b(C), T_b(A))_0 \oplus \mathrm{Ext}(T(C)/T_b(C), A/T(A))$$

*where all the algebraically compact groups on the right can be described by means of invariants of $A$ and $C$.*

We apply (57.1) repeatedly to the pure-exact sequence arising from the monomorphisms $T_b(A) \to T(A)$, $T(A) \to A$ and the same for $C$. If we recall that $\mathrm{Ext}(J, *)$ is divisible for torsion-free $J$ and $\mathrm{Ext}(*, D) = 0$ for divisible $D$, then we get the stated isomorphism with 0 suffix throughout. $T_b(C)$ being a direct sum of cyclic groups, $\mathrm{Pext}(T_b(C), *) = 0$, and in view of (52.3), $\mathrm{Pext}(T(C)/T_b(C), A/T(A)) = 0$, too. The stated isomorphism follows.

In order to make our additional assertion clear, we take into account that $T_b(C)$ is the direct sum of cyclic groups of orders $p^n$, and so $\mathrm{Ext}(T_b(C), G)$ is the direct product of the groups $G/p^n G$. Furthermore, $T(C)/T_b(C)$ is the direct sum of groups of type $p^\infty$, and thus $\mathrm{Ext}(T(C)/T_b(C), T_b(A))_0$ is the direct product of $p$-adic completions of $T_b(A)$ [cf. (56.6)], and the same holds for the last summand in (3) as is shown by **52** (M). $\square$

Next we wish to study $\mathrm{Pext}(C, A)$, the first Ulm subgroup of $\mathrm{Ext}(C, A)$. All that we know about it is contained in (57.4).

**Lemma 57.3** (Harrison [3]). *If $A$ satisfies $A^1 = 0$, then there is an exact sequence*

$$0 \to \mathrm{Pext}(C_0, A) \to \mathrm{Pext}(C, A) \to \mathrm{Hom}(C^1, \hat{A}/A) \to 0.$$

From the exact sequence $0 \to C^1 \to C \to C_0 \to 0$ and from the pure-exact sequence $0 \to A \to \hat{A} \to \hat{A}/A \to 0$ [where $\hat{A}/A$ is divisible], we obtain the commutative diagram

$$
\begin{array}{ccc}
0 \to \operatorname{Hom}(C_0, \hat{A}) & \xrightarrow{\ \kappa\ } & \operatorname{Hom}(C, \hat{A}) \\
\ \ \downarrow{\scriptstyle \mu} & & \ \ \downarrow{\scriptstyle \lambda} \\
0 \to \operatorname{Hom}(C_0, \hat{A}/A) \xrightarrow{\ \nu\ } \operatorname{Hom}(C, \hat{A}/A) \to \operatorname{Hom}(C^1, \hat{A}/A) \to 0 \\
\downarrow \qquad\qquad \downarrow \\
\operatorname{Pext}(C_0, A) \xrightarrow{\qquad} \operatorname{Pext}(C, A) \\
\downarrow \qquad\qquad \downarrow \\
\operatorname{Pext}(C_0, \hat{A}) = 0 \qquad \operatorname{Pext}(C, \hat{A}) = 0
\end{array}
$$

where the rows and columns are exact because of (44.5) and (53.7); the two bottom groups vanish by the algebraic compactness of $\hat{A}$. The first Ulm subgroup of $\hat{A}$ vanishes, and so every homomorphism $C \to \hat{A}$ is induced by some $C_0 \to \hat{A}$, that is, $\kappa$ is an isomorphism. Hence the top row stays exact if we continue it with $\to 0$. Also, we can complete the third row with $\to \operatorname{Hom}(C^1, \hat{A}/A)$; this homomorphism exists in view of both $\operatorname{Pext}(C, A)$ and $\operatorname{Hom}(C^1, \hat{A}/A)$ being epimorphic images of $\operatorname{Hom}(C, \hat{A}/A)$ with kernels $\operatorname{Im} \lambda$ and $\operatorname{Im} \nu$, where $\operatorname{Im} \lambda = \operatorname{Im} \lambda\kappa = \operatorname{Im} \nu\mu \leq \operatorname{Im} \nu$. From the $3 \times 3$-lemma the stated exactness follows [in the proof of (2.4), $\lambda_1$ and $\mu_1$ need not be monic]. $\square$

**Theorem 57.4** (Mac Lane [2], Harrison [3]). *For arbitrary groups $A$ and $C$, $\operatorname{Pext}(C, A)$ has two subgroups $K(C, A) \leq L(C, A)$ such that*

$$\operatorname{Pext}(C, A)/L(C, A) \cong \operatorname{Hom}(C^1, \hat{A}_0/A_0),$$
$$L(C, A)/K(C, A) \cong \operatorname{Pext}(C_0, A_0) \oplus \operatorname{Ext}(C^1, A^1).$$

*If the 0th Ulm factor $C_0$ of $C$ is a direct sum of cyclic groups, then*

$$K(C, A) = 0 \quad and \quad L(C, A) \cong \operatorname{Ext}(C^1, A^1).$$

Starting with the exact sequences $0 \to C^1 \to C \to C_0 \to 0$ and $0 \to A^1 \to A \to A_0 \to 0$, from (51.3), (56.1), and (57.3) one obtains the commutative diagram

$$
\begin{array}{ccc}
\operatorname{Ext}(C_0, A^1) \xrightarrow{\ \alpha\ } \operatorname{Ext}(C, A^1) \to \operatorname{Ext}(C^1, A^1) \to 0 \\
\ \ \downarrow{\scriptstyle \gamma} \qquad\qquad \ \ \downarrow{\scriptstyle \beta} \\
\operatorname{Pext}(C_0, A) \xrightarrow{\ \delta\ } \operatorname{Pext}(C, A) \\
\ \ \downarrow{\scriptstyle \kappa} \qquad\qquad \ \ \downarrow{\scriptstyle \lambda} \\
0 \to \operatorname{Pext}(C_0, A_0) \xrightarrow{\ \mu\ } \operatorname{Pext}(C, A_0) \xrightarrow{\ \nu\ } \operatorname{Hom}(C^1, \hat{A}_0/A_0) \to 0 \\
\downarrow \qquad\qquad \downarrow \\
0 \qquad\qquad 0
\end{array}
$$

with exact rows and columns. Let $K(C, A) = \operatorname{Im} \beta\alpha$ and $L(C, A) = \operatorname{Ker} \nu\lambda$. Then the first isomorphism is trivial. As in (8.3), it results that $L(C, A) = \operatorname{Im} \beta + \operatorname{Im} \delta$. Evidently, $K(C, A) \leq \operatorname{Im} \beta \cap \operatorname{Im} \delta$, and if $x \in \operatorname{Im} \beta \cap \operatorname{Im} \delta$, and say, $x = \delta y$ ($y \in \operatorname{Pext}(C_0, A)$), then $0 = \lambda x = \lambda\delta y = \mu\kappa y$ and $\kappa y = 0$, $y \in \operatorname{Im} \gamma$, showing that $x \in \operatorname{Im} \delta\gamma = K(C, A)$, i.e., $\operatorname{Im} \beta \cap \operatorname{Im} \delta = K(C, A)$. Hence, $L(C, A)/K(C, A)$ is the direct sum of $\operatorname{Im} \beta/\operatorname{Im} \beta\alpha \cong \operatorname{Ext}(C^1, A^1)$ and $\operatorname{Im} \delta/\operatorname{Im} \delta\gamma \cong \operatorname{Pext}(C_0, A_0)$.

If $C_0$ is a direct sum of cyclic groups, then $\operatorname{Pext}(C_0, *) = 0$, whence the second assertion is clear.☐

Unfortunately, (57.4) is not satisfactory for two reasons. First, it does not tell anything about the subgroup $K(C, A)$. Secondly, even if $C_0$ is a direct sum of cyclic groups and so $K(C, A) = 0$, we do not know much about the arising exact sequence

$$0 \to \operatorname{Ext}(C^1, A^1) \to \operatorname{Pext}(C, A) \to \operatorname{Hom}(C^1, \hat{A}_0/A_0) \to 0,$$

so this does not lead to a description of the first Ulm factor of $\operatorname{Ext}(C, A)$. If $C$ is assumed to be a reduced countable torsion group [when all of its Ulm factors are direct sums of cyclic groups by (17.3)], then we can repeat the arguments for $\operatorname{Ext}(C^n, A^n)$ ($n = 1, 2, \cdots$) to obtain a sequence of algebraically compact quotient groups like (3) and $\operatorname{Hom}(C^n, \hat{A}_{n-1}/A_{n-1})$.

EXERCISES

1.  Let $C$ be a $p$-group and $B$ a $p$-basic subgroup of $A$. Prove that

    $$\operatorname{Ext}(C, A)_0 \cong \operatorname{Ext}(C, B)_0 .$$

2.  $\operatorname{Ext}(C, A)_0$ is a direct summand of a direct product of groups of the form $A/p^n A$. [Hint: start with a pure-projective resolution of $C$, then (57.1).]

3.  Give explicitly the cardinal invariants of $\operatorname{Ext}(C, A)_0$ in terms of invariants of $A$ and $C$, by making use of (57.2).

4.  (a) Let $A$ and $C$ satisfy $A^1 = 0$ and $C^1 = $ divisible. Show that

    $$\operatorname{Pext}(C, A) \cong \operatorname{Pext}(C_0, A) \oplus \operatorname{Hom}(C^1, \hat{A}/A).$$

    (b) (Harrison [3]) If $C^1$ is divisible, (57.4) can be simplified as follows:

    $$\operatorname{Pext}(C, A)/K(C, A) \cong \operatorname{Pext}(C_0, A_0) \oplus \operatorname{Ext}(C^1, A^1) \oplus \operatorname{Hom}(C^1, \hat{A}/\hat{A}).$$

5.  Determine the cardinal invariants of the algebraically compact group $\operatorname{Hom}(C^1, \hat{A}/A)$ for $C$ torsion.

6.  Show that if $C$ is a $p$-group and $C_0$ is a direct sum of cyclic groups, $\operatorname{Ext}(C^1, A^1)$ is isomorphic to a group between $\operatorname{Ext}(C, A)^2$ and $\operatorname{Ext}(C, A)^1$.

## 58. INJECTIVE PROPERTIES OF COTORSION GROUPS

Recall that algebraically compact groups can be characterized as groups with the injective property relative to the class of pure-exact sequences. Next, we wish to point out that cotorsion groups can similarly be characterized, furthermore, every group has a "cotorsion hull."

The following result is already known to us for torsion groups $A$; the proof is essentially a repetition of the proof of (55.1).

**Theorem 58.1.** *Every group $A$ can be embedded in a cotorsion group $G$ such that $G/A$ is torsion-free and divisible. If $A$ is reduced, so can $G$ also be chosen.*

Evidently, it suffices to consider the case of reduced $A$. As in (55.1), one gets the exact sequence $0 \to A \to \text{Ext}(Q/Z, A) \to \text{Ext}(Q, A) \to 0$ where $\text{Ext}(Q, A)$ is torsion-free divisible, so that if we define $G = \text{Ext}(Q/Z, A)$—which is, because of (54.6) and (55.3), reduced cotorsion—then the assertion follows.☐

Call an exact sequence $E: 0 \to A \to B \to C \to 0$ *torsion-splitting* if $E\tau$ splits for the injection $\tau: T \to C$, where $T = T(C)$ is the torsion part of $C$. If $C$ is torsion-free, or if $E$ is splitting, then $E$ is trivially torsion-splitting.

**Theorem 58.2.** *A group $G$ is cotorsion if, and only if, it has the injective property relative to torsion-splitting exact sequences.*

First assume $G$ cotorsion. If $0 \to A \to B \to C \to 0$ is exact with $C$ torsion-free, then the induced exact sequence $\text{Hom}(B, G) \to \text{Hom}(A, G) \to \text{Ext}(C, G) = 0$ shows that every homomorphism $A \to G$ is induced by some $B \to G$, i.e., $G$ is injective relative to such exact sequences. If $E$ is any torsion-splitting exact sequence and $\eta: A \to G$, then $\eta$ can be extended to a $\chi: B' \to G$, where $B'$ is the inverse image of the torsion part $T$ of $C$. From the exactness of $0 \to B' \to B \to C/T \to 0$ and what has already been proved, we conclude that $\chi$ can be extended to a homomorphism of $B$ into $G$.

Conversely, suppose $G$ has the injective property relative to torsion-splitting exact sequences. By (58.1), there is an exact sequence $0 \to G \to G^* \to D \to 0$, with $G^*$ cotorsion and $D$ torsion-free divisible. Since $G$ has the injective property relative to this sequence, $G$ is a direct summand of $G^*$ and hence cotorsion.☐

There is an easy way of recognizing within $\text{Ext}(C, A)$ the extensions of $A$ by $C$ that represent torsion-splitting exact sequences.

**Proposition 58.3.** *The exact sequence $E: 0 \to A \to B \to C \to 0$ is torsion-splitting if and only if it is an element of the maximal divisible subgroup of $\text{Ext}(C, A)$.*

From the exact sequence $0 \to T \to C \to C/T \to 0$ [with $T = T(C)$], one infers the exactness of

$$\mathrm{Ext}(C/T, A) \to \mathrm{Ext}(C, A) \to \mathrm{Ext}(T, A) \to 0.$$

The first group is divisible in view of **52** (I), while the third group is reduced by (55.3). $E$ is torsion-splitting exactly if its image in $\mathrm{Ext}(T, A)$ is 0, that is, if it belongs to the image of $\mathrm{Ext}(C/T, A)$, which is just the maximal divisible subgroup of $\mathrm{Ext}(C, A)$. □

On account of (58.2), (58.1) may be given another interpretation. Namely, it shows that for every $A$ there is a torsion-splitting exact sequence $0 \to A \to G \to G/A \to 0$ with $G$ [and $G/A$] cotorsion, i.e, there are enough "torsion-splitting-injectives." Naturally, the problem of minimal injectives arises. For the sake of simplicity, let us confine ourselves to reduced groups $A$, and put

$$A^{\bullet} = \mathrm{Ext}(Q/Z, A).$$

From the proof of (58.1), it is apparent that there is a natural monomorphism $\mu : A \to A^{\bullet}$ under which, if convenient, $A$ may be identified with a subgroup of $A^{\bullet}$, and $A^{\bullet}/A$ is torsion-free divisible. From **54** (H) it is clear that $\mu$ is an isomorphism exactly if $A$ is cotorsion, furthermore $A^{\bullet\bullet} = A^{\bullet}$ for every $A$.

**Lemma 58.4** (Rotman [1]). *Let $A$ and $B$ be reduced groups. Every homomorphism $\eta : A \to B$ has a unique extension $\eta^{\bullet} : A^{\bullet} \to B^{\bullet}$.*

The exactness of $0 \to A \xrightarrow{\ \mu\ } A^{\bullet} \to A^{\bullet}/A \to 0$ implies the exact sequence

$$0 = \mathrm{Hom}(A^{\bullet}/A, B^{\bullet}) \to \mathrm{Hom}(A^{\bullet}, B^{\bullet}) \xrightarrow{\ \mu^{*}\ } \mathrm{Hom}(A, B^{\bullet}) \to \mathrm{Ext}(A^{\bullet}/A, B^{\bullet}) = 0.$$

Since $\mu^{*}$ is surjective, the composite map $A \xrightarrow{\ \eta\ } B \xrightarrow{\ \nu\ } B^{\bullet}$ is induced by some $\eta^{\bullet} \in \mathrm{Hom}(A^{\bullet}, B^{\bullet})$ [this part follows from (58.2), too], and since $\mu^{*}$ is monic, $\eta^{\bullet}$ is unique. □

Notice the commutativity of the diagram

$$
\begin{array}{ccc}
A & \xrightarrow{\ \eta\ } & B \\
\mu \downarrow & & \downarrow \nu \\
A^{\bullet} & \xrightarrow{\ \eta^{\bullet}\ } & B^{\bullet}
\end{array}
$$

where the vertical maps are natural.

**Proposition 58.5.** *If $E : 0 \to A \xrightarrow{\ \alpha\ } B \xrightarrow{\ \beta\ } C \to 0$ is a torsion-splitting exact sequence of reduced groups, then*

$$E^{\bullet} : \quad 0 \to A^{\bullet} \xrightarrow{\ \alpha^{\bullet}\ } B^{\bullet} \xrightarrow{\ \beta^{\bullet}\ } C^{\bullet} \to 0$$

*is splitting exact.*

From the given exact sequence we obtain the exactness of $E^{\bullet}$ by applying (51.3) (3) with $G = Q/Z$. Splittingness will follow from (58.2), if we can show $E^{\bullet}$ torsion-splitting. We have the diagram with exact row

$$\text{Ext}(C, A)$$
$$\downarrow \mu_*$$
$$0 = \text{Ext}(C^{\bullet}/C, A^{\bullet}) \to \text{Ext}(C^{\bullet}, A^{\bullet}) \xrightarrow{\lambda^*} \text{Ext}(C, A^{\bullet}) \to 0$$

where the labeled maps are induced by the natural homomorphisms in the commutative diagram

$$
\begin{array}{ccccccccc}
E: & 0 \to & A & \to & B & \to & C & \to 0 \\
 & & \mu\downarrow & & v\downarrow & & \lambda\downarrow & \\
E^{\bullet}: & 0 \to & A^{\bullet} & \to & B^{\bullet} & \to & C^{\bullet} & \to 0.
\end{array}
$$

$E^{\bullet}$ is obtained from $E$ by applying $\mu_*$ and then the inverse of the isomorphism $\lambda^*$. If $E$ belongs to the maximal divisible subgroup of $\text{Ext}(C, A)$, then $E^{\bullet}$ belongs to that of $\text{Ext}(C^{\bullet}, A^{\bullet})$. $\square$

Let $A$ be a reduced group and assume $G$ is reduced cotorsion such that $A \leq G$. From (58.5) we infer $A^{\bullet} \leq G^{\bullet} = G$, showing that $A^{\bullet}$ is a minimal reduced cotorsion group containing $A$, contained in every reduced cotorsion group containing $A$. Thus $A^{\bullet}$ may be considered as the *cotorsion hull* of $A$. If $A$ is not reduced and $D$ is its maximal divisible subgroup, then $D \oplus A^{\bullet}$ is the cotorsion hull of $A$. Clearly, it is unique up to isomorphism over $A$.

Far-reaching generalizations of certain results in this section were obtained by Irwin, Walker and Walker [1], and especially by Nunke [3].

EXERCISES

1.  Prove that $0 \to A \to B \to C \to 0$ is torsion-splitting exact if and only if the induced sequence

    $$0 \to T(A) \to T(B) \to T(C) \to 0$$

    of torsion subgroups is splitting exact.
2.  (a) Prove that an exact sequence is torsion-splitting exactly if the torsion groups have the projective property relative to it.
    (b) A group has the projective property relative to all torsion-splitting exact sequences if and only if it is a direct sum of a free and a torsion group.
3.  (Irwin, Walker and Walker [1]) Assuming $0 \to A \to B \to C \to 0$ splitting exact, prove the exactness of the sequences (4) and (5) in (53.7) with Pexts replaced by the maximal divisible subgroups of Exts.

4.  (a) The 0th Ulm factor of $A^\bullet$ is the completion of the 0th Ulm factor of $A$. [*Hint*: (57.2).]
    (b) The same fails to hold for the higher Ulm factors.
5.  If $A \leq G$ such that $G$ is reduced cotorsion and $G/A$ is torsion-free divisible, then $G$ is isomorphic to $A^\bullet$ over $A$.
6.  Verify the inequality

$$|A^\bullet| \leq |A|^{\aleph_0}.$$

7.  (C. Megibben) Prove (58.1) for reduced $A$, by considering all groups $H$ such that: (i) $A \leq H$; (ii) $|H| \leq |A|^{\aleph_0}$; (iii) $H/A$ is torsion-free divisible; (iv) $H$ is reduced; and then selecting a maximal one among the $H$.
8.  (Rangaswamy [1]) (a) For a group $A$, $\text{Ext}(C, A)$ is reduced for all $C$ exactly if $A$ is cotorsion.
    (b) For a fixed $C$, $\text{Ext}(C, A)$ is reduced for all $A$ if and only if $C$ is the direct sum of a free and a torsion group.


## NOTES

Of the four basic functors of homological algebra, so far Ext has evoked the greatest attention of group theorists. This is due partly to the fact that—while every group is a Hom and a tensor product, and every torsion group is a Tor—the class of groups which can be Ext is more limited [though every reduced torsion group can be a torsion part of Ext] and has several interesting features. Another reason is that some problems in abelian group theory can be successfully treated if formulated in terms of extensions.

For general rings R, there is a functor $\text{Ext}_R^n$ for every $n \geq 1$, and the exact sequences in (51.3) become infinite with three similar terms for every $n$. (51.3) continues to hold *verbatim* if and only if R is left hereditary. Ext is always an abelian group, it carries an R-module structure if R is commutative; then (52.1) is a very special case of a more general statement on the action of R on Ext.

As already noted, cotorsion groups were discovered independently by Harrison [1], Nunke [1] and Fuchs [9]; the name was coined by Harrison, who also assumed that they are reduced. He introduced the functor Pext and published (53.7). Theorem (53.3) was proved for Dedekind domains by Nunke [1]. The fundamental fact that Ext is always cotorsion is usually derived from the isomorphism

$$\text{Ext}(A, \text{Ext}(B, C)) \cong \text{Ext}(\text{Tor}(A, B), C)$$

which holds over hereditary rings. That $\text{Ext}(C, A)$ is reduced for torsion $C$ follows from the formula

$$\text{Ext}(A, \text{Hom}(B, C)) \oplus \text{Hom}(A, \text{Ext}(B, C)) \cong \text{Ext}(A \otimes B, C) \oplus \text{Hom}(\text{Tor}(A, B), C),$$

valid over commutative hereditary rings. These are consequences of the universal coefficient theorem. In the text, independent proofs are given.

There are various generalizations of (53.6) and (53.7). The most important one is concerned with the case when (1) in **53** represents an element of $S \, \text{Ext}(C, A)$ for a subfunctor $S$ of the identity; cf. Nunke [3]. For another generalization, in the spirit of **45**, see Fuchs [12].

Cotorsion modules over commutative domains were studied by E. Matlis [*Mem. Amer. Math. Soc.* **Nr. 49** (1964)]. For the completion functor corresponding to $G \mapsto \mathrm{Ext}(Q/Z, G)$ over Dedekind domains we refer to Rotman [1].

*Problem 38.* Estimate the number of nonisomorphic extensions of $A$ by $C$ in terms of $A$ and $C$.

*Problem 39.* Determine the invariants of the divisible group $\mathrm{Ext}(C, A)$ for torsion-free $C$.

*Problem 40.* Describe the structure of $\mathrm{Ext}(C, A)$. In particular, determine the Ulm length of $\mathrm{Ext}(C, A)$.

*Problem 41.* Characterize the elements of $\mathrm{Ext}(C, A)$ which represent extensions $G$, such that $A$ belongs to the $\sigma$th Ulm subgroup of $G$, for some given $\sigma$.

*Problem 42.* (R. Baer) Axiomatize $\mathrm{Ext}(C, A)$ as a function of $A$ and $C$.

*Problem 43.* Relate the groups $C$ and $C'$ if $\mathrm{Ext}(C, A) \cong \mathrm{Ext}(C', A)$ for every group $A$ [or, if $\mathrm{Ext}(A, C) \cong \mathrm{Ext}(A, C')$ for every $A$].

*Problem 44.* Investigate groups $A$ with the following property: if $A$ is contained in a direct sum of reduced groups, then there is an integer $n > 0$ such that $nA$ is contained in the direct sum of a finite number of them.

Notice that cotorsion groups, torsion complete groups [see **67**] and direct products of infinite cyclic groups share the stated property. Cf. (39.9)

*Problem 45.* Develop a theory for (countable) direct sums of cotorsion groups.

*Problem 46.* Give conditions on a class of groups to be the class of all injective [projective] objects relative to some class of exact sequences.

# X

## TENSOR AND TORSION PRODUCTS

There is another fundamental bifunctor whose importance is comparable with that of Hom: this is the tensor product. The tensor products play a role that is—in a certain sense—dual to homomorphism groups, as will be apparent from the results of this chapter.

Tensor products are defined in terms of generators and defining relations. They are universal for bilinear functions, and this property makes them of great importance. The exact sequence on tensor products which will be proved in **60** is just as useful as those on Homs. While Hom is left exact, the tensor product turns out to be right exact; exactness can be restored by making use of the functor Tor, the torsion product.

If one of the groups is torsion, then their tensor product can be completely described. In particular, the tensor product of two torsion groups is always a direct sum of cyclic groups. The problem of determining the structure of the torsion product is more difficult, but recently a great deal of information has been obtained for Tor.

### 59. THE TENSOR PRODUCT

Let $A$ and $C$ be arbitrary groups and $g$ a function defined on the set $A \times C$ with values in some group $G$:

$$(1) \qquad\qquad g: A \times C \to G.$$

This $g$ is said to be a *bilinear function* if it satisfies:

$$(2) \qquad\qquad g(a_1 + a_2, c) = g(a_1, c) + g(a_2, c),$$

$$(3) \qquad\qquad g(a, c_1 + c_2) = g(a, c_1) + g(a, c_2),$$

for all $a, a_1, a_2 \in A$, $c, c_1, c_2 \in C$. It follows at once that a bilinear function $g$ satisfies the following conditions, too: $g(a, 0) = 0 = g(0, c)$, $g(-a, c) = -g(a, c) = g(a, -c)$, $g(na, c) = ng(a, c) = g(a, nc)$ for all $a \in A, c \in C$ and

252

$n \in \mathbf{Z}$. Hence, if the $g(a_i, c_j)$ are known for all $a_i$, $c_j$ belonging to generating systems of $A$ and $C$, respectively, then $g(a, c)$ can be determined for all $a \in A, c \in C$.

Given $A$, $C$, we define a group $A \otimes C$ and a bilinear function $e$ of $A \times C$ into $A \otimes C$, such that $e$ is the most general bilinear function in the sense that if (1) is any bilinear function, then there is a unique homomorphism $\phi$: $A \otimes C \to G$ such that $g = \phi e$, i.e., every bilinear function (1) factors through the function $e$.

Define $X$ as the free group whose free generators are all pairs $(a, c)$ with $a \in A, c \in C$, and let $Y$ be the subgroup of $X$ generated by the elements of the form

(4) $(a_1 + a_2, c) - (a_1, c) - (a_2, c)$     and     $(a, c_1 + c_2) - (a, c_1) - (a, c_2)$

for all $a, a_1, a_2 \in A, c, c_1, c_2 \in C$. Define

$$A \otimes C = X/Y;$$

the coset $(a, c) + Y$ will be written as $a \otimes c$. Thus $A \otimes C$ consists of all finite sums $u = \sum k_i (a_i \otimes c_i)$ with $a_i \in A, c_i \in C, k_i \in \mathbf{Z}$, where the elements are subject to the rules corresponding to the generators of $Y$:

(5)                     $(a_1 + a_2) \otimes c = a_1 \otimes c + a_2 \otimes c,$

(6)                     $a \otimes (c_1 + c_2) = a \otimes c_1 + a \otimes c_2.$

Hence

$$na \otimes c = n(a \otimes c) = a \otimes nc$$

for all $a \in A, c \in C, n \in \mathbf{Z}$. We see that the elements $u$ of $A \otimes C$ may be written in the form

$$u = \sum (a_i \otimes c_i) \qquad (a_i \in A, c_i \in C),$$

but an element $u$ will ordinarily have many expressions of the stated form.

Let us point out that the tensor product $A \otimes_R C$ of a right R-module $A$ and a left R-module $C$ is defined in a similar fashion, but in the quotient group $X/Y$ [where $X$ is the abelian group freely generated by all pairs $(a, c)$] the subgroup $Y$ is generated, in addition to (4), by all elements $(a\alpha, c) - (a, \alpha c)$ with $a \in A, c \in C$ and $\alpha \in R$.

Unfortunately, the notation $a \otimes c$ is not unambiguous, since $a \otimes c$ depends on the groups $A$, $C$, too, so that it must always be made clear to which tensor product the element $a \otimes c$ belongs [cf. Ex. 6].

In view of (5) and (6),

$$e: (a, c) \mapsto a \otimes c$$

is a bilinear function $A \times C \to A \otimes C$. If (1) is bilinear, then we prove the existence of a unique homomorphism $\phi: A \otimes C \to G$ making the diagram

$$A \times C \xrightarrow{\ e\ } A \otimes C$$

(7)

$$g \searrow \quad \swarrow \phi$$

$$G$$

commute. If such a $\phi$ exists, then obviously $\phi: a \otimes c \mapsto g(a, c)$, so $\phi$ must necessarily be unique. But the correspondence $a \otimes c \mapsto g(a, c)$ from the generators of $A \otimes C$ into $G$ extends to a homomorphism, since the generators are subject only to the rules (5), (6) [and their consequences] which hold for the respective images, as is shown by (2), (3). This completes the proof of the first part of the following:

**Theorem 59.1.** *Given a pair $A$, $C$ of groups, there exists a group $A \otimes C$ and a bilinear function*

$$e: A \times C \to A \otimes C$$

*such that if $g: A \times C \to G$ is any bilinear function, then there is a unique homomorphism $\phi: A \otimes C \to G$ such that (7) commutes.*

*The stated property determines $A \otimes C$ up to isomorphism.*

To prove the last statement, let us assume that a group $H$ has the same property with the bilinear function $h: A \times C \to H$. Then by what has been proved for $A \otimes C$, we have a homomorphism $\phi: A \otimes C \to H$ such that $h = \phi e$, and by our assumption on $H$, for some $\psi: H \to A \otimes C$ we have $e = \psi h$. Hence $e = \psi \phi e$, $h = \phi \psi h$, and since there is only one $\psi \phi: A \otimes C \to A \otimes C$ with $e = \psi \phi e$, we get $\psi \phi = 1_{A \otimes C}$ and similarly $\phi \psi = 1_H$. $\square$

The group $A \otimes C$ is called the *tensor product* of $A$ and $C$ and the map: $e: (a, c) \mapsto a \otimes c$ the *tensor map*. From the uniqueness assertion of the preceding theorem, there results

$$A \otimes C \cong C \otimes A,$$

a natural isomorphism being induced by the correspondence $a \otimes c \mapsto c \otimes a$.

In order to get acquainted with tensor products, we first establish a number of elementary facts on them. The following, nearly trivial, lemma provides us with useful methods in deriving properties of tensor products.

**Lemma 59.2.** (a) *If $m \mid a$ and $n \mid c$, then $mn \mid a \otimes c$; (b) if $ma = 0$ and $nc = 0$, then $(m, n)(a \otimes c) = 0$; (c) if $m \mid a$ and $mc = 0$, then $a \otimes c = 0$.*

If $a_0 \in A$, $c_0 \in C$ satisfy $ma_0 = a$, $nc_0 = c$, then $mn(a_0 \otimes c_0) = a \otimes c$. If $s, t$ are integers such that $ms + nt = (m, n)$, then $(m, n)(a \otimes c) = msa \otimes c + a \otimes ntc = 0$. Finally, (c) follows from $a \otimes c = ma_0 \otimes c = a_0 \otimes mc = 0$. $\square$

(A) *If either $A$ or $C$ is p-divisible [divisible], then $A \otimes C$ is p-divisible [divisible].*

(B) *For the heights, we have*

$$h_p(a \otimes c) \geq h_p(a) + h_p(c).$$

(C) *If either $A$ or $C$ is a p-group [torsion], then $A \otimes C$ is a p-group [torsion].*

(D) *If $A$ is p-divisible and $C$ is a p-group, then $A \otimes C = 0$.*
In particular, $A \otimes C = 0$ whenever $A$ is a $p$-group and $C$ is a $q$-group, with $p, q$ distinct primes. Also, $A \otimes C = 0$ if $A$ is divisible and $C$ is torsion.

(E) *If, for some $m \in Z$, $a \in mA$ and $c \in C[m]$, then $a \otimes c = 0$ in $A \otimes C$.*

(F) *If $h_p(a) = \infty$ and $C$ is a p-group, then $a \otimes c = 0$ in $A \otimes C$ for every* $c \in C$.
The proofs are evident and may be left to the reader.

(G) *There is a natural isomorphism*

$$Z \otimes C \cong C.$$

The elements of $Z \otimes C$ may be brought to the form $\sum(n_i \otimes c_i) = \sum(1 \otimes n_i c_i) = 1 \otimes \sum n_i c_i = 1 \otimes c$ for some $c \in C$. The map $\phi: c \mapsto 1 \otimes c$ is thus an epimorphism of $C$ onto $Z \otimes C$. Clearly, $(m, c) \mapsto mc$ is a bilinear map $Z \times C \to C$, hence by (59.1) there is a $\psi : Z \otimes C \to C$ such that $\psi : 1 \otimes c \mapsto c$. Hence $\phi$ and $\psi$ are inverse to each other. [Notice that $\phi$ is natural, because $Z$ has a distinguished generator.]

(H) *For every integer $m$, there is a natural isomorphism*

$$Z(m) \otimes C \cong C/mC.$$

Here again, $\phi: c \mapsto 1 \otimes c$ is an epimorphism of $C$ onto $Z(m) \otimes C$. Since $1 \otimes mc = m \otimes c = 0 \otimes c = 0, mC \leq \text{Ker } \phi$. Now $(n, c) \mapsto nc + mC$ is a bilinear map $Z(m) \times C \to C/mC$ which induces an epimorphism $\psi: Z(m) \otimes C \to C/m$ such that $\psi\phi$ is the canonical map $C \to C/mC$. Thus $\text{Ker } \phi = mC$.
In particular, $Z(p^r) \otimes Z(p^s) \cong Z(p^t)$ with $t = \min(r, s)$.

(I) *Let both $A$ and $C$ be direct sums, $A = \bigoplus_{i \in I} A_i$ and $C = \bigoplus_{j \in J} C_j$. Then*

$$A \otimes C \cong \bigoplus_{i,j} (A_i \otimes C_j).$$

It is clearly sufficient to verify the isomorphism $A \otimes C \cong \bigoplus_i (A_i \otimes C)$. If $\pi_i$ are the projections associated with the given direct decomposition of $A$, then the function $g(a, c) = \sum_i (\pi_i a \otimes c)$ is manifestly bilinear on $A \times C$ to $G = \bigoplus_i (A_i \otimes C)$ where $\pi_i a \otimes c \in A_i \otimes C$. Consequently, there is a unique homomorphism $\phi: A \otimes C \to G$ satisfying $g(a, c) = \phi e(a, c)$ with $e$ denoting the tensor map $A \times C \to A \otimes C$. This $\phi$ is epic, for the $\pi_i a \otimes c$ generate $G$. The function $e(\pi_i a, c)$ on $A_i \times C$ is bilinear, thus there is a homomorphism $\psi_i : A_i \otimes C \to A \otimes C$, for every $i$, such that $\psi_i(\pi_i a \otimes c) = e(\pi_i a, c)$. These $\psi_i$ give rise to a homomorphism $\psi: G \to A \otimes C$ which satisfies: $\psi g(a, c) = \psi \sum_i (\pi_i a \otimes c) = \sum_i \psi_i(\pi_i a \otimes c) = \sum_i e(\pi_i a, c) = e(a, c)$. We infer $\psi$ is inverse to $\phi$, so $\phi$ is monic and hence an isomorphism.

A consequence of (I) and (G) is that if $A$ and $C$ are free groups with free generators $\{a_i\}_{i \in I}$ and $\{c_j\}_{j \in J}$, respectively, then $A \otimes C$ is likewise free on the set $\{a_i \otimes c_j\}$, where $i \in I, j \in J$.

(J) *For any three groups $A$, $B$, $C$, there is a natural isomorphism*

$$\phi: \text{Hom}(A \otimes B, C) \to \text{Hom}(A, \text{Hom}(B, C)),$$

*defined as follows: if $\eta: A \otimes B \to C$, then*

$$[(\phi\eta)a](b) = \eta(a \otimes b) \qquad (a \in A, b \in B).$$

Clearly, $\phi\eta$ assigns to $a \in A$ a map $(\phi\eta)a: B \to C$ that sends $b \in B$ into $\eta(a \otimes b) \in C$. It is easy to check the homomorphism properties for $(\phi\eta)a$. Also, $\phi: \eta \mapsto (\phi\eta)a$ is a homomorphism, thus it remains to show that $\phi$ has an inverse $\psi$. Let $\chi: A \to \text{Hom}(B, C)$; then $(a, b) \mapsto (\chi a)(b)$ is a bilinear function $g: A \times B \to C$, hence there exists a unique $\eta: A \otimes B \to C$ such that $\eta(a \otimes b) = (\chi a)(b)$. The mapping $\psi: \chi \mapsto \eta$ is readily seen to be inverse to $\phi$.

EXERCISES

1. (Whitney [1]) Let $A$, $C$, $G$ be additive, but not necessarily commutative groups and $g: A \times C \to G$ a bilinear function. Prove that:
   (a) the values $g(a, c)$ generate an abelian subgroup $G'$ of $G$ [hint: consider $g(a_1 + a_2, c_1 + c_2)$];
   (b) $g(a_1, c) = g(a_2, c)$ if $a_1, a_2$ belong to the same coset mod the commutator subgroup of $A$;
   (c) the hypothesis of commutativity means no restriction on bilinear functions.
2. Characterize the groups $C$ which satisfy in turn the following conditions:
   (a) $A \otimes C \cong A$ for every $A$;
   (b) $A \otimes C \cong A/mA$ for a fixed $m$ and every $A$;
   (c) $A \otimes C$ is divisible for every $A$;
   (d) $A \otimes C = 0$ for every torsion $A$;
   (e) $A \otimes C = 0$ for every $p$-group $A$;
   (f) $A \otimes C = 0$ for every divisible $A$;
   (g) $A \otimes C$ is torsion for every $A$.

3.  Assume $a \in A$ is such that $a \otimes c = 0$ in $A \otimes C$ for all torsion groups $C$ and all $c \in C$. Prove $a \in A^1$.
4.  (a) Find the structure of $A \otimes C$ if $A$ is a direct sum of cyclic groups.
    (b) $A \otimes C \cong \operatorname{Hom}(A, C)$ if both $A$ and $C$ are finite.
5.  Let $A$ have the property that, for every group $C$, $A \otimes C$ is a direct sum of quotient groups of $C$. Show that $A$ is a direct sum of cyclic groups.
6.  (a) If $A' \leq A$ and $C' \leq C$, then the map $a' \otimes c' (\in A' \otimes C') \mapsto a' \otimes c'$ $(\in A \otimes C)$ induces a homomorphism

    $$\phi: A' \otimes C' \to A \otimes C.$$

    (b) Give an example where $\phi$ is not monic. [Hint: $A' = Z(p)$, $A = Z(p^\infty)$.]
7.  (a) If $\sum_{i=1}^{n} (a_i \otimes c_i) = 0$ in $A \otimes C$ for certain $a_i \in A$, $c_i \in C$, then there exist finitely generated subgroups $A' \leq A$, $C' \leq C$ such that $a_i \in A'$, $c_i \in C'$ and $\sum_{i=1}^{n} (a_i \otimes c_i)$ is 0 as an element of $A' \otimes C'$.
    (b) Verify the analog of (a) be replacing "$=0$" by "of order $m$."
8.  Define multilinear functions $g: A_1 \times \cdots \times A_n \to G$, the tensor product $A_1 \otimes \cdots \otimes A_n$ and the tensor map $e$. Prove the analog of (59.1) and the associativity

    $$(A_1 \otimes A_2) \otimes A_3 \cong A_1 \otimes A_2 \otimes A_3 \cong A_1 \otimes (A_2 \otimes A_3).$$

9.  Let $B$ be generated by all epimorphic images of $A$ in $C$. Then there is an epimorphism $A \otimes \operatorname{Hom}(A, C) \to B$.
10. There is a natural homomorphism

    $$A \otimes \prod C_i \to \prod (A \otimes C_i)$$

    which in general fails to be an isomorphism. [Hint: $A = Q$, $C_i$ unbounded torsion.]

## 60. EXACT SEQUENCES FOR TENSOR PRODUCTS

In this section we wish to examine how tensor products behave as bifunctors. Our first concern is therefore to define induced homomorphisms for $\otimes$.

Let $\alpha: A \to A'$ and $\gamma: C \to C'$ be homomorphisms. The mapping $(a, c) \mapsto \alpha a \otimes \gamma c$ of $A \times C$ into $A' \otimes C'$ is evidently bilinear, hence there is a unique homomorphism $\phi: A \otimes C \to A' \otimes C'$ such that $\phi(a \otimes c) = \alpha a \otimes \gamma c$. We shall denote $\phi = \alpha \otimes \gamma$, thus

$$(\alpha \otimes \gamma)(a \otimes c) = \alpha a \otimes \gamma c \qquad (a \otimes c \in A \otimes C, \, \alpha a \otimes \gamma c \in A' \otimes C').$$

To avoid complicated notation, we shall sometimes write simply

$$\alpha_* = \alpha \otimes 1_C, \qquad \gamma_* = 1_A \otimes \gamma,$$

whenever there is no danger of confusion. The following rules are straightforward to check:

$$1_A \otimes 1_C = 1_{A \otimes C}, \qquad\qquad (\alpha \otimes \gamma)(\alpha' \otimes \gamma') = \alpha\alpha' \otimes \gamma\gamma',$$
$$(\alpha + \alpha_1) \otimes \gamma = \alpha \otimes \gamma + \alpha_1 \otimes \gamma, \qquad \alpha \otimes (\gamma + \gamma_1) = \alpha \otimes \gamma + \alpha \otimes \gamma_1,$$

for matching homomorphisms $\alpha$, $\alpha'$, $\alpha_1$, $\gamma$, $\gamma'$, $\gamma_1$. We are therefore led to the
first part of the following result.

**Theorem 60.1** (Cartan and Eilenberg [1]). *The tensor product is an additive
bifunctor on $\mathscr{A} \times \mathscr{A}$ to $\mathscr{A}$, covariant in both variables. It commutes with direct
limits.*

To prove the final assertion, let $A = \{A_i \, (i \in I); \pi_i^k\}$ and $C = \{C_j \, (j \in J);
\rho_j^l\}$ be direct systems with direct limits $A$ and $C$, respectively. The system

$$\{A_i \otimes C_j \, ((i, j) \in I \times J); \pi_i^k \otimes \rho_j^l\}$$

is again direct; let $G$ denote its limit. From the canonical maps $\pi_i : A_i \to A$,
$\rho_j : C_j \to C$, we conclude the existence of homomorphisms

$$\pi_i \otimes \rho_j : A_i \otimes C_j \to A \otimes C$$

such that $\pi_i \otimes \rho_j = (\pi_k \otimes \rho_l)(\pi_i^k \otimes \rho_j^l)$ $(i \leq k; j \leq l)$. By (11.1), there is a
unique $\sigma: G \to A \otimes C$ such that $\sigma\sigma_{ij} = \pi_i \otimes \rho_j$ for all $i, j$, where $\sigma_{ij} : A_i \otimes C_j
\to G$ is the canonical map. Our claim is that $\sigma$ is an isomorphism. Given
$a \in A$, $c \in C$, $a = \pi_i a_i$ and $c = \rho_j c_j$ for some $a_i \in A_i$, $c_j \in C_j$, and the function $g(a, c) = \sigma_{ij}(a_i \otimes c_j)$ is bilinear. Hence some homomorphism $\tau: A \otimes C \to
G$ satisfies $\tau(a \otimes c) = g(a, c)$. Now $a \otimes c = (\pi_i \otimes \rho_j)(a_i \otimes c_j) = \sigma\sigma_{ij}(a_i \otimes c_j) =
\sigma g(a, c)$ shows that $\sigma$ and $\tau$ are inverse to each other on the generators of
$A \otimes C$ and $G$.$\square$

A consequence of this is that if $\sum_{i=1}^n (a_i \otimes c_i)$ *vanishes in $A \otimes C$, then
there are finitely generated subgroups $A'$, $C'$ such that $a_i \in A'$, $c_i \in C'$ and
$\sum_{i=1}^n (a_i \otimes c_i)$ vanishes already in $A' \otimes C'$.* Indeed, this follows from (60.1)
and **11** (c).

In discussing the effect of tensor products on short exact sequences, we
may, in view of symmetry, confine ourselves to tensoring from the right. The
next result shows that $\otimes$ is right exact:

**Theorem 60.2** (Cartan and Eilenberg [1]). *For an exact sequence $A \xrightarrow{\alpha}
B \xrightarrow{\beta} C \to 0$ and for any group $G$, the induced sequence*

$$A \otimes G \xrightarrow{\alpha_*} B \otimes G \xrightarrow{\beta_*} C \otimes G \to 0$$

*is exact.*

The equality $\beta\alpha = 0$ implies $\beta_*\alpha_* = (\beta\alpha)_* = 0$. We need only prove that the homomorphism

$$\phi: H = (B \otimes G)/\operatorname{Im} \alpha_* \to C \otimes G$$

induced by $\beta_*$ is an isomorphism. Given $c \in C$, choose $b \in B$ with $\beta b = c$. The mapping $(c, g) \mapsto (b \otimes g) + \operatorname{Im} \alpha_* \in H$ is clearly well defined and bilinear, hence there is a $\psi: C \otimes G \to H$ such that $\psi(c \otimes g) = (b \otimes g) + \operatorname{Im} \alpha_*$. It is now clear that $\phi$ and $\psi$ are inverse to each other on the generators.☐

A repeated application of the last result yields that, if both $\beta: B \to C$ and $\beta': B' \to C'$ are epimorphisms, then $\beta \otimes \beta': B \otimes B' \to C \otimes C'$ is again an epimorphism. One can with somewhat greater effort prove a little bit more:

**Corollary 60.3.** *If  $A \xrightarrow{\alpha} B \xrightarrow{\beta} C \to 0$  and  $A' \xrightarrow{\alpha'} B' \xrightarrow{\beta'} C' \to 0$  are exact sequences, then the sequence*

$$(A \otimes B') \oplus (B \otimes A') \xrightarrow{\xi} B \otimes B' \xrightarrow{\beta \otimes \beta'} C \otimes C' \to 0$$

*is exact, where $\xi = [\nabla(\alpha \otimes 1_{B'}) \oplus (1_B \otimes \alpha')]$.*
    The diagram

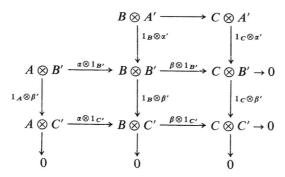

commutes and has exact rows and columns. Therefore the map $\beta \otimes \beta' = (1 \otimes \beta')(\beta \otimes 1)$ is an epimorphism whose kernel is the union of $\operatorname{Ker}(\beta \otimes 1_{B'})$ and $\operatorname{Ker}(1_B \otimes \beta')$, as is easy to verify by diagram chasing [cf. (8.3)]. This union is just $\operatorname{Im} \xi$.☐

For pure-exact sequences, tensor products behave nicely:

**Theorem 60.4** (Fuchs [9]). *If $0 \to A \xrightarrow{\alpha} B \xrightarrow{\beta} C \to 0$ is pure-exact, then for every $G$, the sequence*

(1)                    $$0 \to A \otimes G \xrightarrow{\alpha_*} B \otimes G \xrightarrow{\beta_*} C \otimes G \to 0$$

*is pure-exact.*

If $G = Z$, the assertion follows from **59** (G). If $G = Z(m)$, then the natural isomorphism **59** (H) together with (29.1) implies the pure-exactness of (1). Hence the statement follows for all finitely generated $G$, tensoring being an additive functor. If we represent $G$ as the direct limit of its finitely generated subgroups $G_i$, then the direct limit of the pure-exact sequences $0 \to A \otimes G_i \to B \otimes G_i \to C \otimes G_i \to 0$ yields, by (29.4), a pure-exact sequence $0 \to A \otimes G \to B \otimes G \to C \otimes G \to 0$. A straightforward computation combined with (11.2) shows that the maps must be $\alpha \otimes 1_G$ and $\beta \otimes 1_G$. $\square$

The same proof applies to obtain:

**Corollary 60.5.** *If* $0 \to A \xrightarrow{\alpha} B \xrightarrow{\beta} C \to 0$ *is p-pure-exact and $G$ is a p-group, then* (1) *is pure-exact.*

Indeed, the $p$-pure-exactness of (1) can be replaced by pure-exactness, $B \otimes G$ being a $p$-group. $\square$

A repeated application of (60.4) leads us to the conclusion that if $A'$, $B'$ are pure in $A$, $B$, then $A' \otimes B'$ can actually be considered as a subgroup of $A \otimes B$ under the natural map: $a' \otimes b'\ (\in A' \otimes B') \mapsto a' \otimes b'\ (\in A \otimes B)$. Another case when the same is possible is given by torsion-free $A$, $B$ as follows from:

**Theorem 60.6** (Dieudonné [1]). *If* $0 \to A \xrightarrow{\alpha} B \xrightarrow{\beta} C \to 0$ *is an exact sequence and $G$ is torsion-free, then* (1) *is exact.*

The proof is essentially the same as for (60.4), where $G$ is now the direct limit of [finitely generated] free groups. $\square$

We derive the corollary that the tensor product of torsion-free groups is again torsion-free.

EXERCISES

1.  (a) If $A$ is torsion-free, then $Q \otimes A$ is a divisible hull of $A$, a natural embedding being given by $a \mapsto 1 \otimes a$.
    (b) This method does not apply if $A$ is not torsion-free.
2.  If $\alpha: A \to B$ is a monomorphism such that $\alpha \otimes 1_G : A \otimes G \to B \otimes G$ is monic for every group $G$, then Im $\alpha$ is pure in $B$.
3.  $0 \to A \to B \to C \to 0$ is pure-exact if and only if for every positive integer $m$, the induced sequence

$$0 \to A \otimes Z(m) \to B \otimes Z(m) \to C \otimes Z(m) \to 0$$

    is exact.
4.  If $A'$, $C'$ are pure subgroups of $A$, $C$, respectively, then $A' \otimes C'$ is, in the natural way, a pure subgroup of $A \otimes C$.
5.  $G$ is torsion-free if the exactness of any sequence $0 \to A \to B \to C \to 0$ implies that of $0 \to A \otimes G \to B \otimes G \to C \otimes G \to 0$.

6. If $\alpha$ is an endomorphism of $A$ and $\gamma$ is an endomorphism of $C$, then $\alpha \otimes 1$ and $1 \otimes \gamma$ are commuting endomorphisms of $A \otimes C$.

7. If either $A$ or $C$ is a $p$-adic group, then so is $A \otimes C$. If $\pi$ is a $p$-adic integer, then $\pi(a \otimes c) = \pi a \otimes c$ or $a \otimes \pi c$, whichever makes sense.

8. (a) Assume $A$ and $C$ are torsion-free and $\{a_i\}_{i \in I}$, $\{c_j\}_{j \in J}$ are maximal independent systems of $A$ and $C$, respectively. Show that $\{a_i \otimes c_j\}_{i,j}$ is a maximal independent system in $A \otimes C$.

   (b) Verify the equality

$$r_0(A \otimes C) = r_0(A) \cdot r_0(C)$$

   for any groups $A$ and $C$.

9. (a) Let $A$ and $C$ be torsion-free, and assume $p^t | a \otimes c$ for some $a \in A$, $c \in C$. Then there are nonnegative integers $r$, $s$ with $r + s = t$ such that $p^r | a$ and $p^s | c$. [*Hint*: Ex. 8 (a).]

   (b) Describe the structure of $A \otimes C$, if $A$, $C$ are torsion-free of rank 1.

10. Give the structure of the following tensor products: $J_p \otimes Q$; $J_p \otimes J_q$ with $p$, $q$ any primes; $J_p \otimes C$ with a torsion group $C$.

11. (Head [1]) For a subgroup $A$ of $B$ and for a reduced group $G$, the canonical homomorphism $A \otimes G \to B \otimes G$ is monic exactly if $A \cap p^n B = p^n A$ holds, whenever $G$ has a summand $Z(p^n)$.

12. Prove (60.4) by applying (29.6) with the functor $A \mapsto A \otimes G, \alpha \mapsto \alpha \otimes 1_G$.

## 61. THE STRUCTURE OF TENSOR PRODUCTS

We conclude our study of tensor products with the discussion of the structure of tensor products in case one of the factors is a torsion group. We are then in the favorable situation of having a complete structure theorem.

The following theorem yields a tool for the actual determination of the structure.

**Theorem 61.1** (Fuchs [9]). *If $C$ is a $p$-group and $B$ is a $p$-basic subgroup of $A$, then there is a natural isomorphism*

$$A \otimes C \cong B \otimes C.$$

Starting with the $p$-pure exact sequence $0 \to B \to A \to A/B \to 0$, we get, by virtue of (60.5), the exactness of the sequence

$$0 \to B \otimes C \to A \otimes C \to (A/B) \otimes C \to 0.$$

Here $A/B$ is $p$-divisible and $C$ is a $p$-group, hence $(A/B) \otimes C = 0$, as is seen from **59** (D). The exactness of $0 \to B \otimes C \to A \otimes C \to 0$ amounts to the assertion of the theorem.□

If $B = \bigoplus_i \langle b_i \rangle$ is a direct decomposition of the $p$-basic subgroup into cyclic groups, then we may write more explicitly $A \otimes C = \bigoplus_i (\langle b_i \rangle \otimes C)$ where the summands are isomorphic either to $C$ or to $C/p^k C$ according as $\langle b_i \rangle$ is of infinite order or of order $p^k$.

Theorem (61.1) actually enables us to determine explicitly the tensor product $A \otimes C$ for torsion $C$. If $B_p$ denotes a $p$-basic subgroup of $A$ and $C_p$ the $p$-component of $C$, then

$$A \otimes C \cong \bigoplus_p (A \otimes C_p) \cong \bigoplus_p (B_p \otimes C_p).$$

Notice that this isomorphism shows, in particular, that

$$A \otimes C \cong T(A) \otimes C \oplus (A/T(A)) \otimes C$$

holds for any torsion group $C$. More generally;

**Theorem 61.2** (Harrison [2]). *If $B$ is a pure subgroup of $A$, and $C$ is a torsion group, then there is an isomorphism*

$$A \otimes C \cong B \otimes C \oplus (A/B) \otimes C.$$

Clearly, we need to prove this only for $p$-groups $C$. But if $C$ is a $p$-group, then the statement is an immediate consequence of **34** (G) and the preceding theorem.☐

**Theorem 61.3** (Fuchs [6], Harrison [1]). *The tensor product of torsion groups is the direct sum of cyclic groups.*

Let $A$ and $C$ be torsion groups, and $B$ and $D$ basic subgroups of $A$ and $C$, respectively. Apply (61.1) twice to conclude that $A \otimes C \cong B \otimes D$. Here both $B$ and $D$ are direct sums of cyclic groups, so by **59** (I) and (H) the same must hold for $B \otimes D$. Evidently, an explicit formula may be given for $B \otimes D$ in terms of invariants of $B$ and $D$.☐

There is essentially very little that is known about the structure of the tensor product of torsion-free groups. The main problem we are confronted with is that of finding a satisfactory way of constructing the tensor product out of the components [our knowledge of torsion-free groups being very limited]. Even this cautious formulation yields insurmountable difficulties, as far as the general case is concerned. The best we can do is to find some invariants for $A \otimes C$.

**Corollary 61.4.** *If $A$ and $C$ are torsion-free with $p$-basic subgroups $B$ and $D$, respectively, then $A \otimes C$ is torsion-free and a $p$-basic subgroup of $A \otimes C$ is isomorphic to $B \otimes D$.*

By our final remark in **60**, $A \otimes C$ is torsion-free. The exact sequence $0 \to B \to A \to A/B \to 0$ induces an exact sequence [see (60.6)]

$$0 \to B \otimes C \to A \otimes C \to (A/B) \otimes C \to 0.$$

Here $(A/B) \otimes C$ is torsion-free and $p$-divisible, thus $p$-basic subgroups of $B \otimes C$ are at the same time those of $A \otimes C$. A repetition of this argument shows $B \otimes D$ $p$-basic in $A \otimes C$.☐

We close this section with a result on the torsion and the torsion-free parts of arbitrary tensor products.

**Theorem 61.5** (Fuchs [9]). *For any groups $A$, $C$, there are isomorphisms*

$$T(A \otimes C) \cong [T(A) \otimes T(C)] \oplus [T(A) \otimes C/T(C)] \oplus [A/T(A) \otimes C],$$

$$(A \otimes C)/T(A \otimes C) \cong A/T(A) \otimes C/T(C).$$

Owing to (60.3), the kernel of the natural epimorphism

$$A \otimes C \to A/T(A) \otimes C/T(C)$$

is the subgroup of $A \otimes C$ generated by $A \otimes T(C)$ and $T(A) \otimes C$; these can, in fact, be identified in the natural way with subgroups of $A \otimes C$, as is shown by (60.4). Here manifestly $A \otimes T(C) \cong [T(A) \otimes T(C)] \oplus [A/T(A) \otimes T(C)]$ and $T(A) \otimes C \cong [T(A) \otimes T(C)] \oplus [T(A) \otimes C/T(C)]$ hold because of (61.2). These groups are torsion and have $T(A) \otimes T(C)$ for intersection. The claim is now obvious.☐

EXERCISES

1.  State an explicit form of the tensor product of two $p$-groups if their basic subgroups are given explicitly.
2.  Assume $A/T(A)$ is divisible; prove $A \otimes C \cong T(A) \otimes C$ for every torsion group $C$.
3.  A reduced group $G$ is torsion if and only if for every pure-exact sequence $0 \to A \to B \to C \to 0$, one has $B \otimes G \cong (A \otimes G) \oplus (C \otimes G)$. [*Hint*: choose $B$ free and $C = Q$.]
4.  (a) Let $Z(p^\infty)$ be a subgroup of $A \otimes C$. Then either $A$ or $C$ contains a subgroup $Z(p^\infty)$.
    (b) State conditions on $A$ and $C$ to satisfy $A \otimes C \cong Z(p^\infty)$.
5.  (a) For a pair of groups $A$, $C$, the isomorphism $A \otimes C \cong Z$ implies $A \cong Z \cong C$.
    (b) Show that both $A$ and $C$ are free if $A \otimes C$ is a nontrivial free group. [*Hint*: $Z \otimes C$ is a subgroup.]
6.  If $G$ is adjusted cotorsion, then $Q/Z \otimes G = 0$.
7.  (a) Let $A$ be a reduced torsion-free algebraically compact group. Then

$$A \cong \operatorname{Hom}(Q/Z, A \otimes Q/Z).$$

[*Hint*: compute the tensor product.]
    (b) $\hat{A} \cong \operatorname{Hom}(Q/Z, A \otimes Q/Z)$ for the $Z$-adic completion $\hat{A}$ of torsion-free $A$.

8.  Let $A$, $C$ be unbounded direct sums of cyclic $p$-groups. Show that $\hat{A} \otimes \hat{C}$ is not algebraically compact. [*Hint*: consider the basic subgroup.]
9.  Let $A$ and $C$ have vanishing first Ulm subgroups. Verify the isomorphism $(A \otimes C)^{\wedge} \cong (\hat{A} \otimes \hat{C})^{\wedge}$.
10. Prove that the Ulm length of $A \otimes C$ cannot exceed the Ulm lengths of both $A$ and $C$. [*Hint*: (61.5) implies that it can be, at most, one larger; show that then the last Ulm factor cannot be torsion-free.]
11. (Fuchs [9]) By a *pairing* of the groups $A$, $C$ into a group $G$ is meant a bilinear function $A \times C \to G$. Show that these pairings form a group isomorphic to $\mathrm{Hom}(A \otimes C, G)$, and describe its structure if $G$ is the reals mod 1.

## 62.   THE TORSION PRODUCT

We have seen that the functor Hom does not carry, in general, a short exact sequence into an exact sequence, but the exactitude can be restored by the functor Ext; in fact, (51.3) shows that this can be done at the price of admitting longer exact sequences. Following this line of thought, one might expect a functor that will do the same favor to the tensor product. It is to this functor that we next turn our attention.

Given the groups $A$, $C$, their *torsion product* $\mathrm{Tor}(A, C)$ is defined as the abelian group generated by all triples $(a, m, c)$ with $a \in A$, $c \in C$, $m \in \mathbf{Z}$ $(m > 0)$ such that $ma = mc = 0$, subject to the following defining relations:

$$(1) \begin{cases} (a_1 + a_2, m, c) = (a_1, m, c) + (a_2, m, c) & \text{if} \quad ma_1 = ma_2 = mc = 0, \\ (a, m, c_1 + c_2) = (a, m, c_1) + (a, m, c_2) & \text{if} \quad ma = mc_1 = mc_2 = 0, \\ (a, mn, c) = (na, m, c) & \text{if} \quad mna = mc = 0, \\ (a, mn, c) = (a, m, nc) & \text{if} \quad ma = mnc = 0. \end{cases}$$

Thus the relations (1) are assumed to hold whenever the right hand sides are defined. In view of the apparent symmetry,

$$\mathrm{Tor}(A, C) \cong \mathrm{Tor}(C, A),$$

this being a natural isomorphism $(a, m, c) \leftrightarrow (c, m, a)$.

The first two relations in (1) imply $(0, m, c) = 0 = (a, m, 0)$ and $(-a, m, c) = -(a, m, c) = (a, m, -c)$; furthermore, $(na, m, c) = n(a, m, c) = (a, m, nc)$, whenever each symbol is meaningful. Clearly, the elements of $\mathrm{Tor}(A, C)$ are finite sums of the form

$$\sum_i (a_i, m_i, c_i) \qquad \text{with} \quad m_i a_i = m_i c_i = 0.$$

We now list a number of elementary consequences of the definition.

(A) $\mathrm{Tor}(A, C)$ *is always a torsion group; it is a p-group if either A or C is a p-group.*

(B) *There is a natural isomorphism*

$$\mathrm{Tor}(A, C) \cong \mathrm{Tor}(T(A), T(C)).$$

In fact, in the definition of Tor only elements of finite order are taken into consideration.

Assertions (A) and (B) justify the terminology "torsion product," and at the same time show that we lose nothing by considering torsion groups only. In particular, $\mathrm{Tor}(A, C) = 0$ if either $A$ or $C$ is torsion-free.

(C) *If $nA = 0$, then $n\,\mathrm{Tor}(A, C) = 0$ for every C.*

Clearly, all the generators of Tor are annihilated by $n$.

(D) *If A is a p-group and C is a q-group with p, q distinct primes, then* $\mathrm{Tor}(A, C) = 0$.

For, $\mathrm{Tor}(A, C)$ is now both a $p$-group and a $q$-group.

(E) Tor *commutes with direct sums: if $A = \bigoplus_i A_i$, there is a natural isomorphism*

$$\mathrm{Tor}\left(\bigoplus_i A_i, C\right) \cong \bigoplus_i \mathrm{Tor}(A_i, C).$$

If $(a_{i_1} + \cdots + a_{i_k}, m, c)$ $(a_i \in A_i)$ is a generator of $\mathrm{Tor}(A, C)$, then

$$m(a_{i_1} + \cdots + a_{i_k}) = 0 = mc.$$

This amounts to $ma_{i_1} = \cdots = ma_{i_k} = mc = 0$, hence $(a_{i_1} + \cdots + a_{i_k}, m, c) = (a_{i_1}, m, c) + \cdots + (a_{i_k}, m, c)$ in $\mathrm{Tor}(A, C)$. Obviously, the triples of the form $(a_i, m, c)$ with $a_i \in A_i$ for a fixed $i$ form a subgroup of $\mathrm{Tor}(A, C)$ isomorphic to $\mathrm{Tor}(A_i, C)$. Taking these subgroups for all $i$, they generate their direct sum, which is $\mathrm{Tor}(A, C)$.

(F) *If $A_p$ and $C_p$ denote the p-components of A and C, then*

$$\mathrm{Tor}(A, C) \cong \bigoplus_p \mathrm{Tor}(A_p, C_p).$$

This is an easy consequence of (D) and (E).

In order to examine the functorial behavior of Tor, we start with homomorphisms $\alpha : A \to A'$ and $\gamma : C \to C'$. It is fairly evident that if $(a, m, c)$ is a generator of $\mathrm{Tor}(A, C)$, then $(\alpha a, m, \gamma c)$ is a generator of $\mathrm{Tor}(A', C')$, and the correspondence

$$(a, m, c) \mapsto (\alpha a, m, \gamma c)$$

between generators extends uniquely to a homomorphism

$$\mathrm{Tor}(\alpha, \gamma) : \mathrm{Tor}(A, C) \to \mathrm{Tor}(A', C').$$

As one might expect, these satisfy the usual rules, like $\text{Tor}(\alpha\alpha',\ \gamma\gamma') = \text{Tor}(\alpha,\gamma)\text{Tor}(\alpha',\gamma')$ for matching homomorphisms, and $\text{Tor}(1_A, 1_C) = 1_{\text{Tor}(A,C)}$, which express the functorial character of Tor. Thus we arrive at the first assertion of the following result:

**Theorem 62.1** (Cartan and Eilenberg [1]). *The torsion product is an additive bifunctor on $\mathscr{A} \times \mathscr{A}$ to $\mathscr{A}$, covariant in both variables. It commutes with direct limits.*

It remains to verify the second assertion. Let $A = \{A_i\ (i \in I);\ \pi_i^k\}$ and $C = \{C_j\ (j \in J);\ \rho_j^l\}$ be direct systems of groups whose direct limits we denote by $A$ and $C$, respectively. It is readily checked that

$$\{\text{Tor}(A_i, C_j)\ ((i, j) \in I \times J),\ \text{Tor}(\pi_i^k, \rho_j^l)\}$$

is a direct system; let $T$ denote its direct limit and $\sigma_{ij} : \text{Tor}(A_i, C_j) \to T$ the corresponding canonical maps. The canonical homomorphisms $\pi_i : A_i \to A$, $\rho_j : C_j \to C$ [which satisfy $\pi_i = \pi_k \pi_i^k$ $(i \leq k)$ and $\rho_j = \rho_l \pi_j^l$ $(j \leq l)$] induce homomorphisms

$$\text{Tor}(\pi_i, \rho_j) : \text{Tor}(A_i, C_j) \to \text{Tor}(A, C)$$

such that $\text{Tor}(\pi_i, \rho_j) = \text{Tor}(\pi_k, \rho_l)\ \text{Tor}(\pi_i^k, \rho_j^l)$. From (11.1) we infer the existence of a unique homomorphism $\sigma : T \to \text{Tor}(A, C)$ satisfying $\sigma\sigma_{ij} = \text{Tor}(\pi_i, \rho_j)$. To show $\sigma$ is an isomorphism, let $(a, m, c) \in \text{Tor}(A, C)$, where of course $ma = mc = 0$. Choose $i$, $j$ such that $a = \pi_i a_i$, $c = \rho_j c_j$, and in addition, $ma_i = 0 = mc_j$ which can be achieved by selecting larger $i$ and $j$ if necessary. $\text{Tor}(\pi_i, \rho_j) : (a_i, m, c_j) \mapsto (a, m, c)$ shows $\sigma$ epic. If $x \in \text{Ker}\ \sigma$, then $\sigma_{ij} y = x$ for some $i$, $j$ and $y \in \text{Tor}(A_i, C_j)$. Now $\text{Tor}(\pi_i, \rho_j)y = \sigma\sigma_{ij} y = \sigma x = 0$ implies the existence of indices $k \geq i, l \geq j$, such that $\text{Tor}(\pi_i^k, \rho_j^l)y = 0$. Apply $\sigma_{kl}$ and notice that $\sigma_{kl}\ \text{Tor}(\pi_i^k, \pi_j^l) = \sigma_{ij}$ in order to conclude $x = \sigma_{ij} y = 0$. Hence $\sigma$ is an isomorphism.☐

It is clear that if $\alpha$ is an endomorphism of $A$ and $\gamma$ is one of $C$, then $\text{Tor}(\alpha, \gamma)$ is an endomorphism of $\text{Tor}(A, C)$. The case when the endomorphism is multiplication by an integer is of special interest:

(G) *Multiplication by the integer $n$ in $A$ [or in $C$] induces multiplication by $n$ on $\text{Tor}(A, C)$.*

The mapping $a \mapsto na$ induces the mapping $(a, m, c) \mapsto (na, m, c) = n(a, m, c)$, whence everything is evident.

We return now to the definition of Tor to prove our first explicit isomorphism on Tor:

(H) *For every group $C$, there is a natural isomorphism*

$$\text{Tor}(Z(m), C) \cong C[m].$$

$$0 \to C(p^n) \to Z(p^\infty) \xrightarrow{p} Z(p^\infty) \to 0$$
$$0 \to \operatorname{Tor}(G, C(p^n)) \to G \xrightarrow{p^n} G \to G \otimes C(p^n) \to 0$$

$$\therefore \operatorname{Tor}(G, C(p^n)) = G[p^n]$$

Given $c \in C[m]$, $(1, m, c)$ is a generator of $\operatorname{Tor}(Z(m), C)$. By (1), $\xi : c \mapsto (1, m, c)$ is a homomorphism $C[m] \to \operatorname{Tor}(Z(m), C)$. The equality $(k, n, c) = (1, kn, c) = (1, m, knm^{-1}c)$ guarantees that every element of $\operatorname{Tor}(Z(m), C)$ can be brought to the form $(1, m, c)$ with $c \in C[m]$, hence $\xi$ is surjective. Clearly, $(1, m, c) = 0$ is a consequence of the defining relations only if $c = 0$, thus $\xi$ is monic, and hence the assertion.

Another essential isomorphism is the following which shows, *inter alia*, that every torsion group is a Tor.

(I) *For every group $C$, there is a natural isomorphism*

$$\operatorname{Tor}(Q/Z, C) \cong T(C).$$

Since $Q/Z = \varinjlim_m Z(m)$ with the inclusion maps $\pi_m^n$ $(m \mid n)$ between the cyclic groups $Z(m) \to Z(n)$, by (62.1) we obtain $\operatorname{Tor}(Q/Z, C)$ as the direct limit of the system $\{\operatorname{Tor}(Z(m), C), (m \in Z, m > 0), \operatorname{Tor}(\pi_m^n, 1_C)\}$. By (H), we can replace the groups $\operatorname{Tor}(Z(m), C)$ by $C[m]$; then $\operatorname{Tor}(\pi_m^n, 1_C)$ becomes the inclusion map $C[m] \to C[n]$, as is easily verified. Therefore, the direct limit will simply be the union of all $C[m]$, that is, the torsion part of $C$.

(J) $\operatorname{Tor}(Z(p^\infty), C)$ *is the p-component of $C$.*
This follows by combining (I) with (E).

(K) *For any three groups $A$, $B$, $C$, there is an isomorphism*

$$A \otimes \operatorname{Tor}(B, C) \oplus \operatorname{Tor}(A, B \otimes C) \cong \operatorname{Tor}(A, B) \otimes C \oplus \operatorname{Tor}(A \otimes B, C).$$

We can represent both sides as direct limits of groups where $A$, $B$, $C$ are replaced by their finitely generated subgroups; in fact, direct sums, tensor and torsion products all commute with direct limits. Thus it suffices to verify the stated isomorphism for finitely generated groups $A$, $B$, $C$. These are direct sums of cyclic groups, so we may reduce the whole proof to the case when $A$, $B$, $C$ are all cyclic. The observation that the roles of $A$ and $C$ are symmetric reduces the number of cases to be distinguished.

If $A = Z$, both sides are $\operatorname{Tor}(B, C)$; if $B = Z$, both sides are $\operatorname{Tor}(A, C)$. If $A$ is a $p$-group and $B$ is a $q$-group with distinct primes $p$, $q$, then both sides vanish. The only case left is when the groups are cyclic $p$-groups: $A = Z(p^k)$, $B = Z(p^l)$, $C = Z(p^m)$. It is straightforward to check that in this case all four direct summands are isomorphic to $Z(p^n)$ with $n = \min(k, l, m)$, whence the stated isomorphism follows.

EXERCISES

1. Prove that $(a, m_1 + m_2, c) = (a, m_1, c) + (a, m_2, c)$, provided the right members are meaningful.
2. The group $A$ is torsion-free if and only if (a) $\operatorname{Tor}(A, C) = 0$ for every group $C$. [*Hint*: use (27.3)]; (b) $\operatorname{Tor}(A, A) = 0$.

3. Describe Tor($A$, $C$) for $A$ a direct sum of cyclic groups.
4. Prove the isomorphism Tor($A$, $C$) $\cong A \otimes C$, if both $A$ and $C$ are bounded. [This is not a natural isomorphism.]
5. Verify the isomorphisms

$$\text{Tor}(A \otimes Z(m), C) \cong A \otimes \text{Tor}(Z(m), C)$$

and

$$\text{Tor}(Z(m), A) \otimes \text{Tor}(Z(m), C) \cong \text{Tor}(\text{Tor}(Z(m), A), C).$$

6. If $A$ satisfies Tor($A$, $C$) $\cong T(C)$ for every group $C$, then $A \cong Q/Z \oplus H$ with $H$ torsion-free.
7. If $A$ and $C$ are torsion groups such that Tor($A$, $C$) $\cong Q/Z$, then $A \cong Q/Z \cong C$.
8. If $G$ is adjusted cotorsion, then

$$G \cong \text{Ext}(Q/Z, \text{Tor}(Q/Z, G)).$$

9. Give a counterexample to show that Tor does not commute with direct products and inverse limits, even if the torsion parts are meant throughout. [*Hint:* $A = \prod Z(p^\infty)$.]
10. Give an example in which $n$ Tor($A$, $C$) fails to be isomorphic to Tor($nA$, $C$). [*Hint:* $A = Z(p^\infty)$.]
11. Prove that

$$A \otimes \text{Tor}(B, C) \cong \text{Tor}(A \otimes B, C)$$

if $A$ is torsion-free.

## 63.    EXACT SEQUENCES FOR TOR

The purpose of this section is to investigate more closely the torsion product and to show that Tor is a left exact functor, such that Tor and $\otimes$ are connected in a long exact sequence in a fashion similar to Hom and Ext.

Before entering the discussion of the exact sequence for Tor, we introduce a connecting homomorphism, this time between Tor and $\otimes$, as follows. Let

(1) $$E: 0 \to A \xrightarrow{\alpha} B \xrightarrow{\beta} C \to 0$$

be a short exact sequence and $(c, m, g)$ a generator of Tor($C$, $G$); i.e., $mc = mg = 0$ ($c \in C$, $g \in G$, $m \in Z$, $m > 0$). There is a $b \in B$ with $\beta b = c$, and there is an $a \in A$ with $\alpha a = mb$ [since $mc = 0$]. We set

(2) $$E_* : (c, m, g) \mapsto a \otimes g,$$

defining thereby a map on the generators of Tor($C$, $G$) into the set of generators of $A \otimes G$. In fact, $a \otimes g$ is independent of the choices of $b$, $a$, since if $b'$, $a'$ are other choices, then $b' = b + \alpha a_0$ ($a_0 \in A$), $\alpha a' = \alpha a + \alpha(m a_0)$, and so $a' \otimes g = a \otimes g + m a_0 \otimes g$ where $m a_0 \otimes g = a_0 \otimes m g = 0$. Also, it is easily seen that $E_*$ respects the defining relations in **62**, (1), in the sense that if $E_*$ is extended in the obvious way to the whole of $\text{Tor}(C, G)$, then it carries equal elements of Tor into equal elements of $A \otimes G$. Therefore,

$$E_* : \text{Tor}(C, G) \to A \otimes G$$

is a homomorphism. It is natural, for if $\phi : G \to H$ is any homomorphism, then the diagram

$$\begin{array}{ccc} \text{Tor}(C, G) & \to & A \otimes G \\ \downarrow & & \downarrow \\ \text{Tor}(C, H) & \to & A \otimes H \end{array}$$

with the obvious maps commutes, due to the fact that $(c, m, g)$ is carried into $a \otimes \phi g$ in both ways. $E_*$ is called the *connecting homomorphism* [in view of the symmetry of Tor and $\otimes$ in the variables, we have only one such homomorphism to deal with]. This $E_*$ is needed in the following theorem.

**Theorem 63.1** (Cartan and Eilenberg [1]). *If $E$ in* (1) *is an exact sequence, then for every group $G$, the induced sequence*

(3)
$$0 \to \text{Tor}(A, G) \xrightarrow{\alpha_*} \text{Tor}(B, G) \xrightarrow{\beta_*} \text{Tor}(C, G)$$
$$\xrightarrow{E_*} A \otimes G \xrightarrow{\alpha \otimes 1} B \otimes G \xrightarrow{\beta \otimes 1} C \otimes G \to 0$$

*is exact. Here $\alpha_*$, $\beta_*$ are abbreviations for* $\text{Tor}(\alpha, 1_G)$, $\text{Tor}(\beta, 1_G)$.

Recall that Tor and $\otimes$ commute with direct limits, and the direct limit of exact sequences is exact [this holds for longer exact sequences too, because direct limits preserve kernels and images]. Therefore, it suffices to consider the case when $G$ is finitely generated. Using a direct decomposition of $G$, the proof reduces to the cases $G = Z$ and $G = Z(m)$. The first case is trivial, since then the Tors vanish and the sequence reduces to (1). In the second case, the elements of Tors are of the form $(x, m, 1)$ with $mx = 0$, while those of the tensor products are of the form $x \otimes 1$ with $x$ taken mod $mX$ [where $X$ stands for $A$, $B$, or $C$]. Exactness at the two last places follows from (60.2). Clearly, $\beta_* \alpha_* = (\beta \alpha)_* = 0$; furthermore $E_* \beta_*(b, m, 1) = E_*(c, m, 1) = a \otimes 1 = 0$, because $\alpha a = mb = 0$, and $(\alpha \otimes 1) E_*(c, m, 1) = (\alpha \otimes 1)(a \otimes 1) = \alpha a \otimes 1 = 0$, because $\alpha a \in mB$. Thus, we need only show the inclusions of the kernels in previous images at the first four places.

Let $\alpha_*(a, m, 1) = 0$. Then $(\alpha a, m, 1) = 0$, and $\alpha a = 0$, $a = 0$, i.e., $\alpha_*$ is monic. If $\beta_*(b, m, 1) = 0$, then similarly $\beta b = 0$, and so an $a \in A$ satisfies $\alpha a = b$, and, in addition, $ma = 0$. Hence $\alpha_*(a, m, 1) = (b, m, 1)$, and the exactness at the second Tor follows. If $E_*(c, m, 1) = 0$, then $a \otimes 1 = 0$ [in the notation of (2)], i.e., $a \in mA$. Write $a = ma'$ $(a' \in A)$, $m(b - \alpha a') = 0$ to conclude that $(b - \alpha a', m, 1) \in \text{Tor}(B, Z(m))$ maps onto $(c, m, 1)$. Finally, if $(\alpha \otimes 1)(a \otimes 1) = 0$, then $\alpha a = mb$ for some $b \in B$, and for this $b$, $E_*(\beta b, m, 1) = a \otimes 1$. This establishes exactness at $A \otimes G$.☐

The first part of (3) proves the left-exactness of the functor Tor.

From the theorem just proved, we conclude that if $0 \to H \xrightarrow{\phi} F \to C \to 0$ is an exact sequence with a free group $F$, then $0 \to \text{Tor}(A, C) \to A \otimes H \to A \otimes F \to A \otimes C \to 0$ is exact. Consequently, $\text{Tor}(A, C)$ *can also be defined as the kernel of the map*

$$1 \otimes \phi : A \otimes H \to A \otimes F.$$

This interpretation leads to an alternative set of generators and defining relations for $\text{Tor}(A, C)$ [recall that both $A \otimes H$ and $A \otimes F$ are direct sums of copies of $A$].

If we specialize (1) to a pure-exact sequence, then (3) splits into two short exact sequences. In fact, this follows at once from (60.4). Moreover, we have:

**Theorem 63.2** (Fuchs [9]). *If* (1) *is a pure-exact sequence, then the induced sequence*

$$0 \to \text{Tor}(A, G) \to \text{Tor}(B, G) \to \text{Tor}(C, G) \to 0$$

*is pure-exact, for any group G.*

Owing to (62.1) and (29.6), there is nothing to prove.

EXERCISES

1. If $A'$, $B'$ are pure subgroups of $A$, $B$, then $\text{Tor}(A', B')$ is a pure subgroup of $\text{Tor}(A, B)$.
2. (a) If $G$ is a group such that $0 \to \text{Tor}(A, G) \to \text{Tor}(B, G) \to \text{Tor}(C, G) \to 0$ is exact for every exact $0 \to A \to B \to C \to 0$, then $G$ is torsion-free.
   (b) If the same is assumed only for exact sequences $0 \to A \to B \to C \to 0$ of torsion groups, then the torsion part of $G$ is divisible.
3. An exact sequence $0 \to A \to B \to C \to 0$ such that $0 \to \text{Tor}(A, G) \to \text{Tor}(B, G) \to \text{Tor}(C, G) \to 0$ is exact for every $G$ [every $G = Z(m)$] is pure-exact.
4. Let $A$, $C$ be $p$-groups.
   (a) Apply (63.2) twice to find the basic subgroup of $\text{Tor}(A, C)$ by means of those of $A$, $C$.
   (b) Show that the basic subgroup of $\text{Tor}(A, C)$ is, in general, not isomorphic to the torsion product of the basic subgroups of $A$ and $C$.
5. Write $A = F_1/H_1$ and $C = F_2/H_2$ with $F_1$, $F_2$ free. Prove that

$$\text{Tor}(A, C) \cong [(F_1 \otimes H_2) \cap (H_1 \otimes F_2)]/(H_1 \otimes H_2)$$

where all the groups on the right are regarded as subgroups of $F_1 \otimes F_2$.

6.   (Yahya [2]) Let $r, n$ be integers such that $1 \leq r \leq n - 1$. Given $n$ groups $A_i$, define $S_r^n(A_1, \cdots, A_n)$ as the group generated by symbols $(a_1, \cdots, a_n; m)$ with $a_i \in A_i$, $m \in \mathbf{Z}$, $m > 0$, and $ma_i = 0$ for all $i$, subject to the relations

$$(a_1, \cdots, a_i + a_i', \cdots, a_n; m) = (a_1, \cdots, a_i, \cdots, a_n; m)$$
$$+ (a_1, \cdots, a_i', \cdots, a_n; m)$$
$$(a_1, \cdots, a_r, a_{r+1}, \cdots, a_n; km) = (ka_1, \cdots, ka_r, a_{r+1}, \cdots, a_n; m)$$

whenever the right members make sense. Prove:
(a) $S_r^n$ is an additive functor on $\mathscr{A} \times \cdots \times \mathscr{A}$ to $\mathscr{A}$, covariant in each variable;
(b) $S_r^n$ commutes with direct limits;
(c) $S_{n-1}^n(A_1, \cdots, A_n) \cong \operatorname{Tor}(A_1, \cdots, \operatorname{Tor}(A_{n-2}, \operatorname{Tor}(A_{n-1}, A_n)) \cdots)$.
7.   Prove the associative law for Tor:

$$\operatorname{Tor}(\operatorname{Tor}(A, B), C) \cong \operatorname{Tor}(A, \operatorname{Tor}(B, C))$$

[with a natural isomorphism].

## 64.   THE STRUCTURE OF TORSION PRODUCTS

It follows from the material of 62 that the structure of $\operatorname{Tor}(A, C)$ for arbitrary $A$ and $C$ can be described as soon as we know the same for $p$-groups $A$ and $C$. It is for this reason that we shall restrict ourselves in this section to $p$-groups.

$A$ and $C$ are $p$-groups throughout. The following series of lemmas will lead us to a result that enables us to compute the basic subgroups of the Ulm factors of $\operatorname{Tor}(A, C)$.

**Lemma 64.1** (Nunke [4]).   *If, for some ordinal* $\sigma$, $p^\sigma A = 0$, *then*

$$p^\sigma \operatorname{Tor}(A, C) = 0 \qquad \textit{for every } C.$$

There is an exact sequence $0 \to C \to \bigoplus Z(p^\infty)$ with sufficiently large number of summands. This implies the exactness of the induced sequence

$$0 \to \operatorname{Tor}(A, C) \to \operatorname{Tor}(A, \bigoplus Z(p^\infty)) = \bigoplus \operatorname{Tor}(A, Z(p^\infty)) \cong \bigoplus A.$$

The assumption $p^\sigma A = 0$ implies $p^\sigma(\bigoplus A) = 0$, whence the desired relation follows.☐

**Lemma 64.2** (Nunke [4]).   *For* $p$-*groups* $A$ *and* $C$, *and for every ordinal* $\sigma$,

$$\operatorname{Tor}(p^\sigma A, p^\sigma C) = p^\sigma \operatorname{Tor}(A, C).$$

A repeated application of (63.1) ensures that $\text{Tor}(p^\sigma A, p^\sigma C)$ can be regarded as embedded in $\text{Tor}(A, C)$ in the natural way. First we verify the inclusion $\leq$. To begin with, let $\sigma = 1$. If $(pa, p^k, pc)$ is a generator of $\text{Tor}(pA, pC)$, $p^{k+1}a = p^{k+1}c = 0$, then in $\text{Tor}(A, C)$ we have $(pa, p^k, pc) = (pa, p^{k+1}, c) = p(a, p^{k+1}, c) \in p\,\text{Tor}(A, C)$. We can now proceed by transfinite induction: if the inclusion $\leq$ holds for some $\sigma$, then by virtue of what has been shown

$$\text{Tor}(p^{\sigma+1}A, p^{\sigma+1}C) \leq p\,\text{Tor}(p^\sigma A, p^\sigma C) \leq p^{\sigma+1}\,\text{Tor}(A, C).$$

Suppose $\rho$ is a limit ordinal, and the inclusion has been proved for all $\sigma < \rho$. Then $p^\rho A \leq p^\sigma A$, $p^\rho C \leq p^\sigma C$ imply that $\text{Tor}(p^\rho A, p^\rho C) \leq \text{Tor}(p^\sigma A, p^\sigma C) \leq p^\sigma\,\text{Tor}(A, C)$, the last inclusion being assumed to hold. Therefore

$$\text{Tor}(p^\rho A, p^\rho C) \leq \bigcap_{\sigma < \rho} p^\sigma\,\text{Tor}(A, C) = p^\rho\,\text{Tor}(A, C).$$

To prove the converse inclusion, we start with the commutative diagram

$$
\begin{array}{ccccc}
0 & & 0 & & 0 \\
\downarrow & & \downarrow & & \downarrow \\
0 \to \text{Tor}(p^\sigma A, p^\sigma C) & \longrightarrow & \text{Tor}(A, p^\sigma C) & \longrightarrow & \text{Tor}(A/p^\sigma A, p^\sigma C) \\
\downarrow & & \downarrow & & \downarrow \\
0 \to \text{Tor}(p^\sigma A, C) & \longrightarrow & \text{Tor}(A, C) & \longrightarrow & \text{Tor}(A/p^\sigma A, C) \\
\downarrow & & \downarrow & & \downarrow \\
0 \to \text{Tor}(p^\sigma A, C/p^\sigma C) & \to & \text{Tor}(A, C/p^\sigma C) & \to & \text{Tor}(A/p^\sigma A, C/p^\sigma C)
\end{array}
$$

in which all the rows and columns are exact by (63.1). This yields by (8.3) the exact sequence

$$0 \to \text{Tor}(p^\sigma A, p^\sigma C) \to \text{Tor}(A, C) \to \text{Tor}(A/p^\sigma A, C) \oplus \text{Tor}(A, C/p^\sigma C).$$

By (64.1), the summands are annihilated by $p^\sigma$, hence the last homomorphism has kernel containing $p^\sigma\,\text{Tor}(A, C)$. This yields the desired inclusion $\geq$.☐

**Lemma 64.3** (Nunke [4]). *For every integer* $n$,

$$\text{Tor}(A[p^n], C[p^n]) = \text{Tor}(A, C)\,[p^n].$$

Multiplication by $p^n$ in $A$ yields the exact sequence $0 \to A[p^n] \to A \xrightarrow{p^n} A$ from which Tor gives the exactness of

$$0 \to \text{Tor}(A[p^n], C) \to \text{Tor}(A, C) \xrightarrow{p^n} \text{Tor}(A, C).$$

This implies $\text{Tor}(A, C)\,[p^n] = \text{Ker } p^n = \text{Tor}(A[p^n], C)$. Changing the roles of $A$ and $C$, the same argument leads to the desired result.☐

The following result enables us to determine important structural invariants of $\text{Tor}(A, C)$, namely, the rank of $p^\sigma\,\text{Tor}(A, C)$ and the Ulm–Kaplansky invariants.

**Theorem 64.4** (Nunke [4]). *Let $A$, $C$ be $p$-groups and $\sigma$ any ordinal. Then*

(1) $$r(p^\sigma \operatorname{Tor}(A, C)) = r(p^\sigma A) \cdot r(p^\sigma C),$$

(2) $$f_\sigma(\operatorname{Tor}(A, C)) = f_\sigma(A) f_\sigma(C) + f_\sigma(A) r(p^{\sigma+1}C) + r(p^{\sigma+1}A) f_\sigma(C).$$

Combining (64.2) with (64.3), we obtain

(3) $$(p^\sigma \operatorname{Tor}(A, C))[p] = \operatorname{Tor}((p^\sigma A)[p], (p^\sigma C)[p]).$$

All these groups are direct sums of groups of order $p$, whence **62** (E) and (H) imply that the group on the right is the direct sum of $r(p^\sigma A) \cdot r(p^\sigma C)$ groups of order $p$. This proves (1).

In order to prove (2), let us write $(p^\sigma A)[p] = (p^{\sigma+1}A)[p] \oplus G$ and $(p^\sigma C)[p] = (p^{\sigma+1}C)[p] \oplus H$ where, by definition, $r(G) = f_\sigma(A)$ and $r(H) = f_\sigma(C)$. From (3) it follows

$$(p^\sigma \operatorname{Tor}(A, C))[p] = \operatorname{Tor}((p^{\sigma+1}A)[p], (p^{\sigma+1}C)[p]) \oplus \operatorname{Tor}(G, (p^{\sigma+1}C)[p])$$
$$\oplus \operatorname{Tor}((p^{\sigma+1}A)[p], H) \oplus \operatorname{Tor}(G, H).$$

The first summand on the right is just $(p^{\sigma+1} \operatorname{Tor}(A, C))[p]$, therefore the $\sigma$th Ulm–Kaplansky invariant of $\operatorname{Tor}(A, C)$ is the sum of the ranks of the remaining three summands. This is expressed in formula (2).☐

**Corollary 64.5** (Nunke [4]). *If $A$, $C$ are reduced torsion groups, then the Ulm length of* Tor $(A, C)$ *is the minimum of the lengths of $A$ and $C$.*

This is an immediate consequence of (2).☐

As we shall see in Chapter XII, the direct sums of countable $p$-groups are completely determined by their Ulm–Kaplansky invariants. Consequently, (64.4) renders it possible to determine completely the torsion product of any two countable groups or of direct sums of countable groups.

EXERCISES

1.  Prove
    $$\operatorname{Tor}(A, C)^\sigma = \operatorname{Tor}(A^\sigma, C^\sigma)$$
    for the $\sigma$th Ulm subgroups.
2.  (a) The divisible part of $\operatorname{Tor}(A, C)$ is the torsion product of the maximal divisible subgroups of $A$ and $C$.
    (b) If $\operatorname{Tor}(A, C)$ contains a subgroup of type $p^\infty$, then both $A$ and $C$ do.
3.  (Nunke [4]) Show that the basic subgroups of $\operatorname{Tor}(A, C)$ are always of the same cardinality as $\operatorname{Tor}(A, C)$ itself.
4.  (Nunke [4]) If $A'$, $A''$ are subgroups of $A$, then
    $$\operatorname{Tor}(A' \cap A'', C) = \operatorname{Tor}(A', C) \cap \operatorname{Tor}(A'', C).$$
    [*Hint*: set up a commutative square with these groups and $\operatorname{Tor}(A, C)$.]

5.  (Nunke [4]) Given $u \in \mathrm{Tor}(A, C)$, there are finite subgroups $A_u \leqq A$ and $C_u \leqq C$ such that: (i) $u \in \mathrm{Tor}(A_u, C_u)$; (ii) if $u \in \mathrm{Tor}(A', C')$ for $A' \leqq A$ and $C' \leqq C$, then $A_u \leqq A'$ and $C_u \leqq C'$.
    [*Hint*: let $A_u$, $C_u$ be of minimal orders with (i).]
6.  Let $(a, p^k, c) \in \mathrm{Tor}(A, C)$, where $A$, $C$ are $p$-groups. Prove that the height of $(a, p^k, c)$ is equal to the minimum of the heights of $a$ and $c$ in $A$ and $C$, respectively.
7.* (a) Let $A$, $C$ be $p$-groups, $C$ without elements of infinite height. If $a$ is of order $p$ and of infinite height, then $(a, p, c) \mapsto c$ yields an isomorphism between a subgroup of $\mathrm{Tor}(A, C)$ and $C[p]$ which preserves heights.
    (b) Let $C$ be the torsion part of the $p$-adic completion of $B = \bigoplus_{n=1}^{\infty} Z(p^n)$. Prove that any subgroup of $C[p]$ with elements of bounded heights in $C$ must be finite, and $C[p]$ cannot be embedded by a height-preserving isomorphism in a direct sum of cyclic groups.
    (c) Use (a) and (b) to exhibit a $\mathrm{Tor}(A, C)$ which has no elements of infinite height, but is not a direct sum of cyclic groups.

## NOTES

The Kronecker or tensor product was known a long time ago in linear algebra, it was introduced into group theory by Whitney [1]. The tensor product of modules, algebras, etc. is now fundamental. The use of $\mathrm{Tor}_n^R$ is not so extensive. [Each $\mathrm{Tor}_n^R$ occurs with three terms in the infinite exact sequence which replaces (3) of (63.1) in the general case.]

Nothing is known about the algebraic structure of tensor and torsion products of modules in general, except for trivialities and very special cases.

A. Hattori [*J. Math. Soc. Japan* 9 (1957), 381–385] has shown that over integral domains R, the tensor product of torsion-free R-modules is again torsion-free if and only if R is Prüfer.

Various results have been proved for flat modules. An R-module $F$ is *flat* if every short exact sequence of right R-modules yields an exact sequence of abelian groups when tensored by $F$ over R, or, equivalently, $F$ is flat if $\mathrm{Tor}_1^R(M, F) = 0$ for all right R-modules $M$. Another characterization of flat modules is due to V. E. Govorov [*Sibirsk. Mat. Z.* 6 (1965), 300–304]: they are direct limits of free modules. J. Lambek [*Can. Math. Bull.* 7 (1964), 237–243] proved that an R-module $F$ is flat if, and only if, $\mathrm{Hom}_Z(F, Q/Z)$ is an injective right R-module. Over a left principal ideal domain, the flat modules are exactly the torsion-free modules. All R-modules are flat if, and only if, R is a regular ring in the sense of von Neumann.

The isomorphism in **59** (J) is fundamental: it expresses the fact that the tensor product and Hom are in the adjoint situation.

*Problem 47.* Find the structure $A \otimes C$ for torsion-free [mixed] groups $A$ and $C$.

*Problem 48.* Which torsion groups $G$ can be written in the form $G \cong A \otimes_R C$ for R-modules $A$, $C$ whose additive groups are torsion?

*Problem 49.* Describe [the Ulm factors of] $\mathrm{Tor}(A, C)$ for $p$-groups $A$, $C$.

*Problem 50.* Relate $A$ and $A'$ if $\mathrm{Tor}(A, C) \cong \mathrm{Tor}(A', C)$ for all reduced $C$.

## BIBLIOGRAPHY

Baer, R.
[1] Der Kern, eine charakteristische Untergruppe, *Compositio Math.* **1** (1934), 254–283.
[2] Abelian groups without elements of finite order, *Duke Math. J.* **3** (1937), 68–122.
[3] Abelian groups that are direct summands of every containing abelian group, *Bull. Amer. Math. Soc.* **46** (1940), 800–806.
[4] Die Torsionsuntergruppe einer abelschen Gruppe, *Math. Ann.* **135** (1958), 219–234.

Balcerzyk, S.
[1] Remark on a paper of S. Gacsályi, *Publ. Math. Debrecen.* **4** (1956), 357–358.
[2] On algebraically compact groups of I. Kaplansky, *Fund. Math.* **44** (1957), 91–93.
[3] On factor groups of some subgroups of a complete direct sum of infinite cyclic groups, *Bull. Acad. Polon. Sci.* **7** (1959), 141–142.

Baumslag, G., and Blackburn, N.
[1] Direct summands of unrestricted direct sums of abelian groups, *Arch. Math.* **10** (1959), 403–408.

Birkhoff, G.
[1] Subgroups of abelian groups, *Proc. London Math. Soc.* **38** (1934), 385–401.

Bourbaki, N.
[1] "Algèbre," Chap. VII. Modules sur les anneaux principaux. Paris, 1952.

Boyer, D. L.
[1] On the theory of *p*-basic subgroups of abelian groups, *Topics in Abelian Groups*, 323–330 (Chicago, Illinois, 1963).

Cartan, H., and Eilenberg, S.
[1] "Homological Algebra." Princeton Univ. Press, Princeton, New Jersey, 1956.

Charles, B.
[1] Étude des groupes abéliens primaires de type $\leq \omega$, *Ann. Univ. Sarav., Sciences* **4** (1955), 184–199.
[2] Une caractérisation des intersections de sous-groupes divisibles, *C.R. Acad. Sci. Paris*, **250** (1960), 256–257.
[3] Étude sur les sous-groupes d'un groupe abélien, *Bull. Soc. Math. France* **88** (1960), 217–227.
[4] Méthodes topologiques en théorie des groupes abéliens, *Proc. Colloq. Abelian Groups*, 29–42 (Budapest, 1964).

Cohn, P. M.
[1] The complement of a finitely generated direct summand of an abelian group, *Proc. Amer. Math. Soc.* **7** (1956), 520–521.

Cutler, D. C.
[1] Quasi-isomorphisms for infinite abelian *p*-groups, *Pacific J. Math.* **16** (1966), 25–45.

Dieudonné, J.
[1] Sur les produits tensoriels, *Ann. Sci. École Norm. Sup.* **64** (1948), 101–117.
[2] Sur les *p*-groupes abéliens infinis, *Portugal. Math.* **11** (1952), 1–5.

Dlab, V.
[1] *D*-Rang einer abelschen Gruppe, *Časopis Pěst. Mat.* **82** (1957), 314–334.
[2] The Frattini subgroups of abelian groups, *Czechoslovak. Math. J.* **10** (1960), 1–16.
[3] On cyclic groups, *Czechoslovak Math. J.* **10** (1960), 244–254.

[4] On a characterization of primary abelian groups of bounded order, *J. London Math. Soc.* **36** (1961), 139–144.

Eilenberg, S. and Mac Lane, S.
[1] Group extensions and homology, *Ann. of Math.* **43** (1942), 757–831.

Enochs, E.
[1] Isomorphic refinements of decompositions of a primary group into closed groups, *Bull. Soc. Math. France* **91** (1963), 63–75.

Erdélyi, M.
[1] Direct summands of abelian torsion groups [Hungarian], *Acta Univ. Debrecen* **2** (1955), 145–149.

Fedorov, Yu. G.
[1] On infinite groups all of whose nontrivial subgroups have finite index [Russian], *Uspehi Mat. Nauk.* **6** (1951), 187–189.

Frobenius, G., and Stickelberger, L.
[1] Über Gruppen von vertauschbaren Elementen, *J. Reine Angew. Math.* **86** (1878), 217–262.

Fuchs, L.
[1] The direct sum of cyclic groups, *Acta Math. Acad. Sci. Hungar.* **3** (1952), 177–195.
[2] On the structure of abelian *p*-groups, *Acta Math. Acad. Sci. Hungar.* **4** (1953), 267–288.
[3] On a special kind of duality in group theory. II, *Acta Math. Acad. Sci. Hungar.* **4** (1953), 299–314.
[4] On a property of basic subgroups, *Acta Math. Acad. Sci. Hungar.* **5** (1954), 143–144.
[5] On a useful lemma for abelian groups, *Acta Sci. Math.* (*Szeged*) **17** (1956), 134–138.
[6] Über das Tensorprodukt von Torsionsgruppen, *Acta Sci. Math.* (*Szeged*) **18** (1957), 29–32.
[7] "Abelian groups." Publishing House of the Hungarian Academy of Science, Budapest, 1958.
[8] On character groups of discrete abelian groups, *Acta Math. Acad. Sci. Hungar.* **10** (1959), 133–140.
[9] Notes on abelian groups. I, *Ann. Univ. Sci. Budapest* **2** (1959), 5–23; II, *Acta Math. Acad. Sci. Hungar.* **11** (1960), 117–125.
[10] On algebraically compact abelian groups, *J. Natur. Sci. and Math.* **3** (1963), 73–82.
[11] Note on factor groups in complete direct sums, *Bull. Acad. Polon. Sci.* **11** (1963), 39–40.
[12] Some generalizations of the exact sequences concerning Hom and Ext, *Proc. Colloq. Abelian Groups*, 57–76 (Budapest, 1964).
[13] Note on linearly compact abelian groups, *J. Austral. Math. Soc.* **9** (1969), 433–440.

Gacsályi, S.
[1] On algebraically closed abelian groups, *Publ. Math. Debrecen* **2** (1952), 292–296.
[2] On pure subgroups and direct summands of abelian groups, *Publ. Math. Debrecen* **4** (1955), 89–92.

Golema, K., and Hulanicki, A.
[1] The structure of the factor group of the unrestricted sum by the restricted sum of abelian groups. II, *Fund. Math.* **53** (1963–4), 177–185.

Grätzer G., and Schmidt, E. T.
[1] A note on a special type of fully invariant subgroups of abelian groups, *Ann. Univ. Sci. Budapest* **3–4** (1961), 85–87.

de Groot, J.
  [1] An isomorphism criterion for completely decomposable abelian groups, *Math. Ann.* **132** (1956), 328–332.

Harrison, D. K.
  [1] Infinite abelian groups and homological methods, *Ann. of Math.* **69** (1959), 366–391.
  [2] Two of the problems of L. Fuchs, *Publ. Math. Debrecen* **7** (1960), 316–319.
  [3] On the structure of Ext, *Topics in Abelian Groups*, 195–209 (Chicago, Illinois, 1963).

Harrison, D. K., Irwin, J. M., Peercy, C. L., and Walker, E. A.
  [1] High extensions of abelian groups, *Acta Math. Acad. Sci. Hungar.* **14** (1963), 319–330.

Head, T. J.
  [1] A direct limit representation for abelian groups with an application to tensor sequences, *Acta Math. Acad. Sci. Hungar.* **18** (1967), 231–234.

Hill, P.
  [1] Concerning the number of basic subgroups, *Acta Math. Acad. Sci. Hungar.* **17** (1966), 267–269.

Honda, K.
  [1] Realism in the theory of abelian groups. I, *Comment. Math. Univ. St. Paul* **5** (1956), 37–75; II, *Comment. Math. Univ. St. Paul* **9** (1961), 11–28; III, *Comment Math. Univ. St. Paul* **12** (1964), 75–111.
  [2] From a theorem of Kulikov to a problem of Kaplansky, *Comment. Math. Univ. St. Paul* **6** (1957), 43–48.

Hulanicki, A.
  [1] Algebraic characterization of abelian divisible groups which admit compact topologies, *Fund. Math.* **44** (1957), 192–197.
  [2] Algebraic structure of compact abelian groups, *Bull. Acad. Polon. Sci.* **6** (1958), 71–73.
  [3] The structure of the factor group of an unrestricted sum by the restricted sum of abelian groups, *Bull. Acad. Polon. Sci.* **10** (1962), 77–80.

Irwin, J.
  [1] High subgroups of abelian torsion groups, *Pacific J. Math.* **11** (1961), 1375–1384.

Irwin, J. M., and Richman, F.
  [1] Direct sums of countable groups and related concepts, *J. Algebra* **2** (1965), 443–450.

Irwin, J. M., and Walker, E. A.
  [1] On $N$-high subgroups of abelian groups, *Pacific J. Math.* **11** (1961), 1363–1374.
  [2] On isotype subgroups of abelian groups, *Bull. Soc. Math. France* **89** (1961), 451–460.

Irwin, J. M., Walker, C. P., and Walker, E. A.
  [1] On $p^z$-pure sequences of abelian groups, *Topics in Abelian Groups*, 69–119 (Chicago, Illinois, 1963).

Kaloujnine, L.
  [1] Sur les groupes abéliens primaires sans éléments de hauteur infinie, *C.R. Acad. Sci. Paris* **225** (1947), 713–715.

Kaplansky, I.
  [1] Modules over Dedekind rings and valuation rings, *Trans. Amer. Math. Soc.* **72** (1952), 327–340.
  [2] Infinite abelian groups, Univ. of Michigan Press, Ann Arbor, Michigan, 1954.
  [3] Projective modules, *Ann. of Math.* **68** (1958), 372–377.

Kertész, A.
[1] On fully decomposable abelian torsion groups, *Acta Math. Acad. Sci. Hungar.* **3** (1952), 225–232.
[2] On subgroups and homomorphic images, *Publ. Math. Debrecen* **3** (1953), 174–179.
[3] The general theory of linear equation systems over semisimple rings, *Publ. Math. Debrecen* **4** (1955), 79–86.

Kertész, A., and Szele, T.
[1] On the existence of non-discrete topologies in infinite abelian groups, *Publ. Math. Debrecen* **3** (1953), 187–189.

Khabbaz, S. A.
[1] On a theorem of Charles and Erdélyi, *Bull. Soc. Math. France* **89** (1961), 103–104.
[2] The subgroups of a divisible group *G* which can be represented as intersections of divisible subgroups of *G*, *Pacific J. Math.* **11** (1961), 267–273.

Khabbaz, S. A., and Walker, E. A.
[1] The number of basic subgroups of primary groups, *Acta Math. Acad. Sci. Hungar.* **15** (1964), 153–155.

Kovács, L.
[1] On subgroups of the basic subgroup, *Publ. Math. Debrecen* **5** (1958), 261–264.

Kulikov, L. Ya.
[1] On the theory of abelian groups of arbitrary cardinality [Russian], *Mat. Sb.* **9** (1941), 165–182.
[2] On the theory of abelian groups of arbitrary cardinality [Russian], *Mat. Sb.* **16** (1945), 129–162.
[3] Generalized primary groups. I, [Russian]. *Trudy Moskov. Mat. Obšč.* **1** (1952), 247–326; II, *Trudy Moskov. Mat. Obšč.* **2** (1953), 85–167.
[4] On direct decomposition of groups [Russian], *Ukrain. Mat. Ž.* **4** (1952), 230–275 and 347–372.

Kurosh, A. G.
[1] Zur Zerlegung unendlicher Gruppen, *Math. Ann.* **106** (1932), 107–113.
[2] "Gruppentheorie." Akademie-Verlag, Berlin, 1955.

Levi, F. W.
[1] Abelsche Gruppen mit abzählbaren Elementen, Habilitationsschrift. Leipzig, 1917.

Lewis, P. E.
[1] Characters of abelian groups, *Amer. J. Math.* **64** (1942), 81–105.

Łoś, J.
[1] Abelian groups that are direct summands of every abelian group which contains them as pure subgroups, *Bull. Acad. Polon. Sci.* **4** (1956), 73, and *Fund. Math.* **44** (1957), 84–90.
[2] Linear equations and pure subgroups, *Bull. Acad. Polon. Sci.* **7** (1959), 13–18.
[3] Generalized limits in algebraically compact groups, *Bull. Acad. Polon. Sci.* **7** (1959), 19–21.

Mac Lane, S.
[1] Duality for groups, *Bull. Amer. Math. Soc.* **56** (1950), 485–516.
[2] Group extensions by primary abelian groups, *Trans. Amer. Math. Soc.* **95** (1960), 1–16.
[3] "Homology." Academic Press, New York, 1963.

Mader A.
[1] A characterization of completions of direct sums of cyclic groups, *Bull. Acad. Polon. Sci.* **15** (1967), 231–233.

Maranda, J. M.
[1] On pure subgroups of abelian groups, *Arch. Math.* **11** (1960), 1–13.

Megibben, C.
[1] On subgroups of primary abelian groups, *Publ. Math. Debrecen* **12** (1965), 293–294.

Mitchell, A. R., and Mitchell, R. W.
[1] Disjoint basic subgroups, *Pacific J. Math.* **23** (1967), 119–127.

Mostowski, A. W. and Sąsiada, E.
[1] On the bases of modules over a principal ideal ring, *Bull. Acad. Polon Sci.* **3** (1955), 477–478.

Nunke, R. J.
[1] Modules of extensions over Dedekind rings, *Illinois J. Math.* **3** (1959), 222–241.
[2] A note on abelian group extensions, *Pacific J. Math.* **12** (1962), 1401–1403.
[3] Purity and subfunctors of the identity, *Topics in Abelian Groups*, 121–171 (Chicago, Illinois, 1963).
[4] On the structure of Tor. I. *Proc. Colloq. Abelian Groups*, 115–124 (Budapest, 1964); II, *Pacific J. Math.* **22** (1967), 453–464.

Pierce, R. S.
[1] Homomorphisms of primary abelian groups, *Topics in Abelian Groups*, 215–310 (Chicago, Illinois, 1963).

Pontryagin, L. S.
[1] The theory of topological commutative groups, *Ann. of Math.* **35** (1934), 361–388.

Prüfer, H.
[1] Unendliche abelsche Gruppen von Elementen endlicher Ordnung, Dissertation, Berlin, 1921.
[2] Untersuchungen über die Zerlegbarkeit der abzählbaren primären abelschen Gruppen, *Math. Z.* **17** (1923), 35–61.
[3] Theorie der abelschen Gruppen I. Grundeigenschaften, *Math. Z.* **20** (1924), 165–187; II. Ideale Gruppen, *Math. Z.* **22** (1925), 222–249.

Rado, R.
[1] A proof of the basis theorem for finitely generated abelian groups, *J. London Math. Soc.* **26** (1951), 74–75, 160.

Rangaswamy, K. M.
[1] Extension theory of abelian groups, *Math. Student* **32** (1964), 11–16.
[2] Characterisation of intersections of neat subgroups of abelian groups, *J. Indian Math. Soc.* **29** (1965), 31–36.

Rotman, J.
[1] A completion functor on modules and algebras, *J. Algebra* **9** (1968), 369–387.

Scott, W. R.
[1] The number of subgroups of given index in nondenumerable abelian groups, *Proc. Amer. Math. Soc.* **5** (1954), 19–22.

Specker, E.
[1] Additive Gruppen von Folgen ganzer Zahlen, *Portugal. Math.* **9** (1950), 131–140.

Szele, T.
[1] Über die direkten Teiler der endlichen abelschen Gruppen, *Comment. Math. Helv.* **22** (1949), 117–124.
[2] Ein Analogon der Körpertheorie für abelsche Gruppen, *J. Reine Angew. Math.* **188** (1950), 167–192.

[3] On direct sums of cyclic groups, *Publ. Math. Debrecen* **2** (1951), 76–78.

[4] On groups with atomic layers, *Acta Math. Acad. Sci. Hungar.* **3** (1952), 127–129.

[5] On direct sums of cyclic groups with one amalgamated subgroup, *Publ. Math. Debrecen* **2** (1952), 302–307.

[6] On direct decompositions of abelian groups, *J. London Math. Soc.* **28** (1953), 247–250.

[7] On the basic subgroups of abelian *p*-groups *Acta Math. Acad. Sci. Hungar.* **5** (1954), 129–141.

Szélpál, I.

[1] Die abelschen Gruppen ohne eigentliche Homomorphismen, *Acta Sci. Math. Szeged* **13** (1949), 51–53.

Tellman, S. G.

[1] Images of induced endomorphisms in Ext $(H, G)$, *Acta Sci. Math. Szeged* **23** (1962), 290–291.

Ulm, H.

[1] Zur Theorie der abzählbar-unendlichen abelschen Gruppen, *Math. Ann.* **107** (1933), 774–803.

Walker, C. P.

[1] Relative homological algebra and abelian groups, *Illinois J. Math.* **10** (1966), 186–209.

Walker, E. A.

[1] Cancellation in direct sums of groups, *Proc. Amer. Math. Soc.* **7** (1956), 898–902.

[2] Subdirect sums and infinite abelian groups, *Pacific J. Math.* **9** (1959), 287–291.

[3] Quotient groups of reduced abelian groups, *Proc. Amer. Math. Soc.* **12** (1961), 91–92.

[4] On *n*-extensions of abelian groups, *Ann. Univ. Sci. Budapest* **8** (1965), 71–74.

Whitney, H.

[1] Tensor products of abelian groups, *Duke Math. J.* **4** (1938), 495–520.

Yahya, S. M.

[1] *P*-pure exact sequences and the group of *p*-pure extensions, *Ann. Univ. Sci. Budapest* **5** (1962), 179–191.

[2] Functors $S_r^n(A_1, A_2, \cdots, A_n)$, *J. Natur. Sci. and Math.* **3** (1963), 41–56.

# TABLE OF NOTATIONS

| | |
|---|---|
| $A, B, C, G, H, \cdots$ | groups or subsets of groups |
| $I, J, K, \cdots$ | index sets |
| $k, l, m, n, p, q, \cdots$ | integers ($p, q$ primes) |
| $\alpha, \beta, \gamma, \delta, \eta, \rho, \pi, \phi, \psi, \cdots$ | maps, homomorphisms |
| $\rho, \sigma, \tau, \cdots$ | ordinal numbers |
| $\mathfrak{m}, \mathfrak{n}, \mathfrak{r}, \cdots$ | cardinal numbers |
| $\mathfrak{X}, \mathfrak{Y}, \cdots$ | classes |
| $\mathscr{B}, \mathscr{C}, \mathscr{D}, \cdots$ | categories |
| $R, L, \cdots$ | rings, ideals |
| $\boldsymbol{A}, \boldsymbol{B}, \boldsymbol{C}, \cdots$ | direct or inverse systems |
| $\mathbf{I}, \mathbf{K}, \mathbf{D}, \cdots$ | ideals or dual ideals in a lattice |

## SET THEORY

| | |
|---|---|
| $\in$ | is a member of |
| $\subseteq, \subset$ | is contained, properly contained in |
| $\cup, \cap$ | set union, intersection |
| $\backslash$ | difference set |
| $\varnothing$ | empty set |
| $\times$ | cartesian product |
| $\{a \in A \mid \cdots\}$ | the set of all $a \in A$ with $\cdots$ |
| $\{a_i\}_{i \in I}$ | the set of all $a_i$ with $i \in I$ |
| $\lvert A \rvert$ | cardinality of $A$ |
| $\omega$ | the smallest infinite ordinal |
| $\aleph_0, \aleph_\sigma$ | the smallest, the $\sigma$th infinite cardinal |

## MAPS

| | |
|---|---|
| $\mapsto$ | correspondence |

| $\rightarrow$ | mapping between sets or classes |
| $\alpha \mid A$ | restriction of $\alpha$ to $A$ |
| $1_A$ | the identity map of $A$ |
| Ker $\alpha$ | the kernel of the map $\alpha$ |
| Im $\alpha$ | the image of $\alpha$ |
| $\oplus \alpha_i$, $\Pi \alpha_i$ | direct sum, direct product of maps $\alpha_i$ |
| $\Delta_G$, $\nabla_G$ | diagonal, codiagonal map |

## GROUP THEORY

| $o(a)$ | order of $a$ |
| $e(a)$ | exponent of $a$ |
| $h_p(a)$, $h(a)$, $h^*(a)$ | height, generalized height of $a$ |
| $n \mid a$ | $n$ divides $a$ |
| $\leqq$, $<$ | is a subgroup, a proper subgroup of |
| $\langle \cdots \rangle$, $\langle \cdots \rangle_*$ | subgroup, pure subgroup generated by $\cdots$ |
| $\mid A : B \mid$ | index of $B$ in $A$ |
| $A/B$ | quotient group |
| $B + C$, $\sum B_i$ | subgroup generated by $B$ and $C$, by the $B_i$ |
| $\oplus$, $\underset{m}{\oplus}$ | direct sum |
| $\Pi$, $\underset{m}{\Pi}$, $A^I$ | direct product |
| $\oplus_{\mathbf{K}}$ | K-direct sum |
| $\cong$ | isomorphism |
| $nA$ | the set of all $na$ with $a \in A$ |
| $A[n]$ | the set of all $a \in A$ with $na = 0$ |
| $n^{-1}B$ | the set of all $a \in A$ with $na \in B$ |
| $T(A)$ | torsion part of $A$ |
| $S(A)$ | socle of $A$ |
| $U(A)$, $A^1$ | first Ulm subgroup of $A$ |
| $A^\sigma$, $A_\sigma$ | $\sigma$th Ulm subgroup, $\sigma$th Ulm factor of $A$ |
| $u(A)$ | Ulm length of $A$ |
| $f_\sigma(A)$ | the $\sigma$th Ulm-Kaplansky invariant of $A$ |
| $r(A)$, $r_0(A)$, $r_p(A)$ | ranks of $A$ |
| fin $r(A)$ | final rank of $A$ |
| $\mathbf{L}(A)$ | lattice of subgroups of $A$ |
| $\hat{A}$ | completion of $A$ |
| $A^\bullet$ | cotorsion completion of $A$ |

## PARTICULAR GROUPS, RINGS

| $Z$ | group of integers, infinite cyclic group |
| $Z(m)$ | cyclic group of order $m$ |
| $Z(p^\infty)$ | quasicyclic group |
| $Q$ | group of rationals |
| $Q_p$ | group of rationals with denominators prime to $p$ |
| $Q^{(p)}$ | group of rationals with denominators powers of $p$ |
| $J_p$ | group of $p$-adic integers |
| $\mathsf{Z}$ | ring of integers |
| $\mathsf{Q}$ | field of rationals |
| $\mathsf{Q}_p$ | ring of rationals with denominators prime to $p$ |
| $\mathsf{Q}_p{}^*$ | ring of $p$-adic integers |

| $\mathscr{A}$ | category of abelian groups |
| $\varprojlim, \varinjlim$ | direct, inverse limit |
| Hom | group of homomorphisms |
| Ext, Pext | group of extensions, pure extensions |
| $\otimes$ | tensor product |
| Tor | torsion product |
| Char | character group |

# AUTHOR INDEX

Numbers in italics refer to the pages on which the complete references are listed.

285

# SUBJECT INDEX

# Pure and Applied Mathematics

A Series of Monographs and Textbooks

Edited by

## Paul A. Smith and Samuel Eilenberg

Columbia University, New York

1: ARNOLD SOMMERFELD. Partial Differential Equations in Physics. 1949 (Lectures on Theoretical Physics, Volume VI)

2: REINHOLD BAER. Linear Algebra and Projective Geometry. 1952

3: HERBERT BUSEMANN AND PAUL KELLY. Projective Geometry and Projective Metrics. 1953

4: STEFAN BERGMAN AND M. SCHIFFER. Kernel Functions and Elliptic Differential Equations in Mathematical Physics. 1953

5: RALPH PHILIP BOAS, JR. Entire Functions. 1954

6: HERBERT BUSEMANN. The Geometry of Geodesics. 1955

7: CLAUDE CHEVALLEY. Fundamental Concepts of Algebra. 1956

8: SZE-TSEN HU. Homotopy Theory. 1959

9: A. M. OSTROWSKI. Solution of Equations and Systems of Equations. Second Edition. 1966

10: J. DIEUDONNÉ. Treatise on Analysis. Volume I, Foundations of Modern Analysis, enlarged and corrected printing, 1969. Volume II — in preparation

11: S. I. GOLDBERG. Curvature and Homology. 1962.

12: SIGURDUR HELGASON. Differential Geometry and Symmetric Spaces. 1962

13: T. H. HILDEBRANDT. Introduction to the Theory of Integration. 1963.

14: SHREERAM ABHYANKAR. Local Analytic Geometry. 1964

15: RICHARD L. BISHOP AND RICHARD J. CRITTENDEN. Geometry of Manifolds. 1964

16: STEVEN A. GAAL. Point Set Topology. 1964

17: BARRY MITCHELL. Theory of Categories. 1965

18: ANTHONY P. MORSE. A Theory of Sets. 1965

# Pure and Applied Mathematics

A Series of Monographs and Textbooks